Analytical Calorimetry

Volume 2

Analytical Calorimetry

Volume 2

Proceedings of the Symposium on Analytical Calorimetry at the meeting of the American Chemical Society, held in Chicago, Illinois, September 13-18, 1970

Edited by

Roger S. Porter
Head, Polymer Science and Engineering
University of Massachusetts
Amherst, Massachusetts

and

Julian F. Johnson
Department of Chemistry and Institute of Materials Science
University of Connecticut
Storrs, Connecticut

PLENUM PRESS · NEW YORK–LONDON · 1970

Library of Congress Catalog Card Number 68-8862

SBN 306-30366-3

© 1970 Plenum Press, New York
A Division of Plenum Publishing Corporation
227 West 17th Street, New York, N.Y. 10011

United Kingdom edition published by Plenum Press, London
A Division of Plenum Publishing Corporation, Ltd.
Donington House, 30 Norfolk Street, London W.C. 2, London

All rights reserved

No part of this publication may be reproduced in any form
without written permission from the publisher

Printed in the United States of America

PREFACE

Thermal analysis cuts a broad swath through contemporary science. Within this domain, advances in instrumentation permit the application of quantitative calorimetry to the full spectrum of modern materials. This can be illustrated perhaps no better than by the set of contributions which make up this Volume II of Analytical Calorimetry.

Comprehensive studies are included on an array of polymers, copolymers, and polyblends. Calorimetry is also reported on biopolymers, gels, liquid crystals, alloys, and on a variety of inorganic compounds and materials including moon rocks. Applications of calorimetric techniques to chemical reaction are also detailed. These include the study of decompositions, catalytic reductions, kinetics of dissolution, and the measurement of polymerization heats as well as the curing of thermosetting resins. The techniques employed are principally differential thermal analysis and differential scanning calorimetry. Several contributors have also illustrated the application of important and related techniques such as thermogravimetry, thermal depolarization analysis, and thermetric titrations.

The contributions to this volume represent papers presented before the American Chemical Society Symposium held in Chicago, September 14 and 15, 1970. This symposium on Analytical Calorimetry was sponsored by the Division of Analytical Chemistry.

Roger S. Porter

June 20, 1970

Julian F. Johnson

LIST OF CONTRIBUTORS

Arieh Aviram, International Business Machines Corporation, Yorktown Heights, New York

Eric Baer, Case Western Reserve University, Cleveland, Ohio

H. E. Bair, Bell Telephone Laboratories Incorporated, Murray Hill, New Jersey

Edward M. Barrall II, International Business Machines Corporation, San Jose, California

A. J. Bauman, Jet Propulsion Laboratory, California Institute of Technology, Pasadena, California

R. Becker, Research Laboratories, Eastman Kodak Company, Rochester, New York

Edgar M. Bollin, Jet Propulsion Laboratory, California Institute of Technology, Pasadena, California

Karen E. Bredfelt, International Business Machines Corporation, San Jose, California

Furman H. Brown, Division of Interdisciplinary Studies, College of Engineering, Clemson University, South Carolina

Ross C. Caballero, McDonnell-Douglas Astronautics Company, Santa Barbara, California

Jen Chiu, Plastics Department, E. I. du Pont de Nemours & Company, Incorporated, Wilmington, Delaware

William E. Collins, Photo Products Department, E. I. du Pont de Nemours & Company, Incorporated, Wilmington, Delaware

LIST OF CONTRIBUTORS

James P. Creedon, Photo Products Department, E. I. du Pont de Nemours & Company, Incorporated, Wilmington, Delaware

Robert W. Crossley, Department of Aero-Mechanical Engineering, Air Force Institute of Technology, Wright-Patterson AF Base, Ohio

W. P. Brennan, Department of Chemical Engineering, Princeton University, Princeton, New Jersey

D. J. David, Analytical and Industrial Division, R. L. Stone Company, Austin, Texas

Ronald L. Diggs, Department of Aero-Mechanical Engineering, Air Force Institute of Technology, Wright-Patterson AF Base, Ohio

G. Robert DiMarco, Department of Food Science, Rutgers University, New Brunswick, New Jersey

Ernest A. Dorko, Department of Aero-Mechanical Engineering, Air Force Institute of Technology, Wright-Patterson AF Base, Ohio

George Dugan, Research Center, Hercules Incorporated, Wilmington, Delaware

Stuart M. Ellerstein, Thiokol Chemical Corporation, Trenton, New Jersey

Linda Flannery, Xerox Corporation, Rochester, New York

R. J. Friant, Research Center, Hercules Incorporated, Wilmington, Delaware

Edward Gipstein, International Business Machines Corporation, San Jose, California

Alfred C. Glatz, Voland Corporation, New Rochelle, New York

Stanley E. Gordon, McDonnell-Douglas Astronautics Company, Santa Monica, California

A. H. Guenther, Air Force Weapons Laboratory, Kirtland Air Force Base, Albuquerque, New Mexico

Ian R. Harrison, Case Western Reserve University, Cleveland, Ohio

Howard C. Haas, Polaroid Corporation Research Laboratories, Cambridge, Massachusetts

William J. Hillegas, Xerox Corporation, Rochester, New York

LIST OF CONTRIBUTORS

Stanley A. Hollander, Polaroid Corporation Research Laboratories, Cambridge, Massachusetts

Samuel F. Hulbert, Division of Interdisciplinary Studies, College of Engineering, Clemson University, Clemson, South Carolina

J. F. Jackson, Department of Chemistry, University of Maryland, College Park, Maryland

Bruce Johnson, Department of Chemistry, University of Maryland, College Park, Maryland

A. Kagemoto, Department of Chemistry, Osaka Institute of Technology, Osaka, Japan

F. E. Karasz, Polymer Science and Engineering, University of Massachusetts, Amherst, Massachusetts

John L. Kardos, Department of Chemical Engineering, Washington University, St. Louis, Missouri

Endel Karmas, Department of Food Science, Rutgers University, New Brunswick, New Jersey

Edward W. Kifer, Koppers Company, Incorporated, Pittsburgh, Pennsylvania

H. T. Lee, Polymer Research Branch, Picatinny Arsennel, Dover, New Jersey

Irving Litant, NASA-Electronics Research Center, Cambridge, Massachusetts

W. J. MacKnight, Polymer Science and Engineering, University of Massachusetts, Amherst, Massachusetts

Monis J. Manning, Polaroid Corporation Research Laboratories, Cambridge, Massachusetts

R. T. Marano, Nuclear Materials and Equipment Corporation, Apollo, Pennsylvania

John J. Maurer, Enjay Polymer Laboratories, Linden, New Jersey

K. N. Maycock, Research Institute for Advanced Studies, Martin Marietta Corporation, Baltimore, Maryland

J. D. McCarty, Research Center, Hercules Incorporated, Wilmington, Delaware

G. Meinel, Camille Dreyfus Laboratory, Research Triangle, North Carolina

Robert Mermelstein, Xerox Corporation, Rochester, New York

B. Miller, Department of Chemical Engineering, Princeton University, Princeton, New Jersey

Gerald W. Miller, Instrument and Equipment Division, E. I. du Pont de Nemours & Company, Incorporated, Wilmington, Delaware

Mark B. Myers, Xerox Corporation, Rochester, New York

V. R. Pai Verneker, Research Institute for Advanced Studies, Martin Marietta Corporation, Baltimore, Maryland

A. Peterlin, Camille Dreyfus Laboratory, Research Triangle, North Carolina

S. E. B. Petrie, Research Laboratories, Eastman Kodak Company, Rochester, New York

Roger S. Porter, Polymer Science and Engineering, University of Massachusetts, Amherst, Massachusetts

R. Bruce Prime, International Business Machines Corporation, Endicott, New York

Bernard Rubin, NASA-Electronics Research Center, Cambridge, Massachusetts

John C. Schottmiller, Xerox Corporation, Rochester, New York

John Short, Xerox Corporation, Rochester, New York

E. R. Shuster, Nuclear Materials and Equipment Corporation, Apollo, Pennsylvania

R. J. Tetreault, Polymer Science and Engineering, University of Massachusetts, Amherst, Massachusetts

S. R. Urzendowski, Air Force Weapons Laboratory, Kirtland Air Force Base, Albuquerque, New Mexico

Charles H. Van Dyke, Carnegie, Mellon University, Pittsburgh, Pennsylvania

LIST OF CONTRIBUTORS

J. C. Whitwell, Department of Chemical Engineering, Princeton University, Princeton, New Jersey

William R. Young, International Business Machines Corporation, Yorktown Heights, New York

CONTENTS

A Differential Scanning Calorimetry and Annealing
      Study on Melting of Polyethylene Dilute
      Solution Crystals after Radiation
      Crosslinking   1
  J. F. Jackson and Bruce Johnson

Calorimetry of Drawn and Rolled Linear Polyethylene
      of High and Low Crystallinity   9
  A. Peterlin and G. Meinel

The Effect of Bromination on the Melting Point and
      Heat of Fusion of Polyethylene Crystals 27
  Ian R. Harrison and Eric Baer

Thermal Properties of Partially Hydrolyzed
      Ethylene-Vinyl Acetate Copolymers 41
  R. J. Tetreault, W. J. MacKnight, and
  Roger S. Porter

Thermoanalytic Measurements of Impact Modified Polyblends   51
  H. E. Bair

Thermal Analysis of Plasticized Elastomer Systems 61
  John J. Maurer

Thermodynamical and Statistical Mechanical Implications
      Associated with Grüneisen Ratios Calculated
      by Various Formulations 77
  S. R. Urzendowski and A. H. Guenther

Molecular Weight Determination of Carboxyl-Terminated
      Polystyrene by Thermometric Titration 95
  Robert Mermelstein, John Short, and Linda Flannery

DTA of Polysulfone 103
  H. T. Lee

Scanning Calorimetry of Mesophase Transitions: Marker's Acid 113
William R. Young, Edward M. Barrall II, and Arieh Aviram

Effects of Substituent Chains on the Mesomorphism of the Schiff's Bases of p-Aminocinnamic acid Esters 121
Edward M. Barrall II

Synthesis and Calorimetry of Some Derivatives of Dibenzazepine 127
Edward Gipstein, Edward M. Barrall II, and Karen E. Bredfeldt

Water Binding Index of Proteins as Determined by Differential Microcalorimetry 135
Endel Karmas and G. Robert DiMarco

The Molecular Weight Dependence of the Transition Enthalpy of Poly- -Benzyl-L-Glutamate 147
A. Kagemoto and F. E. Karasz

Derivative Thermogravimetric Analysis of an Epoxy Adhesive and a Phenolic-Silica Prepreg 155
Stanley E. Gordon and Ross C. Caballero

A Dynamic Differential Calorimetric Technique for Measuring Heats of Polymerization 171
Jen Chiu

Thermal Methods for Determination of Degree of Cure of Thermosets 185
James P. Creedon

Dynamic Cure Analysis of Thermosetting Polymers 201
R. Bruce Prime

Differential Thermal Analysis of Thermally Reversible Gels 211
Howard C. Haas, Monis J. Manning, and Stanley A. Hollander

Thermal Behavior of Aqueous Gelatin Solutions 225
S. E. B. Petrie and R. Becker

Some Calorimetric Measurements on Adducts of Organotin Halides with Nitrogen Bases 239
Edward W. Kifer and Charles H. Van Dyke

CONTENTS

Thermal Analysis Studies of the Decomposition of Hydrated and Deuterated Rochelle Salts 255
Alfred C. Glatz, Irving Litant, and Bernard Rubin

Stored Energy Measurements in Apollo 11 Lunar Samples by Differential Thermal Analysis 269
John L. Kardos

Determination of the Heat of Transition of Sodium, Rubidium and Cesium Tetrafluoroborate 281
R. T. Marano and E. R. Shuster

Thermal Analysis of Hydroxylammonium Perchlorate 291
J. N. Maycock and V. R. Pai Verneker

Glass Transition Phenomena in Vapor Quenched Bi-Se Alloys 309
Mark B. Myers, John C. Schottmiller, and William J. Hillegas

Kinetics of Dissolution of Magnesium Oxide in a Sodium Silicate Melt 319
Samuel F. Hulbert and Furman H. Brown

Improvement of Calorimetric Accuracy of DTA by Precision Sample Packing 339
Edgar M. Bollin and A. J. Bauman

The Application of High Pressure DSC to Catalytic Reduction Studies 353
William E. Collins

The Effect of Environment on Quantitative Measurements by Differential Scanning Calorimetry 369
D. J. David

A General Method for Characterizing Thermoanalytical Data 389
Stuart M. Ellerstein

Thermal Analyses of Polymers. VI. Thermal Depolarization Analysis (TDA) 397
Gerald W. Miller

The Application of Combined Differential Scanning Calorimetry-Mass Spectrometry (DSC-MS) to the Study of Thermal and Oxidative Decompositions 417
George Dugan, J. D. McCarty, and R. J. Friant

The Reduction of DSC Curves and the Calculation of
    Refined Autocatalytic Rate Constants by a
    Computer Technique   429
  Robert W. Crossley, Ernest A. Dorko, and
  Ronald L. Diggs

Thermal Resistance Factors in Differential Scanning
    Calorimetry   441
  W. P. Brennan, B. Miller, and J. C. Whitwell

Index   455

A DIFFERENTIAL SCANNING CALORIMETRY AND ANNEALING STUDY ON
MELTING OF POLYETHYLENE DILUTE SOLUTION CRYSTALS AFTER
RADIATION CROSSLINKING[1]

J. F. Jackson and Bruce Johnson

Department of Chemistry

University of Maryland

INTRODUCTION

Studies (2-7) of the dissolution temperature of polyethylene crystals formed in dilute solution have established the value of the interfacial free energy of mature crystals, σ_{ec}. (7) The value observed is independent of the solvent used for crystallization or dissolution. The melting temperatures, T_m^*, for crystals of varying size in the chain direction ζ, are calculated directly from the general equation for melting of lamellar crystals, for high molecular weight polymer. (8)

$$T_m^* = T_m^o \left[1 - \frac{2\sigma_{ec}}{\Delta H_u \, \zeta} \right]$$

The values so obtained are essentially identical to the measured critical annealing temperatures, T^*. (9) The latter quantity is defined as that temperature of annealing above which major changes in the low-angle spacing take place. Early observations of this effect (2, 10-14) led to the proposal that polymer chains migrated within the crystalline lattice, resulting in increased spacings while maintaining an interfacial structure consisting of regular folds of the polymer chains. (15-17) The concordance of T^* with the expected values of T_m^*, (9) supports the contrasting conclusion that partial or complete melting occurs during annealing. (2, 12, 18-20)

Partial melting and rapid recrystallization, as indicated in DSC thermograms by two endothermic peaks separated by an exotherm (19-20) makes determination of T_m^* by this technique impossible. However, radiation crosslinking suppresses the high temperature peak. (21-22) This fact has led to the suggestion that recrystallization is completely suppressed by crosslinking and that the intersection of the trailing edge and the baseline of the resultant thermogram corresponds to the melting temperature of the most stable crystals. (21-23) The melting points so defined exhibit an increase with increasing d spacing in accord with Equation 1. (20-24) However, these melting points are a major function of molecular weight (24) in the high molecular weight range, a fact inconsistent with theory (25) and experimental results. (7,26)

The contention that recrystallization is completely suppressed has not been proven. In addition, the effect of crosslinking on the free energy of fusion of the platelet crystals has not been established. On theoretical grounds, the melting temperature may be expected to increase with crosslinking. (27)

The present work examines the former problem through study of the melting of irradiated dilute solution crystals of polyethylene by DSC, and by critical annealing temperature measurement.

EXPERIMENTAL

One polyethylene fraction (M_η = 350,000) used in this study was prepared by the column elution fractionation technique. (28,29) Marlex 4601, a linear polyethylene supplied to us by Dr. J. E. Pritchard of Phillips Petroleum Company, was selectively precipitated on a Celite 540 base from a solvent-nonsolvent pair of xylene and 2-butoxyethanol above the normal melting point. A column controlled at 127° with n-butyl acetate vapor was filled with the polymer coated Celite powder, and eluent charges of increasing xylene content of a xylene-2-butoxyethanol solution made. The fractions collected were cleaned with methanol, examined by infrared for oxidation, and characterized by measuring intrinsic viscosity in decalin under nitrogen at 135°C in a Cannon-Ubbelohde dilution viscometer. Viscosity average molecular weights were calculated using Chiang's relationship for polyethylene. (29)

Another fraction used was a Hifax fraction with viscosity average molecular weight in the range of 6×10^6, which was kindly supplied by Professor L. Mandelkern.

The samples were dissolved at 125°C and transferred rapidly to a constant temperature bath controlled to ±0.1°C. All crystallizations were done from 0.08% solutions in distilled xylene at 85°C under nitrogen. They remained at 85° for a period ten times longer than that needed for first indication of crystallization to appear. They were filtered by gravity filtration on 1.2μ Gelman Metricel filter paper.

Densities were measured in a toluene-dioxane gradient column. (30) Thermograms were obtained from a Perkin-Elmer DSC-1B differential scanning calorimeter at a scan rate of $10°$/min. The calorimeter was calibrated using Fischer Thermetric standards, and indium, tin and lead standards. Following Bair, et al., (21-24) the trailing edge of the melting peak was extrapolated to the baseline yielding a temperature we designate as T_f, the apparent temperature at which the endotherm is completed.

Samples were irradiated in evacuated tubes at room temperature using cobalt γ-rays to dosages of 26 Mrads at a dose rate of 0.348 Mrad./hr.

The critical annealing temperature, T*, was found using the procedure outlined previously. (9) Separate samples were annealed at a succession of temperatures. Crystallite thickness of each annealed sample was obtained using a Rigaku-Denki low angle X-ray goniometer. The temperature of annealing above which a major change in d-spacing occurred was taken as T*.

RESULTS AND DISCUSSION

Figure 1 presents the variation in d spacing of annealed crystals as a function of annealing temperature. The results for these experiments are quite similar to those for the unirradiated samples. (9) Samples irradiated to 26 Mrads showed no major change in d spacing after annealing for as long as 100 hours at 120°C. A similar sample annealed at 121°C for 48 hours gave a well defined spacing of 156 ± 6A. A sample irradiated to 4 Mrads and annealed at 119°C for 48 hours showed an increase in d spacing to 145 ± 10A. These results taken together with previous data (9) for unirradiated samples suggest a slight increase in critical annealing temperature with increasing dosage. However, the differences observed for the present data are within the limits of error for the measurement. As shown previously, (9) the critical annealing temperature corresponds to the predicted melting temperature for dilute solution polyethylene crystals. The melting point so obtained for crystals prepared at 85°C and irradiated to 26 Mrads is approximately 7°C lower than the T_f reported by Bair, et al., (24) as is shown in Figure 2.

On the other hand, our thermograms show characteristics identical to those previously reported. (24) This is exemplified in Figure 2 by the excellent agreement between our values for T_f and those obtained by Bair, et al. Figure 3 shows similar agreement for unirradiated samples. In addition, temperatures corresponding to the trailing edge of the endotherm decrease as a function of dosage as was shown previously. (24)

Thus, it is clear that complicated thermal properties of dilute solution crystals are reproduced accurately by DSC measurement. However, interpretation of the thermograms in terms of temperatures

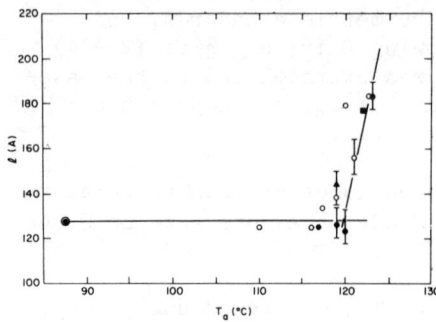

Fig. 1 Plot of low-angle spacing in ångströms against annealing temperature: (●) 26 Mrad; (▲) 4 Mrad; (■) 0 Mrad; (○) 0 Mrad, (Ref. 9).

Fig. 2 Comparison between thermogram measurement of melting, T_f, and critical annealing temperature, T^*: (□), T_f, 25 Mrad, Ref. 24; (●) T_f, 25 Mrad; (●) T^*, 26 Mrad; (○) T^*, 0 Mrad, (Ref. 9).

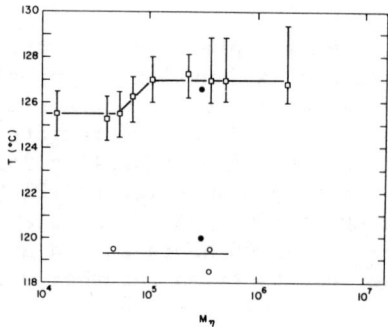

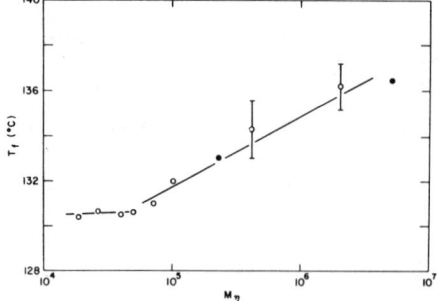

Fig. 3 Reproducibility of DSC thermogram, trailing edge of final endotherm, T_f: (●) present work; (○) Ref. 24.

associated with transitions in platelet polyethylene crystals must be considered carefully. The trailing edge of the endotherm resulting from melting of irradiated crystals does not coincide with the actual melting point of the crystals. Recrystallization probably is not fully suppressed.

CONCLUSION

Differential scanning calorimetric measurements of melting endotherms are reproducible with good precision for very complicated melting processes. Identical results have been obtained in several laboratories with a wide range of polyethylene fractions crystallized from dilute solution. Results agree for both irradiated and unirradiated crystals. Analysis of these endotherms in conjunction with equilibrium melting points determined from critical annealing temperature measurements, has shown that the effect of irradiation is twofold. That is, there is a definite kinetic effect shown in DSC thermograms by a definite displacement toward lower temperature of the trailing edge of the endotherm for irradiated samples. However, the melting point for these low doses is independent of radiation crosslinking, at least to within experimental error.

One of us (JFJ) acknowledges support from the General Research Board, University of Maryland. BJ was an NSF Undergraduate Research Participant Summer 1969.

REFERENCES

1. This research was supported by the Advanced Research Projects Agency, Contract No. SD - 101 under the direction of the Center of Materials Research, University of Maryland.
2. J. B. Jackson, P. J. Flory, and R. Chiang, Trans. Faraday Soc., $\underline{59}$, 1906 (1963).
3. T. W. Huseby and H. E. Bair, J. Appl. Phys., $\underline{39}$, 4969 (1968).
4. A. Peterlin and G. Meinel, J. Polymer Sci., Part B, $\underline{2}$, 751 (1964).
5. D. A. Blackadder and H. M. Schleinitz, Polymer, $\underline{7}$, 603, 1966.
6. V. F. Holland, J. Appl. Phys., $\underline{35}$, 59 (1964).
7. J. F. Jackson and L. Mandelkern, Macromolecules $\underline{1}$, 546 (1968).
8. P. J. Flory, J. Chem. Phys., $\underline{17}$, 223 (1949).
9. L. Mandelkern, R. K. Sharma and J. F. Jackson Macromolecules $\underline{2}$,

644 (1969).

10. W. O. Statton and P. H. Geil, J. Appl. Polymer Sci., 3, 357 (1960).

11. W. O. Statton, J. Appl. Phys., 32, 2332 (1961).

12. E. W. Fischer and G. F. Schmidt Angew. Chem., 1, 488 (1962).

13. M. Takayanagi and F. Nagatashi, Mem. Fac. Eng., Kyushu Univ., 24, 33 (1965).

14. D. H. Reneker, J. Polymer Sci., 59, 539 (1962).

15. N. Hirai, Y. Yamashito, T. Mitshata and Y. Tamura, Dept. Res. Lab. Surface Sci:, Fac. Sci., Okayama U., 2, 1 (1961).

16. A. Peterlin, J. Polymer Sci., Part B, 2, 279 (1963).

17. J. D. Hoffman Soc. Plastic Eng., 4, 315 (1964).

18. L. Mandelkern, A. S. Posner, A. F. Diorio and D. E. Roberts, J. Appl. Physics., 32, 1509 (1961).

19. L. Mandelkern and A. L. Allou, Jr., J. Polym. Sci., Part B, 4, 447 (1966).

20. J. F. Jackson and L. Mandelkern, in "Analytical Calorimetry," R. S. Porter and J. F. Johnson, Ed., Plenum Publishing Corp., New York, New York 1968, p. 1.

21. H. E. Bair, T. W. Huseby and R. Salovey, ibid., p. 31.

22. H. E. Bair, R. Salovey and T. W. Huseby, Polymer 8, 9 (1967).

23. H. E. Bair and R. Salovey, J. Polym. Sci., B5, 429 (1967).

24. H. E. Bair and R. S. Salovey, J. Macromol. Sci. -- Phys. B3, 3 (1969).

25. P. J. Flory, J. Chem. Phys., 17, 223 (1949).

26. L. Mandelkern, A. L. Allou, Jr., and M. R. Gopalan, J. Phys. Chem., 72, 309 (1968).

27. P. J. Flory, J. Am. Chem. Soc., 78, 5222 (1956).

28. P. M. Henry, J. Polym. Sci., 36, 3 (1959).

29. R. Chiang, J. Polymer Sci., 36, 91 (1959).

30. R. K. Sharma and L. Mandelkern, Macromolecules, 2, 266 (1969).

NOTE ADDED IN PROOF: The general conclusions of this work have been corroborated recently by workers using a different analytical technique. See K. Takamizawa, Y, Fukahori and Y. Urabe, Makro. Chem. 128, 236 (1969).

CALORIMETRY OF DRAWN AND ROLLED LINEAR POLYETHYLENE OF HIGH AND LOW CRYSTALLINITY

A. Peterlin and G. Meinel

Camille Dreyfus Laboratory, Research Triangle Institute

The thermal investigation of polymer samples gives very nearly the same information about crystallinity α as the measurement of density. On the basis of a two-component model one has

$$1/\rho = \alpha/\rho_c + (1 - \alpha)/\rho_a \tag{1}$$

$$c = \alpha c_c + (1 - \alpha) c_a$$

$$\Delta h_f = \alpha \Delta h_{fc} - 2\sigma/L$$

where ρ is density, c is specific heat, Δh_f is heat of fusion, σ is surface energy of the folds containing surfaces, L is the thickness of crystal lamellae, and the indices c and a relate to the crystalline and amorphous (supercooled melt) phase, respectively. The expression for the heat of fusion is derived under the assumption that all the polymer is included in crystals and every crystal has on the average the same crystalline (α) and amorphous ($1 - \alpha$) component as the whole sample.

In drawn or rolled polymer, e.g., polyethylene, the crystalline density is higher and the amorphous density is lower than that of a completely relaxed sample, e.g., of a slowly cooled or thoroughly annealed sample.[1] The deviations increase with decreasing temperature of drawing and increasing draw ratio. In first approximation, however, one may neglect this effect and interpret the observed ρ, c, and Δh_f in terms of Eq. 1. As a rule the mass crystallinity

$$\alpha_\rho = (1/\rho_a - 1/\rho)/(1/\rho_a - 1/\rho_c) \tag{2}$$

derived from density or heat capacity is smaller than

$$\alpha_f = (\Delta h_f + 2\sigma/L)/\Delta h_{fc} \tag{3}$$

derived from the heat of fusion. This is a consequence of the heat content defect of the highly strained amorphous component in the sample drawn at low temperature[2] (Fig. 1). Annealing of such a sample or drawing at a high temperature completely removes this difference.[3]

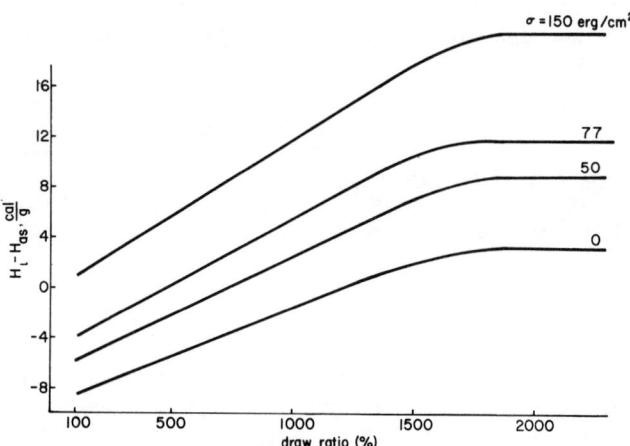

Fig. 1 - Heat content defect of the amorphous component of linear polyethylene drawn at 60°C as function of draw ratio λ. The absolute values depend on the choice of surface energy σ (Peterlin, Meinel[2])

The melting curves of plastically deformed polyethylene which has been treated with fuming nitric acid (FNA) for a sufficiently long time show a conspicuous difference between the first and second run.[4,5] The single melting peak of the first run of the etched material separates into two maxima during the second run indicating the presence of essentially two molecular weights. The ratio of the areas under the two maxima does not change appreciably with prolonged etching time. The interpretation of these data is based on the assumption that the acid preferentially attacks the folded chains and the chains in relaxed amorphous conformation but does not attack the very nearly crystallized strained tie molecules connecting either the subsequent folded chain blocks of the same microfibril (intrafibrillar tie molecules) or in adjacent microfibrils (interfibrillar t. m.). The thoroughly intermixed location of folded and tie molecules in the drawn sample yields, after FNA treatment, a distribution of chains in the debris as shown in Fig. 2a and hence a single melting peak in the first run. During crystallization from the melt the chains segregate according to their length and form two kinds of crystals with different thickness (Fig. 2b) showing up in two melting peaks in the second run. In support of this explanation the smaller molecular weight can indeed be correlated with the

thickness of the crystalline core of a single lamella. The larger
molecular weight corresponds to a length slightly greater than twice
this value and is interpreted as the thickness of the sandwich formed
by two superimposed lamellae with the intervening amorphous layer.

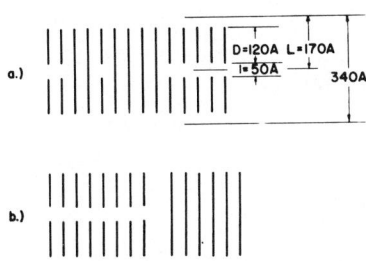

Fig. 2 - Schematic diagram of chain arrangement in a FNA treated
drawn PE sample with T_d = 60°C, λ = 10: (a) as etched and
(b) after first melting and recrystallization (Meinel,
Peterlin[4]).

In that which follows the combined analysis of density, heat
of fusion and calorimetry of FNA treated sample will be applied to
plastically deformed polyethylene (PE) of high and low crystallinity.
The difference in crystallinity is the consequence of the large
difference of viscosity average molecular weight being M_v = 56000
in the former and 1.4×10^6 in the latter case (Table I). The high
molecular weight enhances the chances for the molecule to be involved simultaneously in crystallization in more than one location
which as a rule prevents the crystallization of the connecting chain
section. As a consequence the solidified sample contains an
unusually large number of tie molecules and switch board chain loops.
Both contribute to the amorphous component which may be as high as
44% as derived from the low density. In contrast to that the amorphous component of the lower molecular weight sample is only about
25 to 32% depending on thermal history. These values agree rather
well with those obtained on single crystals precipitated from dilute
solution and annealed at the same temperatures as the bulk samples
were crystallized.[6] One hence concludes that the tie molecules in
the high crystallinity PE do not constitute a very large fraction
of the amorphous component.

Drawing and rolling transforms the unoriented more or less
spherulitic starting material into the highly ordered and aniso-
tropic fiber structure which depends on the temperature of drawing
draw rate, and draw ratio, but it is very nearly completely independent
of the thermal and mechanical history of the starting material.[7]

TABLE I

Properties of PE Samples before Plastic Deformation

	M_n	M_v			ρ		α
FF	6000	56000	quenched in ice water		0.949	g/cm^3	68%
			annealed at	127°C	0.960		75
				128°C	0.965		78
ACX	35000	1400000	quenched in ice water		0.930		56
			annealed		0.938		61

The spherulites are fully developed if the number of primary nuclei is sufficiently small as in the case of crystallization at very slow cooling in absence of a large number of heterogeneous nuclei. With rapid cooling and particularly with quenching in ice water the number of nuclei is so large that only the embryonic central part of the spherulite can develop. It consists of multilayered single crystals. Hence the whole "microspherulitic" sample is composed of randomly oriented stacks of closely packed parallel lamellae. During plastic deformation the lamellae of such a stack undergo a nearly identical sequence of transformation depending on the original orientation: lamella slip and tilt, chain slip and tilt inside the lamella, final fracture of lamella into small blocks with partial chain unfolding at the cracks and the incorporation of the folded chain blocks into microfibrils of the new fiber structure (micronecking). The last discontinuous step occurs for each stack of lamellae in a thin destruction zone which has the same lateral dimensions as the stack of lamellae.[8] The distribution of destruction zones in the sample undergoing plastic deformation, i.e., in the neck at cold drawing[9] (Fig. 3) and in the whole sample at hot drawing, is more or less at random depending on the local fluctuation of the stress and strain field caused by the morphological inhomogeneity of the sample. The random distribution yields a gradual transformation of the initial microspherulitic into the final fiber structure as observed by small-angle and wide-angle X-ray scattering, density measurement or calorimetric investigation although the transformation on molecular scale (micronecking) is discontinuous as observed by electron microscopy.

With small draw ratio the transformation into the new fiber structure is not complete. In such a case the drawn or rolled sample still contains substantial remains of the original microspherulitic structure. Hence its properties, e.g., density and heat of fusion, will depend on those of the starting material and its thermal history (slowly cooled, quenched or annealed).

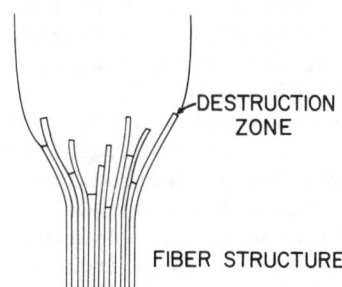

Fig. 3 - Schematic model of random distribution of destruction zones in the neck and the ensuing macroscopically gradual transformation of the initial microspherulitic into the final fiber structure (Peterlin, Baltá-Calleja[9]).

Experimental

A. Samples

The measurements were made on unfractionated samples of Fortiflex (FF, trademark of the Celanese Corporation) and ACX polyethylene (trademark of the Allied Chemical Corporation). FF is a low pressure, linear PE with M_n = 6000 and a viscosity average molecular weight M_v = 56000. ACX is a very nearly linear PE with about 1.8 CH_3 groups per 1000 CH_2, M_n = 35000 and M_v = 1.4×10^6. The 0.5 mm thick film samples were pressed and cast from the melt at 155°C and subsequently quenched in ice water. Annealing was performed for 1 hour at 127 and 128°C. The properties of the quenched and annealed samples are summarized in Table I. The films were cut into 5 mm wide strips and drawn in a thermostated nitrogen atmosphere at a rate of 0.1 cm/min at 60°C in a Table Model Instron TM-M tester. Some FF films were drawn to a draw ratio between 8 and 9 at 110°C. Rolling was performed at room temperature (25°C) on 4 cm wide strips by means of a calendar with 5 cm diameter rolls, both rotating at the same speed. In order to obtain higher draw ratio, the rolling was repeated a great many times. Some shrinkage was observed even at room temperature. Therefore the draw ratio was determined 24 hours after the drawing or rolling from the displacement of ink marks printed on the sample before deformation.

B. FNA Treatment

The plastically deformed samples were treated with FNA at 80°C for 50 hours. After etching the material was filtered, washed in distilled water, rinsed with acetone in a soxhlet extractor and dried in vacuum for a few hours. Thermograms of the degraded material were obtained by a Perkin Elmer Differential Scanning Calorimeter at a heating rate of 2.5°C/min. The degraded material was not used as etched but was first melted and then recrystallized by cooling from the melt at 2.5°C/min. This "second run" curve allows an estimate of a molecular weight distribution obtained after a preferential acid attack and removal of the noncrystalline regions. An estimate of the molecular weight can be derived from the relationship between melting point and crystal thickness obtained from nitric acid treated samples of narrow molecular weight distribution.[10] The values used are plotted in Fig. 4 and are taken from our own experiment[11] and from the literature.[12-14] The crystal thickness D was calculated from molecular weights of the etched samples, from the long periods derived from SAXS of nitric acid treated single crystal, or from long periods L of untreated samples reduced by subtraction of the amorphous layer $D = (1 - \alpha)L$. In the case of a double maximum in the following results only the low melting peak will be used for calculation of the thickness, since the position of the high temperature peak is influenced by the presence of low melting material and the previous rate of cooling.

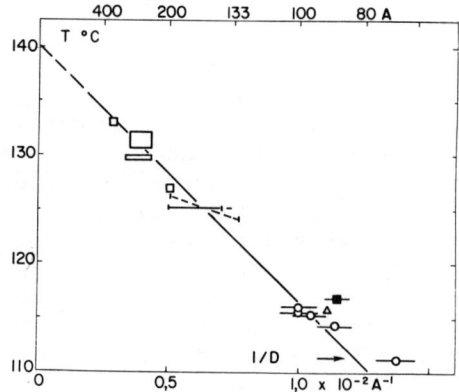

Fig. 4 - Melting temperatures of FNA treated PE as function of the reciprocal crystallite thickness (□ Palmer, Cobbold,[12] ■ Winslow et al.,[13] △ Holdsworth et al.,[14] Meinel[11])

Experimental Results

    A. Density and Heat of Fusion

        1. High Crystallinity PE (FF)

During drawing, the density of the material changes (Fig. 5). The density of annealed FF drops from $\rho = 0.963$ g/cm^3 and approaches the density of the quenched and drawn material of 0.951 g/cm^3 around a draw ratio of $\lambda = 4$. The density of the

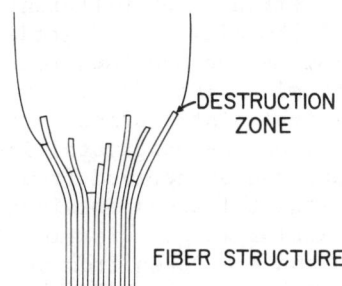

Fig. 3 - Schematic model of random distribution of destruction zones in the neck and the ensuing macroscopically gradual transformation of the initial microspherulitic into the final fiber structure (Peterlin, Baltá-Calleja[9]).

Experimental

A. Samples

The measurements were made on unfractionated samples of Fortiflex (FF, trademark of the Celanese Corporation) and ACX polyethylene (trademark of the Allied Chemical Corporation). FF is a low pressure, linear PE with M_n = 6000 and a viscosity average molecular weight M_v = 56000. ACX is a very nearly linear PE with about 1.8 CH_3 groups per 1000 CH_2, M_n = 35000 and M_v = 1.4x10^6. The 0.5 mm thick film samples were pressed and cast from the melt at 155°C and subsequently quenched in ice water. Annealing was performed for 1 hour at 127 and 128°C. The properties of the quenched and annealed samples are summarized in Table I. The films were cut into 5 mm wide strips and drawn in a thermostated nitrogen atmosphere at a rate of 0.1 cm/min at 60°C in a Table Model Instron TM-M tester. Some FF films were drawn to a draw ratio between 8 and 9 at 110°C. Rolling was performed at room temperature (25°C) on 4 cm wide strips by means of a calendar with 5 cm diameter rolls, both rotating at the same speed. In order to obtain higher draw ratio, the rolling was repeated a great many times. Some shrinkage was observed even at room temperature. Therefore the draw ratio was determined 24 hours after the drawing or rolling from the displacement of ink marks printed on the sample before deformation.

B. FNA Treatment

The plastically deformed samples were treated with FNA at 80°C for 50 hours. After etching the material was filtered, washed in distilled water, rinsed with acetone in a soxhlet extractor and dried in vacuum for a few hours. Thermograms of the degraded material were obtained by a Perkin Elmer Differential Scanning Calorimeter at a heating rate of 2.5°C/min. The degraded material was not used as etched but was first melted and then recrystallized by cooling from the melt at 2.5°C/min. This "second run" curve allows an estimate of a molecular weight distribution obtained after a preferential acid attack and removal of the noncrystalline regions. An estimate of the molecular weight can be derived from the relationship between melting point and crystal thickness obtained from nitric acid treated samples of narrow molecular weight distribution.[10] The values used are plotted in Fig. 4 and are taken from our own experiment[11] and from the literature.[12-14] The crystal thickness D was calculated from molecular weights of the etched samples, from the long periods derived from SAXS of nitric acid treated single crystal, or from long periods L of untreated samples reduced by subtraction of the amorphous layer $D = (1 - \alpha)L$. In the case of a double maximum in the following results only the low melting peak will be used for calculation of the thickness, since the position of the high temperature peak is influenced by the presence of low melting material and the previous rate of cooling.

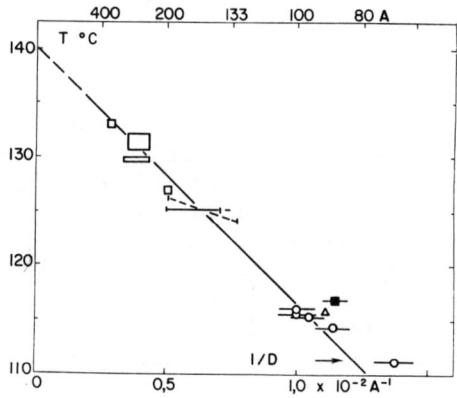

Fig. 4 - Melting temperatures of FNA treated PE as function of the reciprocal crystallite thickness (□ Palmer, Cobbold,[12] ■ Winslow et al.,[13] △ Holdsworth et al.,[14] Meinel[11])

Experimental Results

A. Density and Heat of Fusion

1. High Crystallinity PE (FF)

During drawing, the density of the material changes (Fig. 5). The density of annealed FF drops from $\rho = 0.963$ g/cm^3 and approaches the density of the quenched and drawn material of 0.951 g/cm^3 around a draw ratio of $\lambda = 4$. The density of the

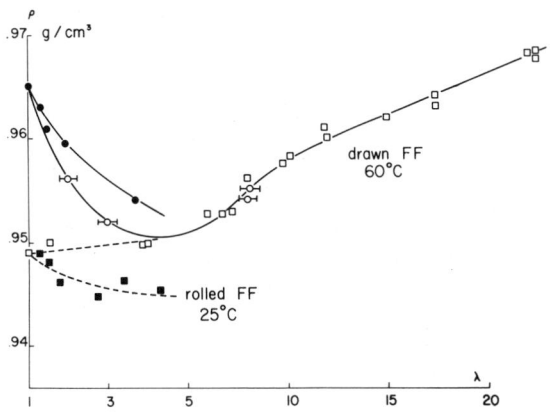

Fig. 5 - Density ρ of drawn (T_d = 60°C open signs) and rolled (T_r = 25°C full signs) FF as function of draw ratio λ. Starting material: □ quenched; ○ annealed, ◻ slowly cooled from melt. Note the change of abscissa scale at λ = 5.

quenched material does not change initially during drawing; it increases above λ = 5 and soon reaches the same values as were obtained with the drawn annealed FF. A behavior similar to that of the density is found in the heat of fusion (Fig. 6). The heat of fusion of the quenched sample increases steadily from 42 cal/g at λ = 1 to 52 cal/g at λ = 20. The heat of fusion of the annealed sample, however, decreases until it reaches the value of the quenched sample around a draw ratio of $\lambda \simeq 5$. Beyond this draw ratio, the values are independent of the thermal history of the film used for drawing. From the density and heat of fusion curves in Figs. 5 and 6 one may derive a similar dependence of crystallinity on draw ratio and thermal history of the sample. The density and heat of fusion of samples of draw ratio 8 are partly affected by the drawing temperature as shown in Fig. 7. The density increases slightly from 0.950 g/cm³ to 0.959 g/cm³ if the draw temperature goes from T_d = 9° to 110°C, the increase being stronger at low draw temperature. The data were all obtained on translucent, drawn samples either as drawn (open circles) or after slight pressing (full dots). The pressing did not affect the crystallinity, as indicated by results at draw temperatures above 80°C, where some translucent samples were additionally pressed. Parallel to this behavior is that of the heat of fusion which also increases from 43 to 49 cal/g between 9 and 110°C. Other properties like elastic modulus and ultimate tensile strength both measured at room temperature are practically constant and independent of the temperature of drawing, as already mentioned before.[5] At higher draw ratios above λ = 10, however, the elastic modulus rapidly increases with decreasing temperature of drawing.

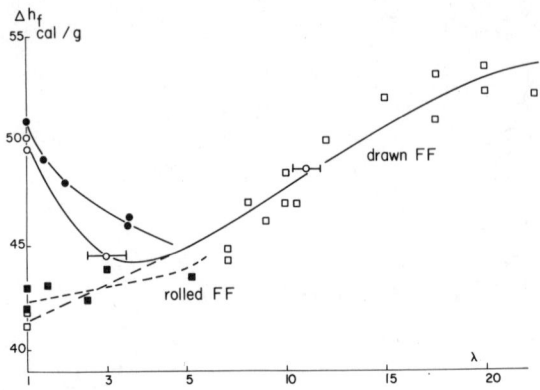

Fig. 6 - Heat of fusion Δh_f of drawn and rolled FF as function of λ (see Fig. 5).

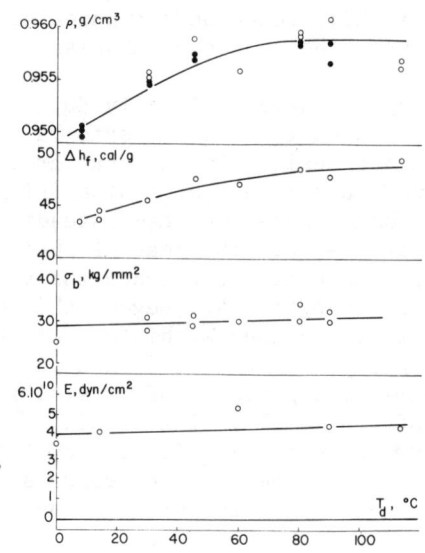

Fig. 7 - Density ρ, heat of fusion Δh_f, stress to break and elastic modulus of FF drawn to $\lambda = 8$ as function of temperature of drawing T_d. Full dots are data obtained with drawn samples after slight pressing.

During rolling of PE quenched in ice water the density decreases from 0.949 at $\lambda = 1$ to 0.945 g/cm^3 at $\lambda = 4$ (Fig. 5). A large change in the density, however, is observed after rolling of annealed films. It decreases from 0.965 g/cm^3 at $\lambda = 1$ to 0.954 g/cm^3 at $\lambda = 3.6$. Very similar to the behavior of the density is that of the heat of fusion (Fig. 6) which stays nearly constant at about 43 cal/g during rolling of quenched material but decreases from 50.5 to 46 cal/g between λ = 1 and 3.6 when annealed FF is rolled.

The mechanical properties of the rolled material, measured at -196°C are shown in Fig. 8. The modulus E increases from 3×10^{10} dynes/cm^2 at λ = 1 to about 7×10^{10} dynes/cm^2 at λ = 4. For comparison, the elastic modulus measured at room temperature is added in the drawing. In this case one observes a drop of the modulus from 1.1×10^{10} dynes/cm^2 to a minimum of 0.6×10^{10} dynes/cm^2 at λ = 2, and a subsequent increase up to 1.2×10^{10} dynes/cm^2 at λ = 5. The tensile strength σ_b at -196°C increases linearly with increasing draw ratio from 1.3 kg/mm^2 to about 3.5 kg/mm^2 between λ = 1 and 4.

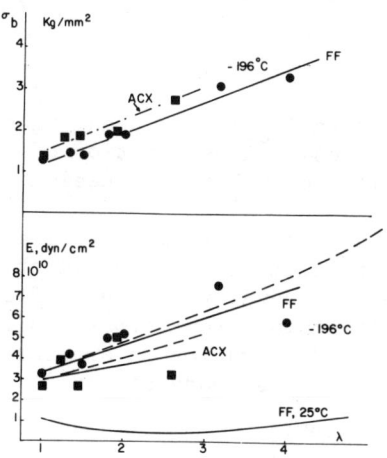

Fig. 8 - Tensile strength σ_b and elastic modulus E of quenched FF (●) and ACX (■) at -196°C after rolling at 25°C as function of the draw ratio λ. Broken lines refer to drawn samples.

2. Low Crystallinity PE (ACX)

Drawing of a quenched film slightly decreases the density (Fig. 9) and increases the heat of fusion (Fig. 10). But both changes are very nearly within the error limits of the experiment so that one may safely conclude that the crystallinity of the quenched film remains nearly constant up to the highest draw ratio λ = 5 attainable with quenched ACX film. During drawing of annealed film, however, the density falls from 0.937 to 0.932 g/cm and the heat of fusion from 38 to 29 cal/g between λ = 1 and 3.

During rolling of ACX, similar changes in crystallinity occur as in FF. The density drops slightly from 0.930 to 0.929 g/cm (Fig. 9) and the heat of fusion from 29 to 28 cal/g (Fig. 10)

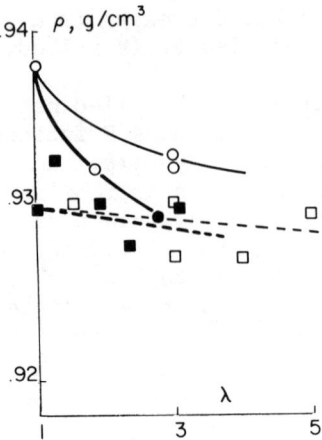

Fig. 9 - Density ρ of drawn (T_d = 60°C open signs) and rolled (T_r = 25°C full signs) ACX as function of draw ratio λ. Starting material: □ quenched, ○ annealed.

between λ = 1 and 3, when a quenched film is rolled. Both changes are very close to the error limit of the experiment. After rolling an annealed film, however, the density decreases from 0.938 to 0.929 g/cm³ and the heat of fusion from 38.8 to 33.2 cal/g between λ = 1 and 2.6. The range of draw ratios is smaller than in FF, because the ACX sample fractures already at $\lambda \approx 3$.

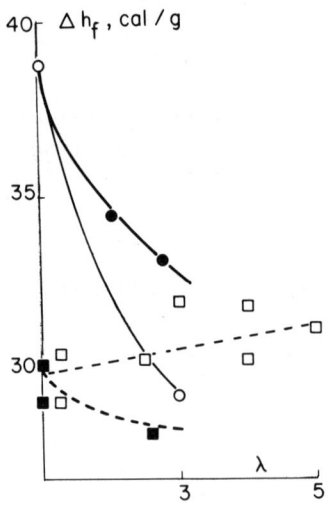

Fig. 10 - Heat of fusion Δh_f of drawn and rolled ACX as function of λ (see Fig. 9).

The mechanical properties of the rolled material, measured at -196°C are shown in Fig. 8. At this temperature, the modulus increases from 3×10^{10} dynes/cm^2 to 4×10^{10} dynes/cm^2 with the draw ratio going from 1 to 2.5. Similarily, the tensile strength at -196°C increases linearly from 1.5 kg/mm^2 at $\lambda = 1$ to 2.6 kg/mm^2 at $\lambda = 2.5$. These changes are quite similar to those observed with FF.

B. Fuming Nitric Acid Treatment

1. High Crystallinity PE (FF)

The thermograms of the degraded samples after first melting (second run) are shown in Fig. 11. After treatment of an undrawn FF a single melting peak is obtained with a maximum at 124.0°C and 130.0°C and a half width of 3.0°C and 2.6°C for the quenched and the annealed sample, respectively. The melting peak drops some 1°C to 2°C and increases slightly in half width, when the sample is drawn. At a draw ratio of 4.5 an additional change can be detected: in the sample drawn from the quenched film an additional maximum below the melting temperature of the original sample seems to occur. With further drawing two well separated peaks are found with their maxima at 120°C and 127°C when $\lambda = 10$, and at 120°C and 129°C when $\lambda = 20$. The area of the low temperature peak, which is of similar magnitude as that of the high temperature peak at $\lambda = 10$, decreases between $\lambda = 10$ and 20 whereas that of the higher peak increases with further drawing. In the material drawn from the annealed film a small maximum appears around 118-120°C, and increases in area between $\lambda = 4$ and $\lambda = 10$. No further changes in the thermogram are seen with further drawing.

Rolled FF samples of draw ratio $\lambda = 1, 2$, and 4.5 yield second run thermograms as plotted in Fig. 11. The single maximum of the quenched film at 124°C shifts to a slightly lower temperature after rolling, i.e., by about 1°C at $\lambda = 2$ and 2°C at $\lambda = 4.5$. The width of the maximum seems to increase slightly with λ. At a draw ratio of 4.5 an additional maximum appears at the high temperature side of the melting peak, i.e., at about 125°C. When annealed FF is rolled, the single maximum of the second run thermogram also shifts to a lower temperature with increasing draw ratio, i.e., from 130°C to 129°C between $\lambda = 1$ and 4.5. The intensity of the low temperature side slightly increases. In addition, a small, very broad, new maximum appears at about 118°C, when $\lambda = 4.5$.

2. Low Crystallinity PE (ACX)

The second run thermograms of HNO_3 etched ACX samples of different draw ratio are shown in Fig. 12. Neither the quenched nor the annealed undrawn material show a sharp melting peak after the acid treatment as observed with FF. In quenched ACX two maxima

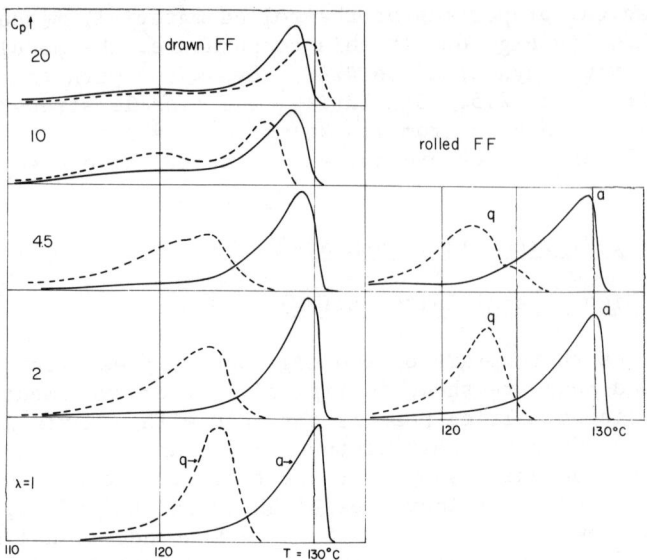

Fig. 11 - Second run thermograms of FNA treated FF drawn at 60°C or rolled at 25°C to the specified draw ratio λ. — Starting material, --- quenched, —— annealed.

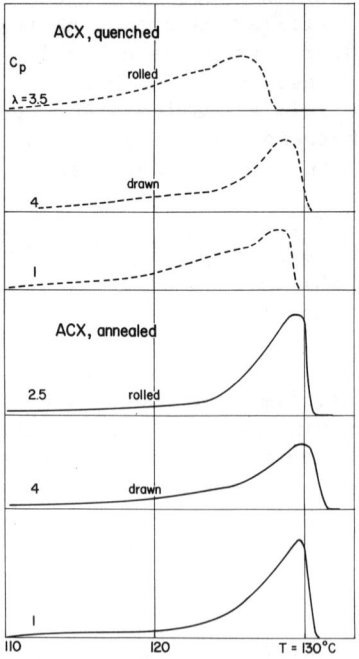

Fig. 12 - Second run thermograms of FNA acid treated ACX (quenched and annealed) drawn at 60°C to λ = 4 and rolled at 25°C to λ = 3.5 and 2.5

with slightly different melting points at 124 and 128°C seem to be superimposed. In the annealed material a rather broad maximum at about 129°C and a relatively large low melting tail can be observed. After drawing the thermograms show again two peaks, with the maxima at similar temperatures as before. In the sample drawn from the annealed film a much larger portion of the material melts between 110 and 125°C than in the annealed undrawn sample. The main peak at 130°C is slightly broadened.

The thermograms of the etched ACX samples before and after rolling with a maximum draw ratio of 2.5 and 3.5 for the annealed and quenched sample, respectively, are shown in Fig. 12. After rolling of the annealed sample a small increase in the width of the maxima is observed. After rolling of the quenched sample the whole melting curve is shifted by about 1°C to lower temperatures. The amount of low melting material is slightly increased.

The similarity of the thermograms from drawn and from rolled samples supports the interpretation of small-angle and wide-angle X-ray scattering data about the far going similarity of deformation during drawing and rolling.

Discussion

A. Mechanical and Thermal Properties

Above a draw ratio $\lambda = 5$, the density and heat of fusion and hence the crystallinity as derived from these data are independent of the thermal history of the film used for drawing (Figs. 5 and 6). This observation supports the previous claims[7] derived from electron microscopy,[8] small-angle and wide-angle X-ray scattering,[15] and crystal size determination[16] that the new fiber structure and its basic element, the microfibril, as produced by the micronecking of lamellae within the destruction zones depends on the conditions of drawing (temperature, draw rate and draw ratio) but not on the morphology of the starting material.

The situation is more complicated in the range of low draw ratios between 1 and 5. Here the transformation from the old microspherulitic into the new fiber structure is not yet completed. The sample still contains a significant amount of more or less modified original structure. This superposition of old and new elements yields a conspicuous dependence of density and heat of fusion on the properties of the starting material, e.g., the differences between the drawn or rolled quenched and annealed material. Since in our investigations the draw ratios obtained by rolling never went beyond 5 all these experiments belong to the case of incomplete transformation into fiber structure. The same applies to the drawing experiments with ACX.

The initial decrease of the crystallinity of annealed samples observed at low draw ratios can be the consequence of gradual spherulite destruction and microfibril formation. The existence of the new fiber structure can indeed be observed by small-angle X-ray scattering (SAXS) already at $\lambda = 1.1$ for highly branched,[17] and at $\lambda = 1.3$ for linear PE.[18] Since a finite concentration is needed for detection by SAXS, one might conclude that microfibrils are formed extremely early in the drawing process. Most probably, they have a lower density than the original annealed film for two possible reasons. First, the crystals of the fiber structure exhibit larger paracrystalline disorder,[16] and secondly, the decrease of long period from 300 to 170Å increases the surface to volume ratio and hence reduces the density. With drawing at room temperature, the density of microfibrils must be similar to that of the quenched sample since no changes in density are observed during drawing of quenched PE. A small change in density may also be attributed to the appearance of the monoclinic phase. Because of its presence, no quantitative analysis of the density or the heat of fusion is made in terms of crystallinity.

As indicated in Fig. 7, the crystallinity depends on the draw temperature. It is lower at lower draw temperature. Since the long period does not change appreciably in this temperature range, one has to assume that more imperfections are incorporated in the microfibrils if they are formed at lower temperatures.

A surprising fact is the steady increase in crystallinity between $\lambda = 10$ and 20 (Fig. 5). An explanation for this might be the increasing number of tie molecules, as derived from the molecular weight distribution of FNA treated samples,[5] from the elastic modulus[5] and from the crystal size determined by wide-angle and small-angle X-ray scattering.[16]

A conspicuous similarity between drawing and rolling can be found in the mechanical and thermal properties. In both cases the density and heat of fusion of the annealed samples decrease at small λ and approach the values of the quenched samples (Figs. 5 and 8). The mechanical properties, i.e., ultimate tensile strength and elastic modulus of rolled and drawn material (Fig. 8), can be compared in ACX only, since in FF sufficiently large homogeneous drawn pieces of low draw ratio are not available. The comparison is done with data obtained at low temperature where plastic deformation is negligible. With ACX the elastic moduli of rolled and drawn material are in good agreement; the ultimate tensile strength, however, is slightly higher in the rolled sample than in the drawn material. A similar behavior is found in FF, when the values of the drawn material of high draw ratio extrapolated to low λ are compared to those of the rolled FF. The disagreement in the tensile strength indicates that, despite a great many similarities in the basic deformation process during drawing and rolling, some differences in the structure and properties exist between the two cases. They manifest themselves also in the different ultimate elongations of the samples. They are smaller in rolled samples than in the drawn material of the same draw ratio.

B. Nitric Acid Treatment

In etched samples, drawn from <u>annealed</u> FF film, no large change of the melting peak at 130°C is found below a draw ratio of 3. With inqreasing draw ratio a second maximum appearing near 120°C indicates the presence of a crystalline core thickness D around 120Å (see Fig. 4). With an addition of the etched away non-crystalline surface layer of roughly 50Å this can account for the SAXS long period of ca. 170Å in the final fiber structure.

A substantial amount of material melts at higher temperature, i.e., around 129°C (Fig. 11). This may be partly due to the fact that eventually not all material is transformed into the new lamellae, even at $\lambda = 10$. More important, however, seems to be the influence of tie molecules connecting the crystalline blocks of every microfibril and not vulnerable to HNO_3 attack.[5] As a consequence of higher molecular length (at least twice the crystallite thickness) the tie molecules contribute to the high temperature peak. Its area increases with increasing draw ratio.

The thermograms of the nitric acid treated samples, drawn from the <u>quenched</u> FF film, also show a new maximum around 120°C. It corresponds to a crystal core thickness of 120Å. The second maximum of the thermograms at high temperature is due to the presence of material of higher molecular weight, i.e., to tie molecules invulnerable to HNO_3 attack. Its presence can be detected first at $\lambda = 2.4$ as a tail at the high temperature side of the melting peak. In all cases, therefore, the thermograms can be interpreted in agreement with data on SAXS to support the contention of the destruction of the original lamellae and the formation of new thinner crystallites. There are indications, however, that this process persists to higher values of λ than SAXS suggests. This interpretation finds support in the line-broadening measurements of the wide-angle X-ray scattering.[16]

The transformational behavior of PE of low crystallinity is slightly different from that of Fortiflex. This is not too unexpected, since the low crystallinity material has not only a different density but also a different microstructure. In contrast to Fortiflex the light scattering of ACX with crossed polarizers shows no spherulites.[19]

The evaluation of the results from nitric acid treatment is rather difficult, since the melting peaks are very broad already with undrawn and etched ACX in contrast to narrow peaks on undrawn FF. The broad crystal size distribution in the starting material severely limits the information one can derive from the thermograms. Quenched and etched ACX even shows a double peak (Fig. 12) which indicates that two types of crystallites with different thickness are present in the quenched ACX. At least in the annealed sample one notices, that the amount of low melting material increases with draw ratio similarly as in FF. This points to a destruction of the thick original lamellae and the formation of thinner platelets. The amount of the transformed material will be small, since the ratio of the area of low temperature peak to that of the high temperature peak is small. The relative amount of the new material, however, is not directly proportional to this ratio, since the high temperature maximum does not necessarily consist of the original material only but contains also material from the new tie molecules invulnerable to the acid.

The thermograms of rolled ACX at a draw ratio of 2 show a single maximum similar to that of the undrawn etched samples. The half width is increased slightly and the maximum, at least with the quenched and rolled sample, are shifted to slightly lower temperatures. This indicates that the crystal size distribution has not changed very much. This agrees with the results of small-angle X-ray scattering which show changes in orientation and possibly changes in chain tilt, but hardly the complete destruction of original crystallites.[11] At $\lambda = 4.5$, however, such a destruction is already very conspicuous. At this draw ratio an additional very broad maximum at 116°C is observed in the annealed rolled sample, which indicates the formation of new crystallites with a thickness D of the crystalline core of about 100Å. The very little changed original maximum at 129°C should be connected with the remaining old lamellae which were not yet destroyed by rolling. The increase of intensity of the low temperature shoulder centering about 125°C is very likely caused by a new maximum characteristic for tie molecules invulnerable to HNO_3 attack and generated during rolling. Such a higher temperature maximum corresponds to tie molecules. Tie molecules are also formed in quenched rolled samples as seen by the new maximum at the high temperature side of the thermogram at 125°C.

REFERENCES

1. E. W. Fischer, H. Goddar and G. F. Schmidt, Makromol. Chem. 118, 144 (1968).
2. A. Peterlin and G. Meinel, J. Polymer Sci. B3, 783 (1965); J. Appl. Phys. 36, 3028 (1965); Appl. Polymer Symp. 2, 85 (1966).
3. G. Meinel and A. Peterlin, J. Polymer Sci. B5, 613 (1967).
4. G. Meinel and A. Peterlin, J. Polymer Sci. B5, 197 (1967); J. Polymer Sci. A2, 6, 587 (1967).
5. G. Meinel, A. Peterlin and K. Sakaoku, Analytical Calorimetry, Ed. R. S. Porter and J. F. Johnson, Plenum Press (New York 1968) p. 15.
6. E. W. Fischer and G. F. Schmidt, Angew. Chem. 74, 551 (1962).
7. A. Peterlin, J. Polymer Sci. C9, 61 (1965); C15, 427 (1966); C18, 123 (1967); Kolloid-Z. & Z. Polymere 216-217, 129 (1967); 233, 857 (1969).
8. A. Peterlin and K. Sakaoku, J. Appl. Phys. 38, 4152 (1967).
9. A. Peterlin and F. J. Baltá-Calleja, J. Appl. Phys. 40, 4238 (1969).
10. K. H. Illers, Makromol. Chem. 118, 88 (1963).
11. G. Meinel, unpublished data.
12. R. P. Palmer and H. T. Cobbold, Makromol. Chem. 74, 174 (1964).
13. F. H. Winslow, M. Y. Hellman, W. Matreyek and R. Salovey, J. Polymer Sci. B5, 89 (1967).
14. P. T. Holdsworth, A. Keller, I. M. Ward and T. Williams, Makromol. Chem. 125, 70 (1969).
15. R. Corneliussen and A. Peterlin, Makromol. Chem. 105, 193 (1967).
16. G. Meinel, N. Morosoff and A. Peterlin, J. Polymer Sci. A2 (in press).
17. A. Peterlin and F. J. Baltá-Calleja, Kolloid-Z. & Z. Polymere (in press).
18. A. Peterlin and G. Meinel, Makromol. Chem. (in press).
19. W. Glenz (private communication).

THE EFFECT OF BROMINATION ON THE MELTING POINT AND HEAT OF FUSION
OF POLYETHYLENE SINGLE CRYSTALS

Ian R. Harrison and Eric Baer

Case Western Reserve University

Cleveland, Ohio 44106

The demonstration that the chain axis of a polymer is approximately perpendicular[1] to the lamella face initiated controversy on the nature of the chain fold. Use of primarily physical methods has not been successful in determining the degree of surface roughness, which is still in question[2]. Chemical attack of the fold particularly in PE has mainly resulted in degradation of the polymer[3]. However chemical modification of a polymer fold followed by the application of physical methods has been reported[4]. It is felt that this kind of work will prove useful not only as an extension of analytical techniques but also as a means of modifying those properties of the polymer which are sensitive to the nature of the lamella surface. In addition the selective attack of the fold could produce novel block copolymers which may themselves be the precursors of other copolymers, gels or rubbers.
The halogenation of single crystals of PE has been reported[5] and the suggestion made that a substructure on the scale of 30-200Å exists, units of which either react completely or remain unreacted on exposure to chlorine. This led to the conclusion that in PE halogenation is not primarily a surface phenomena[6]. Research in progress in our laboratory is concerned with the elucidation of the nature of chain-folded crystals by the controlled bromination of the chain-folded surfaces. This work has produced single crystals containing varying amounts of bromine whose thermal behavior is reported here prior to the main experimental material.

Experimental

Single crystals of Marlex 6015 polyethylene were prepared by crystallization from 0.05% xylene solution at 85°C. These crystals were washed with fresh xylene at the crystallization temperature then solvent exchanged to carbon tetrachloride. The resulting slurry was brominated using liquid bromine and U.V. light at room temperature. The brominated crystals were washed in distilled carbon tetrachloride and dried in vaccuo at room temperature for several days. The crystals were then analyzed chemically for bromine content.

The heats of fusion of the dried brominated single crystals were recorded using the Perkin Elmer DSC 1B. Sample weights ranged from 2 to 6 µg and were weighed to 1% accuracy. The heating rate was 20°C/minute; calibration of the instrument was performed using a sample of indium. (ΔH_f = 6.79 cals/gram.) Areas of the power-time recordings were measured by planimetry; the areas involved ranging from 4 to 10 square inches. The graphically reported values are the average of at least two runs at the appropriate bromine content. After fusion had occurred the samples were further heated to 150°C and then cooled in the instrument from the melt to room temperature at 40°C/minute. The heat of fusion of these melt recrystallized samples was then measured at the same heating rate as the brominated single crystals. Additional melting and recrystallization of the samples lowered the heat of fusion of the melt recrystallized material by no more than 3%.

The melting points of the dried brominated single crystals were recorded on a DuPont 900 Differential Thermal Analyzer. The melting point was defined as the minimum in the endothermic reaction and was reproducible to +0.5° using different thermocouples. Acetanilide (114.5°C) and phenacetine (134.5°C) were used as calibration materials. Samples were run in 2 mm. tubes and weighed 1 to 2µg; the heating rate was 20°C/minute. The reported melting points were again the average of at least two runs and differences between runs were generally less than one degree. The samples were heated through their melting points to 150°C and then cooled from the melt to room temperature in the instrument at approximately 35 to 40°C/minute. The melting points of these melt recrystallized materials were then measured at the same heating rate as the single crystals. Further melting followed by recrystallization depressed the melting point of the melt recrystallized materials by no more than 0.5°C.

Results and Discussion

The heat of fusion of single crystals PE has been measured as a function of crystallization temperature[7-10] (or lamellae thickness) and molecular weight[8-11]. Equations have been developed to allow for surface effects and defects in the crystals[12] in an effort to determine the degree of crystallinity from the measured heat content of a specimen. Literature values of the degree of crystallinity from heat of fusion measurements are, however, still open to question.

There exists conflicting evidence on the effect of heating rate on the rearrangement of single crystals[7,13] It has been reported that rearrangement takes place at heating rates of 15°/min. or less, but no superheating of folded chain crystals takes place from 15°/min. up to 3000°/min.[13] However the majority of heats of fusion reported are measured at 5 or 10°/min. Lamellae thickening may also be prevented by irradiation induced cross-linking.[7,11] The melting point depression observed on increased irradiation may also be due in some part to a decrease in entropy of the solid, especially since the greater degree of the cross-linking takes place at the folds[14,15]. The effect of molecular weight[8-11] and presumably molecular weight distribution on the shape of the endotherm is well documented. The reliability of calibrating an instrument with a metal or organic crystal with a sharp melting point and then using the instrument on a polymer with a 15° melting range may also be questioned. These points lead one to question the values of 15 to 20% amorphous content of single crystals, arrived at via heat of fusion measurements.

With these problems in mind we report the heats of fusion (ΔH_f) in cals./gram of PE for solution grown single crystals and melt recrystallized material. The heats are shown in fig. 1 as a function of bromine content. The bromine is recorded as mole fraction (X_B) of ($\sim CH_2$-$CHBr \sim$) in ($\sim CH_2 CH_2 \sim)_n$, since the assumption is made that the bromine is attached to alternate carbon atoms[16]. The figure has two main points of interest. The heat of fusion of the brominated as prepared single crystals is apparently unchanged on addition of up to 4 bromines per fold. In contrast the heat of fusion of the melt recrystallized material decreases with addition of bromine. This decrease shows a maximum change of slope at 1 bromine atom per fold. The position of the maximum change of slope at 1 bromine per fold is emphasized by a plot of the rate of change of slope of this curve as a function of bromine content, as shown in figure 2.

The interpretation of the constant value for the heat of fusion of the as prepared crystals is that the crystalline portion of the single crystals is unchanged on bromination. If bromination takes place on alternate carbon atoms at the surface of the crystals then the average fold will contain at least eight carbon atoms, since no disruption of the crystal occurs. Naturally if adjacent carbons are brominated, the fold may contain only 4 carbon atoms.

The depression of the heat of fusion of the melt crystallized material is consistent with the inclusion of defects (bromine atoms) within the crystal thereby lowering the crystallinity. Assuming a regular fold model with selective attack of the folds, the addition of up to 1 bromine per fold is shown in the melt as the presence of isolated bromine atoms spaced a decreasing distance apart along the chain. At 1 bromine per fold the minimum separation of bromine atoms, namely one fold length of the original crystal, is achieved. The addition of more than one bromine per fold does not decrease this minimum distance noticeably but rather increases the size of the defect involved. At 1 bromine per fold then there is a transition from increasing numbers of defects to increasing size of defects, which may give rise to the change in slope observed in figure 1. On crystallization of melts with up to one bromine per fold the number of defects is proportional to the bromine content and the crystallinity might be expected to decrease linearly with bromine

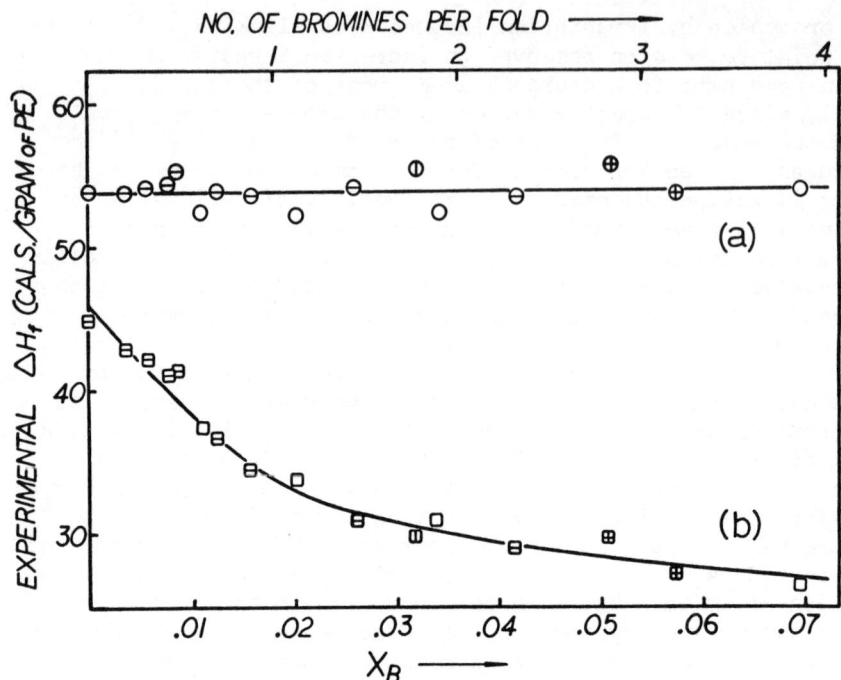

Figure 1 Measured heat of fusion (ΔH_f) in cals/gram of PE, as a function of mole fraction (X_B) of $+CH_2$-CHBr $\sim$) for (a) the brominated as prepared crystals and (b) these same materials after recrystallization from the melt.

content. Since above one bromine per fold the number of defects remains essentially the same, while their size increases, the crystallinity should have different relationship to bromine content.

A slightly different approach could be used which assumes that defects the size of an isolated bromine atom would be acceptable in a crystallite, whereas larger defects would be rejected from the crystal. This too would lead to a similar conclusion concerning the effect of bromine on crystallinity. Needless to say, it is very difficult to see how any model other than that of a regular type of fold could explain these results, although some surface roughness can be tolerated.

The problems involved in the determination of melting point or melting range of polymers by DTA have been well documented[17]. In our own study of heating rate on melting point it was found for this polymer that the melting point was essentially independent of heating rate over the range 1°C/min. to 100°C/min.[18] On the other hand the extrapolated onset temperature and the extrapolated final temperature were greatly influenced by the heating rate. All the melting points reported at 20°C/min. resulted from single peaked endotherms.

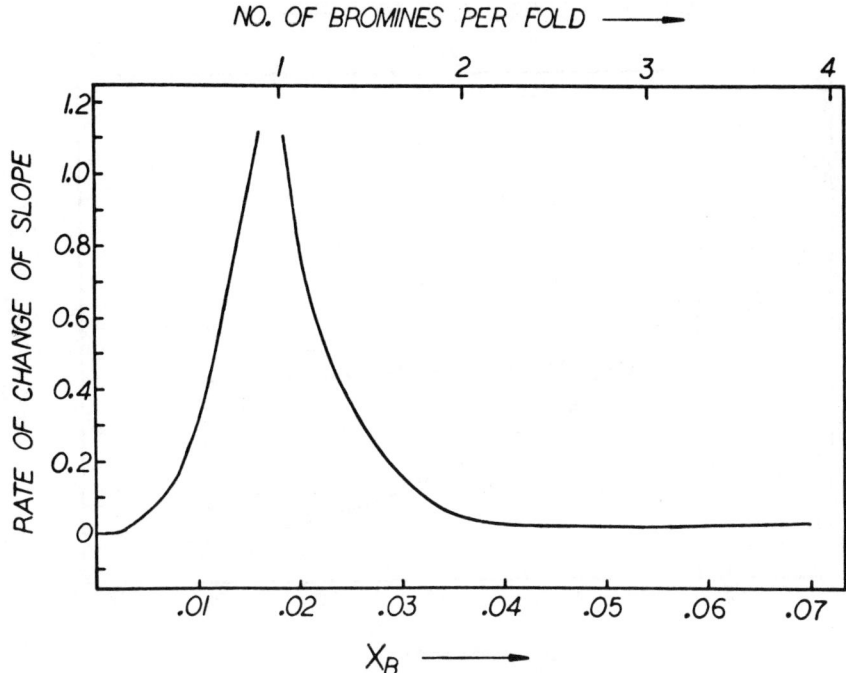

Figure 2   Rate of change of slope of curve (b) from figure (1) as a function of (X_B)

The melting points(T_m) of single crystals and melt recrystallized materials as a function of mole fraction of ($\sim CH_2$-CHBr$\sim$) in the polymer ($\sim CH_2$-$CH_2\sim)_n$ are shown in Fig. 3. The melting point of the single crystals decreases continuously up to the addition of 1 bromine per fold and then stays constant out to 4 bromines per fold. Low bromine content melt recrystallized material shows a smaller initial depression of melting point. However as more bromine is introduced the melting point continues to be depressed further and does not reach a constant value.

This melting point data in conjunction with the heat of fusion data may be analyzed in a variety of ways. A plot of the entropy of fusion ($\Delta S_f = \Delta H_f/T_m$) versus mole per cent bromine using data from figures 1 and 3 is shown in Figure 4 for both single crystals and melt recrystallized material. Note the slight increase in ΔS_f for the as prepared single crystals up to 1 bromine per fold, then an essentially constant value out to 4 bromines per fold. The as prepared single crystals show no reduction in enthalpy, yet the melting point has dropped. This means that the entropy difference (ΔS_f) between melt and crystal has increased the most likely explanation being a decrease in configurational entropy at the fold surface of the crystal. In the regular fold model the

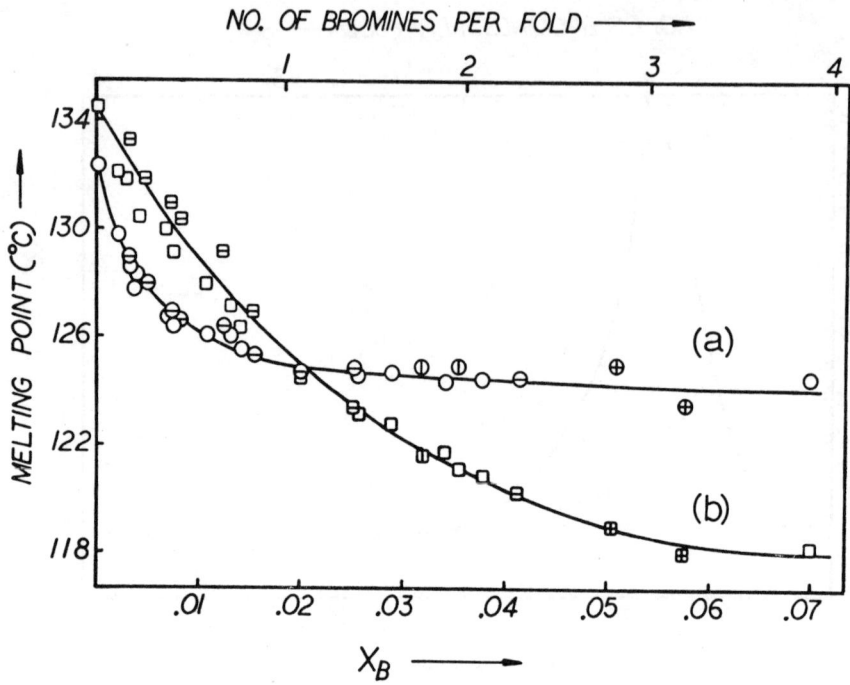

Figure 3 Melting point (T_m) as a function of X_B for (a) the brominated as prepared crystals and (b) these same materials after recrystallization from the melt.

structure of a fold is assumed to be fixed. The introduction of bromine onto such a fold will not "free" it, but will restrict its mobility still further, the net result being an entropy decrease. Presumably the major restrictions come with the addition of the first bromine to the fold.

More dramatic is the decrease in ΔS_f for the melt recrystallized material, which undergoes a change in slope at about 1 bromine per fold (Fig. 4). This change may be readily explained in terms of the previously mentioned transition from increasing numbers of defects to increasing size of defects, which takes place at 1 bromine per fold. The melt recrystallized materials show both a reduction in enthalpy and a decrease in melting point. This combination gives rise to a decrease in the difference in entropy between melt and crystal, i.e., the crystal becomes more disordered. Presumably on recrystallization the bromine containing groups can accommodate themselves more comfortably in the crystal than in the forced low entropy state of a fixed fold.

The value of ΔS_f for unbrominated single crystals (0.132 eu/grm.) is in agreement with those values reported for high density

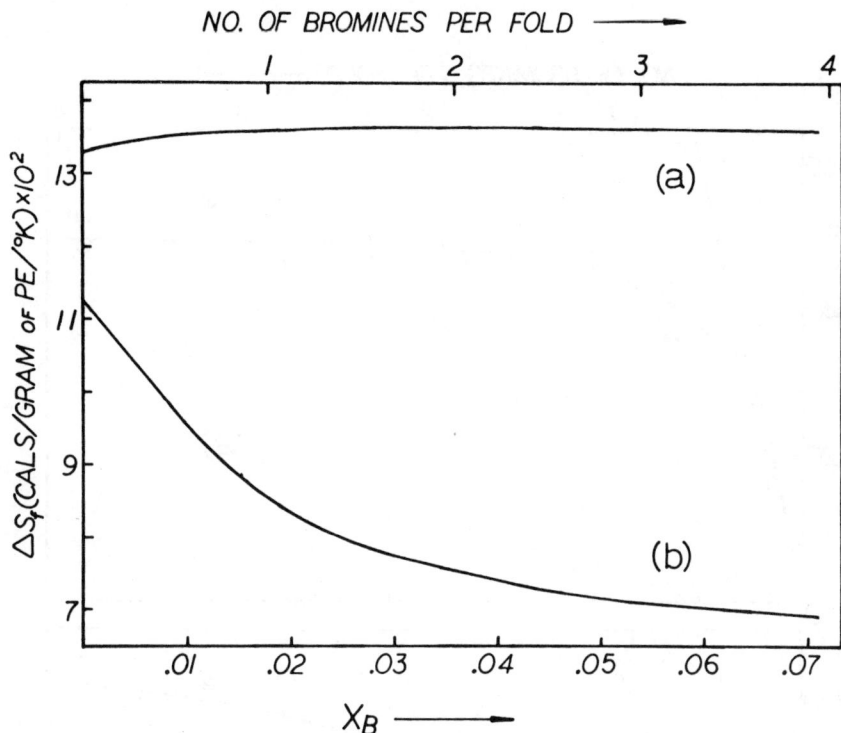

Figure 4 Entropy of fusion (ΔS_f) as a function of X_B using ΔH_f and T_m data from figures 1 and 3 respectively; (a) the brominated as prepared crystals and (b) these same materials after recrystallization from the melt.

polyethylene (0.130-0.144 eu/grm.)[19,20] and close to the value predicted for an ideal CH_2 chain crystal from n-paraffin data (0.171 eu/grm.)[21].

The brominated single crystals and the melt recrystallized materials are copolymers, and as such their melting points can be compared to those predicted by the Flory equation[22], in an effort to determine the copolymer type. A modified equation was used: -
$$1/T_m - 1/T_m^\circ = [R/\Delta H_f] (X_B)$$
where T_m and T_m° are the melting points of the copolymer and homopolymer respectively. A plot of $(1/T_m - 1/T_m^\circ)$ versus X_B should be a straight line with slope $R/\Delta H_f$ for random copolymers. The melting point data for the brominated as prepared crystals and the melt recrystallized materials have been treated in this way and are shown in Figures 5(a) and 5(b) respectively. Also shown on these figures is the behavior predicted by the Flory equation and that reported for PE containing presumably random methyl and ethyl side groups[23].

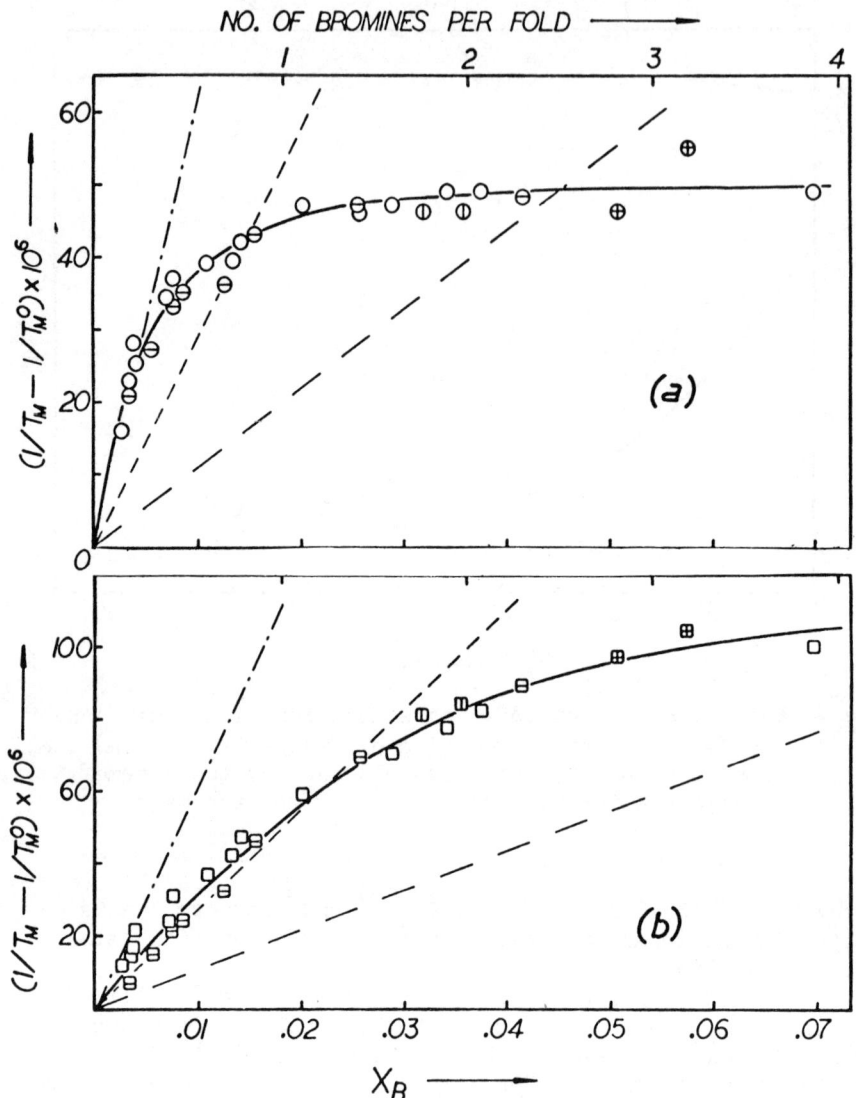

Figure 5 A plot of the Flory melting point depression as a function of X_B
— — —Predicted by equation using $\Delta H_f \sim 65$ cals/gram
— — — — —Experimental data for methyl branched PE.(23)
—·—·—·—Experimental data for ethyl branched PE. (23)
(a) the brominated as prepared crystals and (b) these same materials after recrystallization from the melt.

The data for the melt recrystallized materials Figure 5(b) are discussed first. With bromine contents of less than 1 bromine per fold the melting point depression is close to that reported for a polyethylene with methyl side groups, presumably randomly distributed along the chain. It is tempting to draw the comparison between a bromine atom and a methyl group since they have approximately the same volume, whereas the larger ethyl group produces a much greater melting point depression for the same mole fraction. Note that since the equation does not take into account the size of the non-crystallizable group, the observed melting point depression is usually much greater than that predicted for a random copolymer. When the bromine content exceeds 1 bromine per fold, one observes deviations from random copolymer behavior shown by an increasing flattening out of the curve, behavior expected of a block copolymer. If the model of a step-wise bromination at the folds is correct, then those materials containing less than one bromine per fold will have isolated bromine atoms. On recrystallization from the melt and subsequent remelting these materials should behave as random copolymers. However when these materials contain more than 1 bromine per fold a block copolymer results. The equation predicts that materials containing 1 or more bromines per fold should have the same melting point depression, since the mole fraction of non-crystallizable units stays constant. The equation is however not sensitive to comonomer size. This results in deviations from predicted values both for random copolymers (methyl and ethyl) and presumably in the case reported here for block copolymers where the block size changes.

The data for brominated single crystals is shown in Figure 5(a). An initial interpretation in terms of the Flory theory would be that the materials show a transition from random to blocky behavior at 1 bromine per fold. However since it is believed that the non-crystallizable units are outside the crystallite for the brominated single crystals, the application of this type of equation to the melting points is not justified, since the requirements of the theory have not been met.

In an effort to further analyze the melting point data of the brominated as prepared single crystals, the Hoffman-Weeks equation was[24] employed: -

$$T_m = T_m°(1 - 2\sigma e/\Delta H_f \ell)$$

where T_m and $T_m°$ are now the melting points of a folded chain and extended chain crystal respectively, σe is the surface energy, and ℓ the lamellae thickness. Since ΔH_f and ℓ are essentially constant during bromination, a drop in T_m on bromination implies an increase in surface energy. A plot of σe as a function of mole fraction of (CH_2-CHBr) for the brominated as prepared single crystals is shown in Figure 6. The surface energy rises rapidly on addition of bromine until 1 bromine per fold and then is essentially constant out to 4 bromines per fold. Although a direct correlation is not

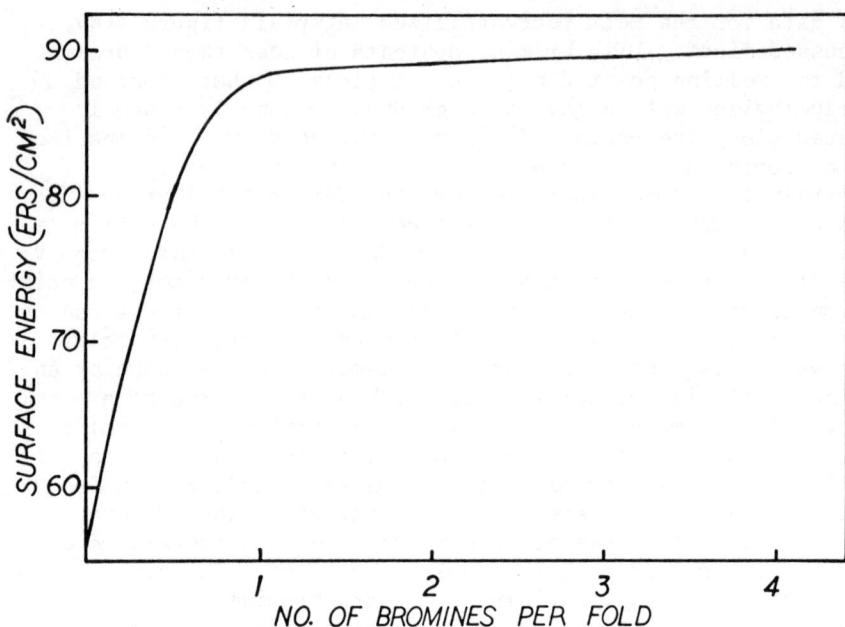

Figure 6 Surface energy σe from the Hoffman-Weeks equation applied to melting point data of the brominated single crystals, shown as a function of numbers of bromines per fold.

proposed, the changes in σe are consistent with "surface free energy" changes as measured by wettability. It has been shown that on addition of increasing numbers of the same bulky chemical group to a surface unit, the major change in "surface free energy" takes place on addition of the first group[25]. Also the nature of a surface (i.e., high or low energy) is usually determined by that group closest to the surface[25].

More recently this work has been extended to single crystals grown at 75°C. Although the data are not as complete as that for crystals grown at 85°C the same general effects are observed. The heat of fusion of the as prepared crystals grown at 75°C stays constant with bromine content out to approximately 2.5 bromine atoms per fold. In addition the heat of fusion of the melt recrystallized materials shows a change of slope at approximately one bromine per fold for the 75°C crystals as did the 85°C crystals. However there is one major difference since the fusion curve changes from a doublet to a singlet with increasing bromine content for the 75°C crystals. Figure 7(a)

It has been suggested that the melting point data of the 85°C as prepared crystals may be explained as a decreasing ability to anneal to a longer fold period, as more bromine is attached to the surface[26]. Since only a single melting point was observed with

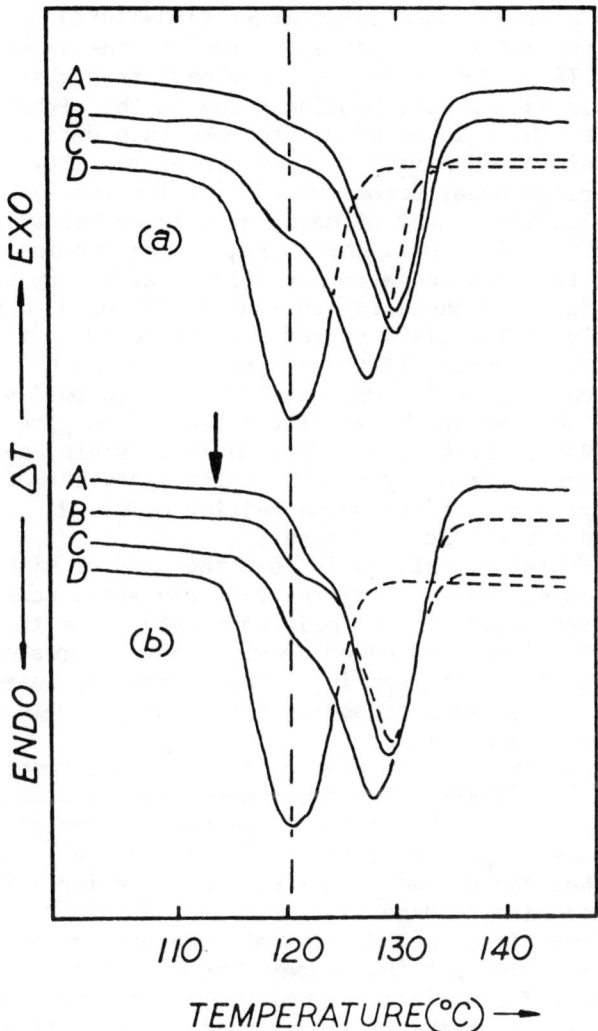

Figure 7   DTA scans for samples A-D containing approximately 0, 1/4, 1/2 and 1 bromine per fold. (a) for brominated as prepared single crystals. (b) these same materials after annealing at the temperature shown by the arrow, and subsequent cooling to room temperature before the scan was made.

the 85°C crystals this proposition could not be immediately tested without some small angle x-ray or electron microscopy work. On the other hand the doublet melting point in the DTA observed for the 75°C crystals showed an increased development of the lower peak on addition of bromine; Figure 7(a). Annealing of these single crystals at the same temperature (113-114°C) also demonstrates that the materials with less bromine anneal more. This is shown in the DTA as motion of the lower peak to higher temperatures, those materials having less bromine going to the highest temperatures. Figure 7(b). In an effort to make this data more quantitative similar experiments were run on the DSC. Measurements of the appropriate areas demonstrated that for the 75°C as prepared single crystals, the amount of material melting below 124°C increased from 22% to 95% of the total endotherm, as the bromine content was raised from 0 to 1 bromine atom per fold. Further on taking the as prepared crystals and annealing them at 112°C for 30 mins. the material melting below 124°C showed similar behavior to that observed in the DTA. For example, if the polymer contained no bromine, then, of that portion of it which originally melted below 124°C approximately 60% showed the ability to anneal to give a melting point higher than 124°C. However on addition of 1 bromine per fold only 5% of the polymer originally melting below 124°C showed the ability to anneal to give a material whose melting point was higher than 124°C. These data are summarized in Figure 8.

Still another possibility is that the polymer without bromine on it super heats. Considering the 75°C crystals, some melt out at lower temperatures (the "true" melting point) while the remainder super heat. The introduction of bromine on the surface then acts as a site from which melting can initiate. Hence the introduction of bromine again causes an increase in the amount of material melting out at the lower temperature.

It is apparent from the heat of fusion measurements of the 85°C crystals that bromination does not disrupt the crystal lattice. While the data on the 75°C crystals are not as complete, they do seem to indicate similar results. This is somewhat strange in view of the fact that the doublet endotherm observed for 75°C crystals has been associated with an annealing process. This could indicate that annealing taking place in 75°C crystals containing less than one bromine per fold does not incorporate bromine atoms in the lattice. Presumeably only folds without bromine atoms on them can anneal. This suggestion arises from the fact that melt recrystallization of those materials containing less than one bromine atom per fold, does lead to a decrease in the heat of fusion on increasing the bromine content. The decrease in presumeably due to incorporation of the bromine atoms in the lattice. In addition single crystals with one or more bromine atoms per fold show no doublets in the DTA or DSC which could be associated with an annealing process.

The melting point depression observed for the 85°C and 75°C single crystals seems consistent with the models proposed. A surface energy increase, a surface entropy decrease or loss of the ability to anneal would all explain the observed depression for the 85°C crystals. The annealing experiments on the 75°C crystals would indicate that the explanation lies in a decrease ability to anneal on increased bromination. We are at present working on some small

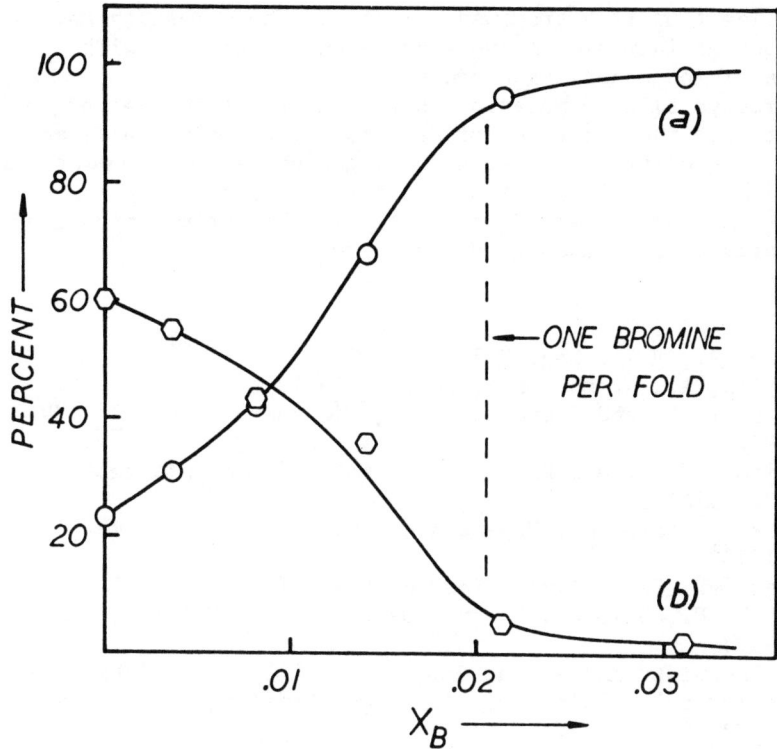

Figure 8 (a) That percentage of the brominated as prepared crystals which melt below 124°C as a function of bromine content.
(b) That percentage of those materials originally melting below 124°C which on annealing at 112°C raise their melting points higher than 124°C.

angle x-ray experiments to help solve this problem. However even if there is no annealing of brominated samples there must also be an additional melting point decrease due to the introduction of a higher energy surface onto the polymer. Hence the melting point depression on bromination is probably a two fold effect.

Conclusions

As a result of this work the following conclusions may be stated:-
1) Bromination of polyethylene single crystals takes place primarily at the surface, as evidenced by the constant heat of fusion with increased bromine content. The observed effects can be readily explained by postulating selective bromination at a regular type of chain folded crystal surface.
2) Bromination leads to a reduction in melting point of the brominated as prepared single crystals on addition of up to 1 bromine per fold. Beyond this amount of bromine no major reduction of melting point takes

place, at least up to 4 bromines per fold. These results may be explained by an increase in the surface energy coupled with prevention of the annealing process.
3) Melt recrystallized materials show a decrease in heat of fusion, consistent with the inclusion of defects in the recrystallized material. The nature of these defects changes at one bromine per fold. At this point in bromination, the number of defects has reached a constant value. Subsequently further bromination leads predominently to increases in defect size.

References

1. Keller, A., Phil. Mag. 8 1753 (1963).
2. Peterlin, A., J. Macromol. Sci. B3(1) 19 (1969).
3. Palmer, R. P. and Cobbold, A. J., Die Makro. Chem. 74 174 (1964).
4. Harrison, I. R. and Baer, E., J. Colloid and Interface Science 31 176 (1969).
5. Keller, A. Matreyek, W. and Winslow, F. H., J. Polymer Sci. 62 291 (1962).
6. Bassett, D. C., Polymer, London, 5 457 (1964).
7. Bair, H. E., Salovey, R. and Huseby, T. W., Polymer, London, 8 9 (1967).
8. Mandelkern, L. and Allou, Jr., A. L., Polymer Letters 4 447 (1966).
9. Mandelkern, L., Allou Jr., A. L. and Gopalan, M., J. Physical Chem. 72 309 (1968).
10. Hamada, F., Wunderlich, B., Sumida, T., Hayashi, S. and Nakajima, A., J. Physical Chem. 72 178 (1968).
11. Bair, H. E. and Salovey, R., J. Macromol.Sci. B3(1) 3 (1969).
12. Hoffman, J. D., Society of Plastics Engineers Trans. 4 315 (1964).
13. Hellmuth, E. and Wunderlich, B., J. Appl. Phys. 36 3039 (1965).
14. Salovey, R., J. Polym. Sci. 61 463 (1962).
15. Salovey, R. and Bassett, D. C., J. Appl. Phys. 35 3216 (1964).
16. Nambu, K., J. Appl. Polymer Sci. 4 69 (1960).
17. Double, J. S., Trans. J. Plastics Inst. 34 73 (1966).
18. Harrison, I. R., Baer, E. and Stolz , T., Unpublished work
19. Ke, B., J. Polym. Sci. 42 15 (1960).
20. Dole, M. and Wunderlich, B., Die Makro. Chem. 34 29 (1959).
21. Broadhurst, M. G., J. Res. N.B.S. 67A 233 (1963).
22. Flory, P. J., J. Chem. Phys. 17 223 (1949).
23. Ke, B., J. Polym. Sci. 61 47 (1962).
24. Hoffman, J. D. and Weeks, J. J., J. Res. N.B.S. 66A 13 (1962).
25. Shafrin, E. G. and Zisman, W. A., J. Physical Chem. 64 519 (1960).
26. Bair, H. E., Private communication.

Acknowledgements

The authors wish to express their sincere appreciation to the Manufacturing Chemists' Association for partial financial support of this work.

THERMAL PROPERTIES OF PARTIALLY HYDROLYZED ETHYLENE-VINYL ACETATE COPOLYMERS

R. J. Tetreault, W. J. MacKnight and Roger S. Porter

Polymer Science and Engineering and Chemistry Department, University of Massachusetts, Amherst 01002

INTRODUCTION

Recently there has been interest in the properties of ethylene-vinyl alcohol copolymers formed by the hydrolysis of ethylene-vinyl acetate copolymers and studies of the mechanical (1,2) and thermal (2,10) properties of such copolymers have been carried out over considerable ranges of composition. Such copolymers exhibit the phenomenon of co-crystallization with the hydroxyl groups replacing CH_2 groups in the polyethylene crystal lattice at low hydroxyl group concentration and the CH_2 groups replacing hydroxyl groups at high hydroxyl group concentration in the polyvinyl alcohol unit cell (3,4,10).

In the present work we report the results of a Differential Scanning Calorimetry (DSC) study on two series of partially hydrolyzed ethylene-vinyl acetate copolymers. These two series are based on starting ethylene-vinyl acetate copolymers containing 75 and 59 mole percent ethylene. As might be expected, the properties of the two series of terpolymers show dramatic changes in going from the starting copolymers which are amorphous, rubbery materials to the highly hydrolyzed copolymers which are semi-crystalline, hard plastics. These changes are the result of the substitution of the hydroxyl group for the acetate group and are principally due to the fact that the hydroxyl group can cocrystallize with methylene sequences as discussed above and to the hydrogen bonding capabilities of the hydroxyl group. Neither cocrystallization (5) nor hydrogen bonding can occur in the case of the acetate group.

EXPERIMENTAL

Materials

The starting copolymers were obtained through the courtesy of Monsanto Company. They were experimentally prepared by a free radical, high pressure method similar to the techniques applied in the homopolymerization of low density polyethylene. Elemental analysis showed the starting copolymers to contain 75.4 and 59.1 mole percent ethylene. These two copolymers are referred to as A1 and B1 respectively. Vapor phase osmometry yielded $\overline{M}_n$ for A1 of 12,000 and $\overline{M}_n$ for B1 of 14,000. Analysis of 100 Mc NMR data according to the method of Schaefer (6) indicated that both A1 and B1 were random copolymers and that A1 and B1 both contained less than 1.5 methyl groups per 100 carbon atoms.

Acidic hydrolysis of A1 and B1 was carried out in a toluene-methanol mixture at 66°C, according to

$$\sim(CH_2CH_2)_x-(CH_2CH)_y^\sim \atop | \atop O \atop | \atop C=O \atop | \atop CH_3} \quad + \quad CH_3OH \xrightarrow{H_2SO_4}$$

$$\sim(CH_2CH_2)_x-(CH_2-CH)_y^\sim \atop | \atop OH} \quad + \quad CH_3-O-\overset{O}{\overset{\|}{C}}-CH_3$$

The desired chemical composition was achieved by cooling and neutralizing the reaction mixture after specific reaction times. The hydroxyl content was determined by saponification. Table I presents the chemical compositions of all polymers studied.

After hydrolysis the polymers were precipitated, washed and vacuum dried. The dried polymers were pressed into films at temperatures 40 to 50°C above the melting points (T_m's) in the case of the crystalline polymers and at temperatures of 100 to 125°C above their glass transition temperatures (T_g's) in the case of the amorphous polymers. Samples B4-B6 were molded just above their T_m's inasmuch as chemical crosslinking occurred in these materials at higher temperatures. All of the films were slowly cooled in the press to room temperature. The cooling rate was ca. 0.5°C/min.

TABLE I

CHEMICAL COMPOSITION OF ETHYLENE-VINYL ACETATE-VINYL ALCOHOL TERPOLYMERS PREPARED FOR STUDY

Composition (Mole %)

Sample No.	Ethylene	Vinyl Acetate	Vinyl Alcohol
A1	75.4	24.0	0
A2	75.4	20.6	4.0
A3	75.4	13.5	11.1
A4	75.4	8.92	15.7
A5	75.4	5.03	19.6
A6	75.4	0.45	24.2
B1	59.1	40.9	0
B2	59.1	35.2	5.7
B3	59.1	20.9	20.0
B4	59.1	11.1	29.8
B5	59.1	4.67	36.2
B6	59.1	0.57	40.3

Measurements

T_g's and T_m's were determined on a Perkin-Elmer Differential Scanning Calorimeter, Model DSC-1B. Temperature calibration was accomplished using a series of pure organic compounds (TherMetric standards). Indium was used as the standard for the heat of fusion (ΔH_f) measurements.

In the polymers under investigation both the degrees of crystallinity and the T_m's are strong functions of thermal history. It was attempted to minimize such effects by treating all samples in the following manner. In the case of the T_g measurements, approximately 10 mg film samples were heated at 20°/min in the DSC from room temperature to 30° above their T_m's in the case of crystalline samples or 100° above their T_g's in the case of amorphous samples and then cooled to below their T_g's at 10°/min. The samples were then heated at 10°/min and the T_g's were taken as the midpoint in the step observed in the DSC trace. In the case of the T_m and ΔH_f determinations, 5 mg samples were heated at 10°/min in the DSC to 30° above their T_m's and then cooled at 2.5°/min to below the crystallization temperature. ΔH_f's were calculated from the areas under the crystallization exotherms. Samples were heated at 5°/min and T_m's were taken to be the temperatures at which the endotherms returned to the baselines.

X-ray results were obtained using a flat plate film camera. A beam of Cu Kα radiation, filtered through nickel, was passed through 20 mil film samples.

RESULTS AND DISCUSSION

Table II summarizes the DSC results for all samples studied

TABLE II

SUMMARY OF DSC DATA

Sample No.	Composition E/VA/VOH (Mole %)	T_g (°K)	T_m (°K)	ΔH_f (Cal/gm)
A1	75.4/24.6/0	232	None	None
A2	75.4/20.6/4.0	236	None	None
A3	75.4/13.5/11.1	243	346	2.11 ± .18
A4	75.4/8.92/15.7	261	360	3.59 ± .33
A5	75.4/5.03/19.6	266	373	5.12 ± .23
A6	75.4/0.45/24.2	280	382	10.50 ± .25
B1	59.1/40.9/0	239	None	None
B2	59.1/35.2/5.7	252	None	None
B3	59.1/20.9/20.0	260	None	None
B4	59.1/11.1/29.8	269	379	0.682 ± .023
B5	59.1/4.67/36.2	283	387	3.53 ± .31
B6	59.1/0.57/40.3	304	393	6.14 ± .03

Glass Transitions

T_g's for both ethylene-vinyl acetate and ethylene-vinyl alcohol copolymers have been previously determined by dynamic mechanical (1,2,5,7) and thermal (2,8) methods. It has been found that the T_g's of the ethylene-vinyl acetate copolymers are insensitive to vinyl acetate concentration so long as they remain crystalline or up to about 20 mole percent vinyl acetate, remaining constant at ca -20°C. As the vinyl acetate concentration is further increased, T_g assumes intermediate values between -20°C and the T_g of polyvinyl acetate, 30°C. Ethylene-vinyl alcohol copolymers, on the other hand, show a sharp increase in T_g at low vinyl alcohol contents, a value of 30°C having been reported (2) for a copolymer containing 20 mole percent of vinyl alcohol.

Table II shows that the terpolymers studied behave in a manner similar to the ethylene-vinyl alcohol copolymers inasmuch as T_g is a monotonically increasing function of vinyl alcohol content for both series whether or not the materials are crystalline or amorphous.

To a first approximation the T_g of many random vinyl copolymers is a linear function of the composition (9). Such an approximation is obviously invalid for both the ethylene-vinyl acetate and ethylene-vinyl alcohol copolymers. A simple extension of the empirical equation relating composition to T_g for copolymers to a terpolymer system results in (9)

$$T_g = v_1 T_{g_1} + v_2 T_{g_2} + v_3 T_{g_3} \qquad (1)$$

In the present terpolymer, v_1 is the volume fraction of polyvinyl alcohol with T_{g_1}, v_2 is the volume fraction of polyethylene with T_{g_2} and v_3 is the volume fraction of polyvinyl acetate with T_{g_3}. In view of the inapplicability of the linear approximation to the copolymers, it would not be expected that Eq. (1) would be valid for the terpolymer system. Table III shows that this is the case. The calculated T_g for polyethylene which should be independent of terpolymer composition is an increasing function of vinyl alcohol content. The values used in Eq. (1) to calculate T_{g_2} are: $T_{g_1} = 358$ (11), $T_{g_3} = 301$ (11), $\rho_1 = 1.303$ (10), $\rho_2 = 0.920$, and $\rho_3 = 1.179$ (10).

TABLE III

T_g CALCULATED FOR POLYETHYLENE

Sample No.	$(T_g)_{obs}$ (°K)	$(T_{g_2})_{calc}$ (°K)
A1	232	178
A2	236	186
A3	243	199
A4	261	227
A5	266	234
A6	280	252
B1	239	136
B2	252	171
B3	260	188
B4	269	205
B5	283	229
B6	304	263

It may be that the deviations of the terpolymer series from the predictions of Eq. (1) can be ascribed to the strong intermolecular interactions introduced as a result of the presence of the hydrogen bonding vinyl alcohol groups. That such an explanation may be valid is indicated by Fig. 1 which is a plot of the difference between the calculated polyethylene T_g's for the terpolymers given in Table III and the polyethylene T_g calculated for the starting ethylene vinyl acetate copolymers as a function of hydroxyl content. It can be seen from Fig. 1 that this difference is directly proportional to hydroxyl content with points for both series falling on the same line.

Crystallinity

Samples A1, A2 and B1-B3 showed no detectable crystallinity by DSC under the thermal histories investigated. This is, of course, a reflection of the inability of the acetate group to be accommodated in the polyethylene crystal lattice. All of the amorphous samples have vinyl acetate concentrations greater than that at which the last trace of crystallinity disappears in ethylene-vinyl acetate copolymers (7). At sufficiently low vinyl acetate contents the terpolymers crystallize with the inclusion of the hydroxyl groups in the crystal lattice. Figs. 2 and 3 show schematic DSC traces for the crystalline members of series A and B, respectively. Besides the increase noted in the melting points and degrees of crystallinity, the rather diffuse melting endotherm of the low crystallinity members of both series becomes very sharp in samples A6 and B6, both of which are essentially ethylene-vinyl alcohol copolymers. There remains, however, an extremely broad "tail" on the melting endotherm apparently extending all the way to T_g. Similar results have been reported previously on comparable copolymers (2). It has been shown (2) that a copolymer of polyethylene and vinyl alcohol containing 23 mole percent vinyl alcohol exhibits two melting points and an x-ray diffraction pattern corresponding to a superposition of the polyethylene reflections and the polyvinyl alcohol reflections. Thus, apparently two crystal phases are present in this copolymer. X-ray examination of samples A6, B5 and B6 revealed a crystalline reflection at 4.6Å characteristic of the (101) reflection of polyvinyl alcohol. The characteristic polyethylene reflection at 3.7Å assigned to the (200) crystal plane was absent. All the terpolymers exhibited only a single melting point in the range up to 210°C. The T_m quoted for the 23 mole percent copolymer is 110°C for the polyethylene phase and 160°C for the polyvinyl alcohol phase. Table II indicates that sample A6 with a 24 mole percent vinyl alcohol content has a melting point of 109°C. Thus, although this melting point was previously assigned to a polyethylene crystal phase in a copolymer of the same vinyl alcohol content, present evidence indicates it to be the melting point of a polyvinyl alcohol crystal phase into which methylene units are incorporated as defects. The reason for the difference in crystallization between our materials and those of Illers is not firmly established. One possibility is that the ethylene-vinyl alcohol copolymer studied previously (2) was compositionally inhomogeneous. If some of the chains were largely composed of polyethylene and others were largely composed of polyvinyl acetate, this would account for the presence of two distinct crystal phases. The melting point of 160°C observed for the polyvinyl alcohol phase compared to the 109°C observed in the present work for a comparable copolymer would then be rationalized on the basis of the incorporation of fewer methylene units into the polyvinyl alcohol unit cell for the high melting point material.

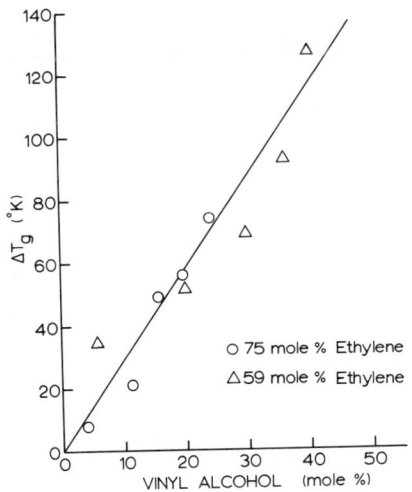

Figure 1. $\Delta T_g = T_{g2(calc)} - T_{g2(calc)}$ (starting copolymer) vs. hydroxyl concentration for both series.

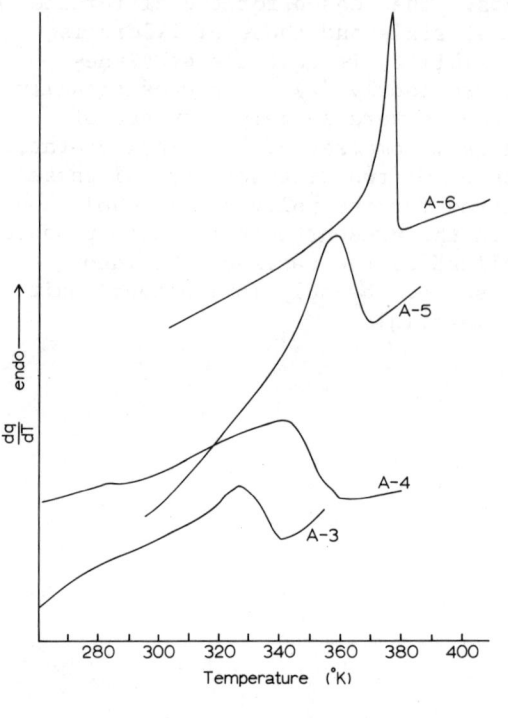

Figure 2. DSC traces for series A showing melting endotherms.

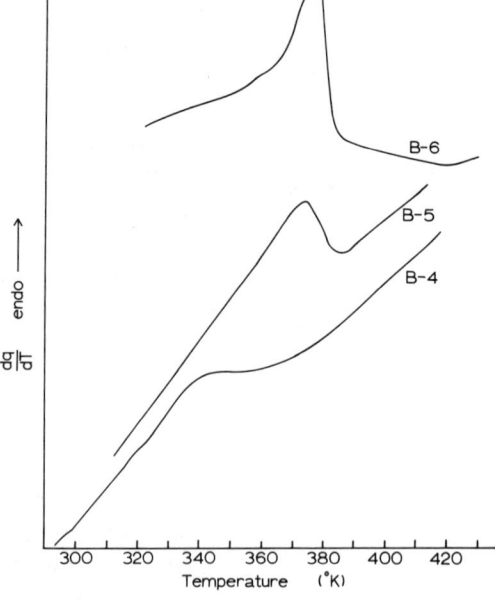

Figure 3. DSC traces for series B showing melting endotherms.

The sharp melting endotherms of samples A6 and B6 (Figs. 2 and 3) indicate a narrow distribution of crystallite sizes. The reason for this is not clear at present but an examination of the data of Bodily and Wunderlich (10) shows that a series of ethylene-vinyl alcohol copolymers exhibit similar sharp melting points.

SUMMARY

The two series of ethylene-vinyl acetate-vinyl alcohol terpolymers studied exhibit composition dependent T_g's which are governed by hydrogen bonding introduced by the presence of hydroxyl groups. Crystallinity starts to occur in the hydroxyl group concentration range between 4 and 11 mole percent in the case of series A and 20 and 30 mole percent in the case of series B. X-ray evidence indicates that the basic structural unit is the polyvinyl alcohol unit cell in all cases with methylene units present as defects.

ACKNOWLEDGEMENTS

We are grateful to Dr. M. Wisotsky for the 100 Mc NMR spectra and to the National Science Foundation for partial support of this research.

REFERENCES

1. T. Fujiki, M. Saito, M. Uemura and Y. Kosaka, J. Polym. Sci. A-2, 8, 153 (1970).
2. K. H. Illers, European Polym. J. Supplement, p. 133 (1969).
3. C. W. Bunn and H. S. Peiser, Nature 159, 161 (1947).
4. B. Wunderlich, Polymer 5, 611 (1964).
5. L. E. Nielsen, J. Polym. Sci. 42, 357 (1960).
6. J. Schaefer, J. Phys. Chem. 70, 1975 (1966).
7. F. P. Reding, J. A. Faucher and R. D. Whitman, J. Polym. Sci. 57, 483 (1962).
8. R. K. Tubbs, J. Polym. Sci. A3, 4181 (1965).
9. L. A. Wood, J. Polym. Sci. 28, 319 (1950).
10. D. Bodily and B. Wunderlich, J. Polym. Sci. A-2, 4, 25 (1966).
11. L. E. Nielsen, Mechanical Properties of Polymers, Reinhold, New York, 1962.

THERMOANALYTIC MEASUREMENTS OF IMPACT MODIFIED POLYBLENDS

H. E. Bair

Bell Telephone Laboratories, Incorporated

The production of multiphase polymer systems is reportedly the fastest growing field in the entire chemical industry. One type of these multicomponent materials is the rubber reinforced plastics where small, micron sized, rubber particles are blended or grafted into a rigid polymer matrix. By this technique, a brittle copolymer such as styrene-acrylonitrile (SAN) may be toughened by the addition of a rubber-like polybutadiene (PB). In this case, the resulting combination of plastic and rubber is an acrylonitrile-butadiene-styrene (ABS) resin with high impact strength. The elucidation of the composition and structure of these polyblends may be accomplished by a combination of phase separation methods, spectroscopic analysis[1] and electron microscopy.[2,3]

Recently a rapid quantitative thermal analysis technique was developed to characterize the structure of impact modified polyblends.[4] In this method the type and amount of each component within an amorphous polyblend may be identified thermally by a component's glass transition temperature and the magnitude of the increase in specific heat occurring at the transition. It is the purpose of this paper to describe the quantitative thermal analysis of several impact modified poly(vinyl chloride) (PVC) and Noryl polyblends.

Four commercial polyblends, two impact modified PVC resins (I, II) and two Noryl resins (III, IV) were selected for analysis. In addition, the transition behavior of six ABS resins were studied. Three polyblends of polystyrene (PS) and polyphenylene oxide (PPO) were made by mixing in chloroform and drying under reduced pressure at 50°C for five days.

The specific heats, C_p, of the resins were determined with a Perkin-Elmer Differential Scanning Calorimeter (DSC-1) by an intermittent heating technique with temperature intervals of 30°C.[5] The heating rate was 40°C/min.

The glass transition temperature, T_g, in this work is defined as the temperature which corresponds to the onset of the discontinuity in C_p. Although the C_p of an unknown sample is normally measured by comparison to a standard reference material (Al_2O_3), the magnitude of the increase in the specific heat, ΔC_p, at a glass transition, may be determined directly in the DSC without reference to any standard from the following relationship.

$$\Delta C_p = \frac{\Delta S}{MR} \qquad (1)$$

where Δ is fractional increase in the ordinate at the transition, S is full scale value of the power (ordinate) in millical./sec., M is the mass of the sample and R is the heating rate of the calorimeter.

This technique affords a rapid and accurate measurement of ΔC_p to within 1%. However, a few precautionary rules should be remembered in order that an accurate measurement of ΔC_p may be obtained: first, a heating rate should be selected which coincides with the effective cooling rate for the formation of the glass;[4] second, any stored energy due to orientation effects must be released by heating prior to evaluation of the material; third, if any plasticizer exists in the material it will not only lower T_g but also spread the transition over a much wider temperature interval and make the measurement of ΔC_p more difficult. Thus, the plasticizer should be extracted, if possible, or a similar material without plasticizer should be evaluated in order to understand the role of the plasticizer within the resin.

The fraction, χ, of a component in a polyblend is the ratio of the observed change in C_p for a particular component in the polyblend, ΔC_p^{blend}, to the known increase in C_p of the pure parent of the component in the blend, ΔC_p^{parent}

$$\chi = \frac{\Delta C_p^{blend}}{\Delta C_p^{parent}} \qquad (2)$$

A 1 mm thick film of the Noryl resin III was prepared by compression molding at 205°C and exposed to osmium tetroxide vapor at room temperature. In this manner, the rubber phase was selectively stained and hardened. Then, ultrathin sections were microtomed from the film and studied by transmission electron microscopy.

In Figs. 1 and 2 the specific heat of the two impact modified PBC resins (I and II) are compared with the C_p values of an ABS resin and pure PVC against temperature. The C_p curve of PVC* increases linearly from -100° to 75°C. An abrupt increase in C_p occurs at about 80°C, which is a manifestation of the glass to liquid transformation, T_g, occurring in the homopolymer, and then, once again, C_p increases smoothly from 95 to -190°C. The latter phenomenon is an indication of the absence of any crystals within the sample.

The C_p measurements of the ABS resin reveal two second order transitions at -85°C and 103°C.[4] The T_g at -85° coincides with the glass temperature of pure polybutadiene, and, therefore, is assigned to the rubber within the polyblend. In addition, since there is no shift in the transition temperature of the rubber in the ABS resin, it may be inferred that the grafting of the SAN to PB has occurred at the interface between the rubber and copolymer. It will be shown later that, if the polymer diffuses into the rubber particles before grafting occurs, the T_g of the resulting copolymer will be between the T_g's of the two individual components. From a comparison of the observed ΔC_p (0.014 cal.°C^{-1}g.$^{-1}$) for the rubber in the ABS resin to ΔC_p (0.092 cal.°C^{-1}g.$^{-1}$) for pure polybutadiene it is estimated the resin contains 15% rubber. This amount of rubber agrees closely with the value of 18% calculated for the same resin by a phase separation technique.[1] The T_g at 103°C is due to the styrene-acrylonitrile copolymer within the ABS. The T_g of the SAN which has been extracted from this resin is 104°C. From Eq. 2 it is estimated that the fraction of SAN in the resin is 75%. This value is in good agreement with 78% as calculated by the phase separation technique.

The thermal behavior of several additional ABS resins, two methacrylate-butadiene-styrene (MBS) resins and a methacrylate-acrylonitrile-butadiene-styrene (MABS) system, have been investigated. The percentage of rubber in the MBS and MABS resins ranges from 45 to 56%. Not only is this a far greater amount of rubber than found in any of the ABS materials in this study, but also the T_g's at -57° and -52°C indicate grafting has occurred inside the rubber particles. The results are summarized in Table 1.

The specific heat behavior of the two high impact PVC resins (I, II) (the C_p of resin II is calculated from the measured C_p of

---

*The C_p values are the averaged values of Alford and Dole,[6] and Dunlap.[7]

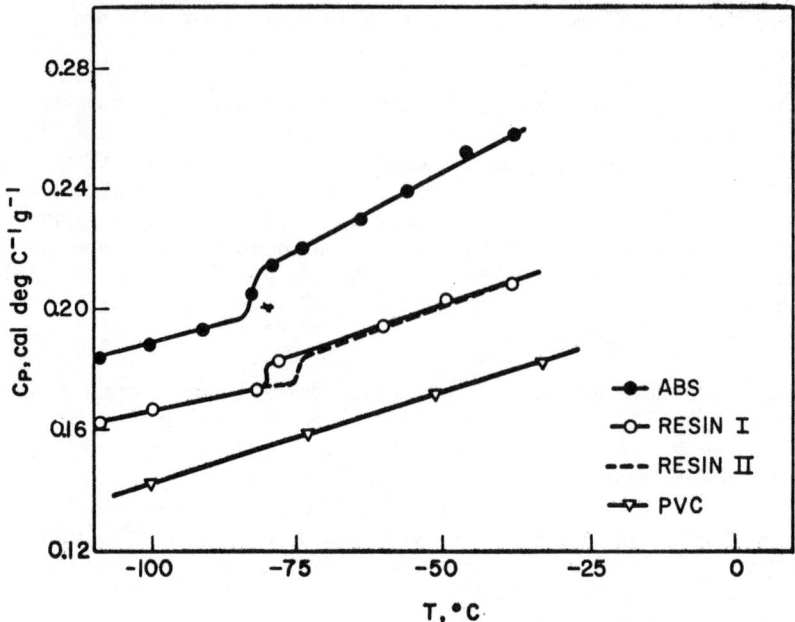

Fig. 1 - Comparison of the Low Temperature Specific Heats of ABS, High Impact PVC, and Pure PVC

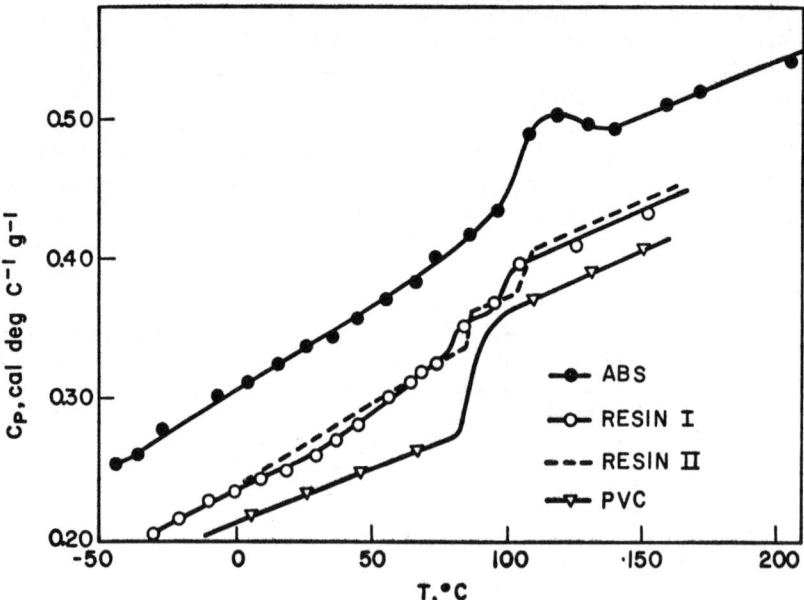

Fig. 2 - Comparison of the High Temperature Specific Heats of ABS, High Impact PVC, and Pure PVC

TABLE 1

	T_g, °C	ΔC_p cal.°C^{-1}g.$^{-1}$
Polyacrylonitrile	93	0.092
Polybutadiene	-85	0.092
Polystyrene	106	0.069
PVC	84	0.055
SAN	104	0.093
ABS I	103	0.069
	-85	0.014
ABS II	99	0.066
	-85	0.016
ABS III	103	0.061
	-85	0.029
MBS I	97	0.029
	-58	0.052
MBS II	98	0.032
	-57	0.041
MABS I	88	0.023
	-52	0.052

resin I and the ΔC_p measurements of resin II) indicates the materials are composed of three separate phases. Resins I and II have T_g's at -85°, 71°, and 95°C, and -80°, 80°, and 104°C, respectively. The low temperature transition is attributed to polybutadiene; the intermediate T_g is due to PVC, and the highest glass temperature is associated with the SAN which has been incorporated within the polyblends. The 5 to 9 degree lowering of the T_g's in resin I is due to the addition of a plasticizer which facilitates the processing of this material.

Resin II, which has no plasticizer, appears to be composed of approximately 48% ABS and 45% PVC and 7% additives. X-ray fluorescence measurements indicate the resin has about 55% PVC. Table 2 lists the complete analysis of each resin.

The infrared spectrum of the Noryl resin III has absorptions characteristic of a 50-50 blend of polystyrene and polyphenylene oxide.[8] The C_p curve of a PS-PPO mixture prepared from solution by blending 50% by weight of both PS and PPO is compared in Fig. 3 with the C_p data of Karasz, Bair and O'Reilly for amorphous PPO[9] and PS[10], which was determined by adiabatic calorimetry. The T_g of the blend (140°C) is about midway between the T_g's of the PS (90°C) and the PPO (200°C). The T_g's of PS-PPO alloys of

TABLE 2

	T_g, °C	ΔC_p cal.°C^{-1}g.$^{-1}$	Composition
Resin I	95	0.024	26% SAN
	71	0.020	37% PVC
	-85	0.006	7% PB
Resin II	104	0.030	32% SAN
	80	0.025	45% PVC
	-80	0.004	5% PB

If no free SAN has been added to either resin I or II, it is estimated resin I is a blend of 38% ABS and 37% PVC and 25% additives, while resin II is a mixture of 45% PVC, 48% ABS, and 7% additives.

different compositions were found to increase as the concentration of PPO was increased (Fig. 4). Thus, the PS-PPO polyblends behave as a single phase.[11,12,13] This unusual solution compatability of PPO-PS alloys has been studied spectroscopically. The results indicate that no complex has been formed which could account for the single phase behavior. The NMR spectrum shows only a very weak interaction has occurred, perhaps dipolar in nature.[14]

In Fig. 5 the C_p against T is plotted for Noryl resins III and IV (the specific heat of resin IV is calculated from the measured C_p of resin III and ΔC_p measurements of resin IV) and exhibits: 1) T_g's at -61° and 60°C with ΔC_p of resin IV more than twice as large as ΔC_p of resin III, 2) a first order transition between 83° to 103°C with an apparent heat of fusion of 0.2 cal./g., 3) a glass transition at 141° and 111°C for resins III and IV, respectively, and 4) a linear increase in C_p between 155 and 260°C. The T_g at -61 is due to styrene-butadiene (SBR) rubber. In this case styrene monomer has penetrated the polybutadiene particles before grafting occurred. The polymerization which resulted inside the rubber particle produced the cellar structure[2] which is shown in the electron micrograph of the ultrathin section taken from Noryl resin III (Fig. 6).

The rubber particles are comparatively large, measuring 10 to 15μ across. The orientation of the particle may have happened during compression molding at 205°C or during the microtoming operation. Resin III has 5% SBR and resin IV 13%. The small endotherm found in both resins is presumed to be the melting of 1% of low density polyethylene (PE), which probably acts as a lubricant. Lastly, the T_g at 141°C indicates (Fig. 4) that resin III is composed of 47% by weight of both PS and PPO, whereas the T_g at 111°C for resin IV is due to a mixture of 69% PS to 17% PPO (Table III). Therefore, Noryl resins III and IV appear to be blends of high impact strength PS and PPO.

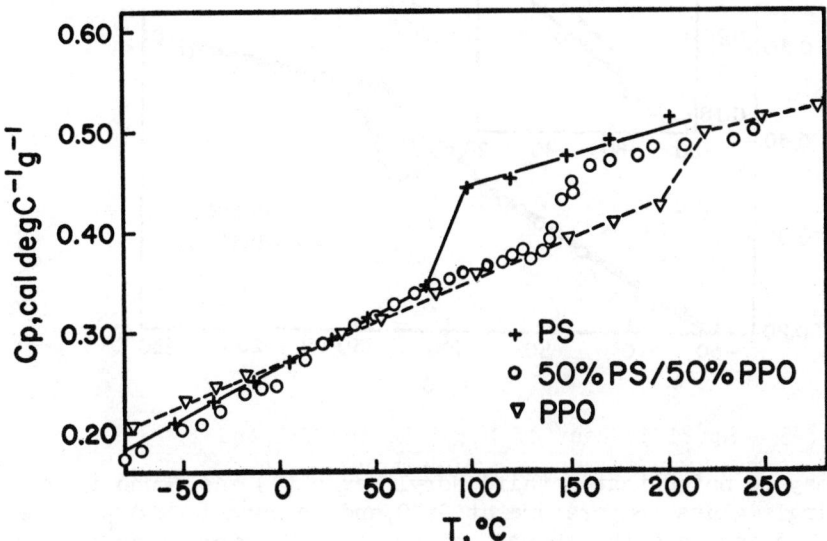

Fig. 3 - Comparison of the Heat Capacity of an Amorphous Polystyrene and Polyphenylene Oxide with a Polyblend of 50% (by weight) PS and 50% PPO

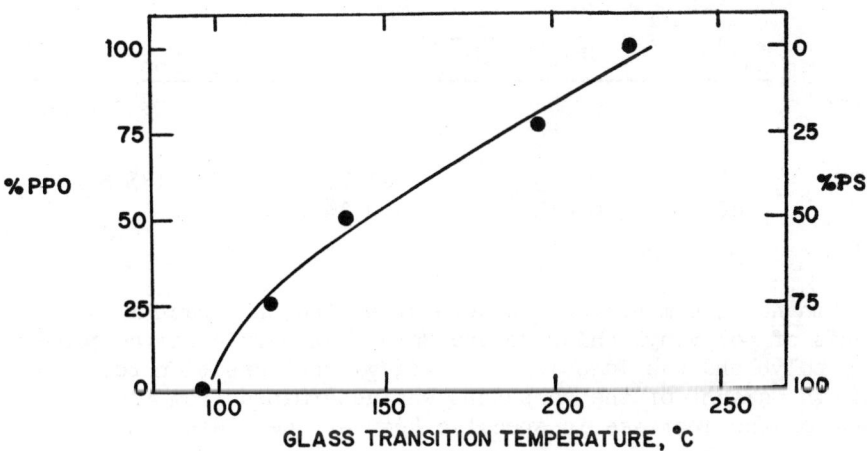

Fig. 4 - The T_g of PS/PPO Polyblends as a Function of Concentration

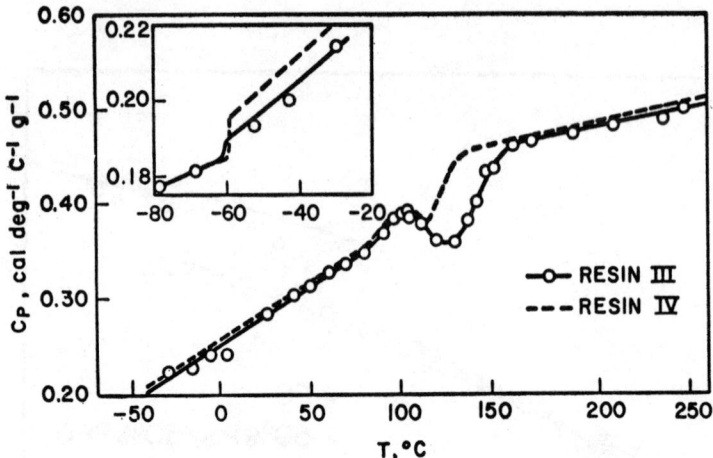

Fig. 5 - Specific Heat of Noryl Resins III and IV

It may be noted that a third Noryl resin (V) was found to have a single glass temperature at 95°C and an anomalous ΔC_p. The low T_g would indicate the absence of any PPO; however, recent chemical analysis indicates the presence of a fire retardant additive which apparently acts as a plasticizer. In order to analyze this type of Noryl the effect of the plasticizer on the T_g of the PS-PPO blends must be evaluated.

TABLE 3

	T_g, °C	ΔC_p cal.°C^{-1}g.$^{-1}$	Composition
Resin III	141 -61	0.044 0.005	47% PS, 47% PPO, 5% SBR and 1% PE
Resin IV	111 -60	0.043 0.012	69% PS, 17% PPO, 13% SBR and 1% PE
Resin V	95	0.072	

Thermoanalytic measurements were attempted on impact modified polyblends of polyvinyl chloride and Noryl. A particular component within a polyblend was identified by its glass transition temperature and the amount of the component was determined from the magnitude of the increase in specific heat at the glass temperature.

Several commercial impact modified PVC resins were determined to be blends of ABS and PVC, while two Noryl resins were found to be blends of polyphenylene oxide and high impact polystyrene.

Fig. 6 - Transmission Electron Micrograph of an Ultrathin Section of Noryl III

Polyblends of PS and PPO yield single sharp glass transitions which are a function of concentration. Polystyrene appears to be cosoluble with PPO without the formation of any complex.

ACKNOWLEDGMENTS

The author would like to acknowledge helpful discussions with M. Matsuo, T. W. Huseby and P. G. Kelleher. In addition, I wish to thank M. Matsuo for making the ultrathin specimen of the Noryl resin.

REFERENCES

1) B. D. Gesner, Polymer Sci., 3A, 3825 (1968).

2) M. Matsuo, C. Nozaki and Y. Jyo, Polymer Eng. and Sci., 9, 197 (1969).

3) M. Matsuo, Polymer Eng. and Sci., 9, 266 (1969).

4) H. E. Bair, SPE Technical Papers, 16, 115 (1970); Polymer Eng. and Sci., in press.

5) J. A. Currie and M. Dole, in Analytical Calorimetry, R. S. Porter and J. F. Johnson, eds., Plenum Press, New York, 1968, p. 51.

6) S. Alford and M. Dole, J. Am. Chem. Soc., 77, 4774 (1955).

7) L. H. Dunlap, J. Polymer Sci., Part A-2, 4, 673 (1966).

8) B. D. Gesner, BTL Memorandum for Record (1966).

9) F. E. Karasz, H. E. Bair and J. M. O'Reilly, J. Poly. Sci., A2, 6, 1141 (1968).

10) F. E. Karasz, H. E. Bair and J. M. O'Reilly, J. Phys. Chem., 69, 2657 (1965).

11) H. E. Bair, F. E. Karasz and J. M. O'Reilly, G.E. Memo Report (1964).

12) U.S. Patent No. 3,383,435, May 14, 1968 by E. P. Cizak, Assigned to General Electric Co.

13) J. Stoelling, F. E. Karasz and W. J. MacKnight, Polymer Preprints, 10, No. 2, 629 (1969).

14) E. P. Otocka, BTL Memorandum for Record (1968).

THERMAL ANALYSIS OF PLASTICIZED ELASTOMER SYSTEMS

John J. Maurer

Enjay Polymer Laboratories

Linden, New Jersey

INTRODUCTION

Differential Thermal Analysis (DTA) has broad utility in the study of elastomer systems[1]. The major area of application consists of qualitative studies of glass transition temperature (Tg) or melting points (Tm) of the elastomer or other ingredients in the formulation. This report, however, will be concerned with quantitative considerations of plasticized elastomer systems both from the standpoint of the influence of the plasticizer on Tg and vulcanizate properties, and the analysis of plasticizer content via DTA or Differential Scanning Calorimetry (DSC). In particular, the relationship between Tg and the low temperature dynamic properties of butyl rubber formulations will be examined. Implications regarding optimum plasticizer type and level are also considered. As a general theme, the paper seeks to illustrate the scope and depth of investigations which may be conducted based on consideration of basic thermograms.

EXPERIMENTAL

<u>DTA and DSC</u>. Commercial units (DuPont Instrument Products) were employed using DTA conditions described in detail elsewhere[1]. Essential features of DTA procedure: ~ 30-35 mg sample; glass bead reference, Chromel/Alumel thermocouples. DSC: see text for details; empty pan used as reference. Heating rate: 20°/min for both methods.

<u>Butyl Compounds</u>. Prepared by conventional procedures and conditions for butyl rubber (i.e., Banbury mixer, mills, curing press).

Formulation (phr) of laboratory dynamics compounds: Enjay Butyl 268-100, HAF-35, FEF-20, MT-25, Plasticizer-25, Stearic Acid-1, ZnO-5, Ethyl Cadmate-2, Altax-0.5, sulfur-1.

<u>Rubber-Plasticizer Mixtures</u>. Blended for 10-15 minutes on 6" x 12" mill.

Analysis of Butyl Dynamics Compounds

Plasticizers are sometimes used in elastomer formulations to improve specific physical properties of the end-product vulcanizate. The present investigation is concerned with a series of laboratory compounds developed during such a study of the Low Temperature Efficiency (LTE) of Butyl dynamics compounds. LTE is defined as the ratio of the deflection at -20°F divided by that at +75°F. These dynamics compounds enable comparison of the following plasticizers: a process oil (F845:Flexon 845); two ester plasticizers (BCP:Butyl Cellosolve Pelargonate and PLGN:Plastogen); and a 1:1 mixture of F845 and PLGN. The initial DTA objective was to establish the relationship between Tg and LTE and thus determine if LTE of new compounds could be rapidly estimated by this means.

DTA Characteristics of Accelerated Compounds

The thermograms for these systems, as illustrated in Figure 1, reveal several interesting characteristics. First, Tg differs; second, an endotherm appears between Tg and 0°C; third, endotherms are noted between 80 and 150°C. The latter information, while incidental to the present investigation, was included to indicate a type of qualitative analysis of curatives (and interaction products thereof) which can readily be detected in some formulations, thus possibly enabling quality control checks during the manufacturing process. The endotherm between Tg and 0°C is believed due to the process oil or plasticizer, as discussed in detail below. To test this idea, thermograms of several types of process oils and plasticizers were compared. Figure 2 reveals a strong endotherm in BCP near -33°C, and a diffuse endotherm in F845 (-45 to 0°C). These are assumed to represent melting regions, but this has not been rigorously checked. The alternative possibility, that they represent glass transition behavior, is considered unlikely. Even if it were the case it would not seriously impair the arguments presented here except to describe certain regions of the thermogram in terms of glassy behavior rather than in terms of crystallinity development. It seems reasonable, in any event, to attribute the low temperature endotherms in Figure 1 to F845 and BCP. For comparison, the scan for F640 shows that not all process oils resemble F845, and that of Plgn shows that not all plasticizers resemble BCP.

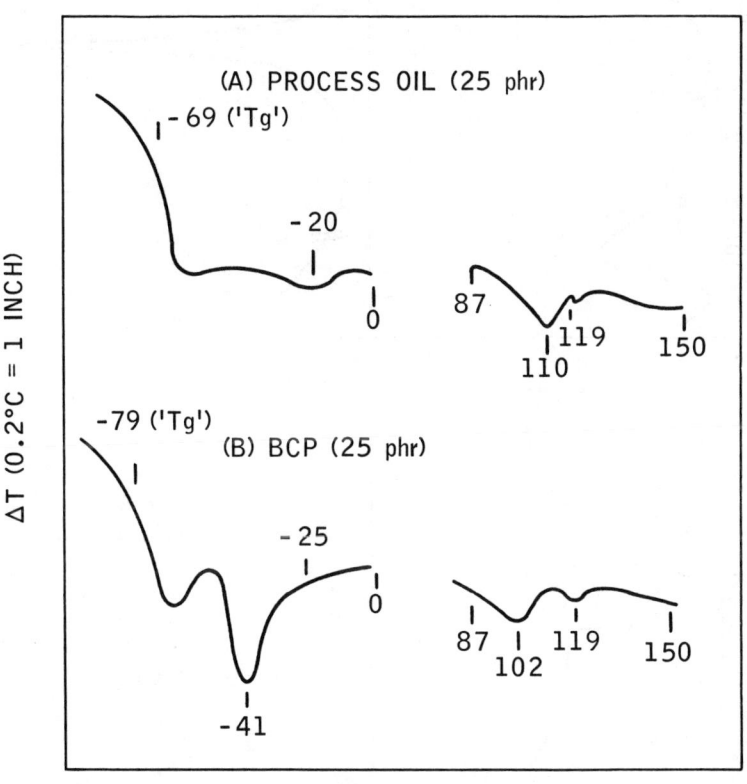

Figure 1
PLASTICIZED BUTYL RUBBER COMPOUNDS

Figure 2
PLASTICIZER CHARACTERISTICS

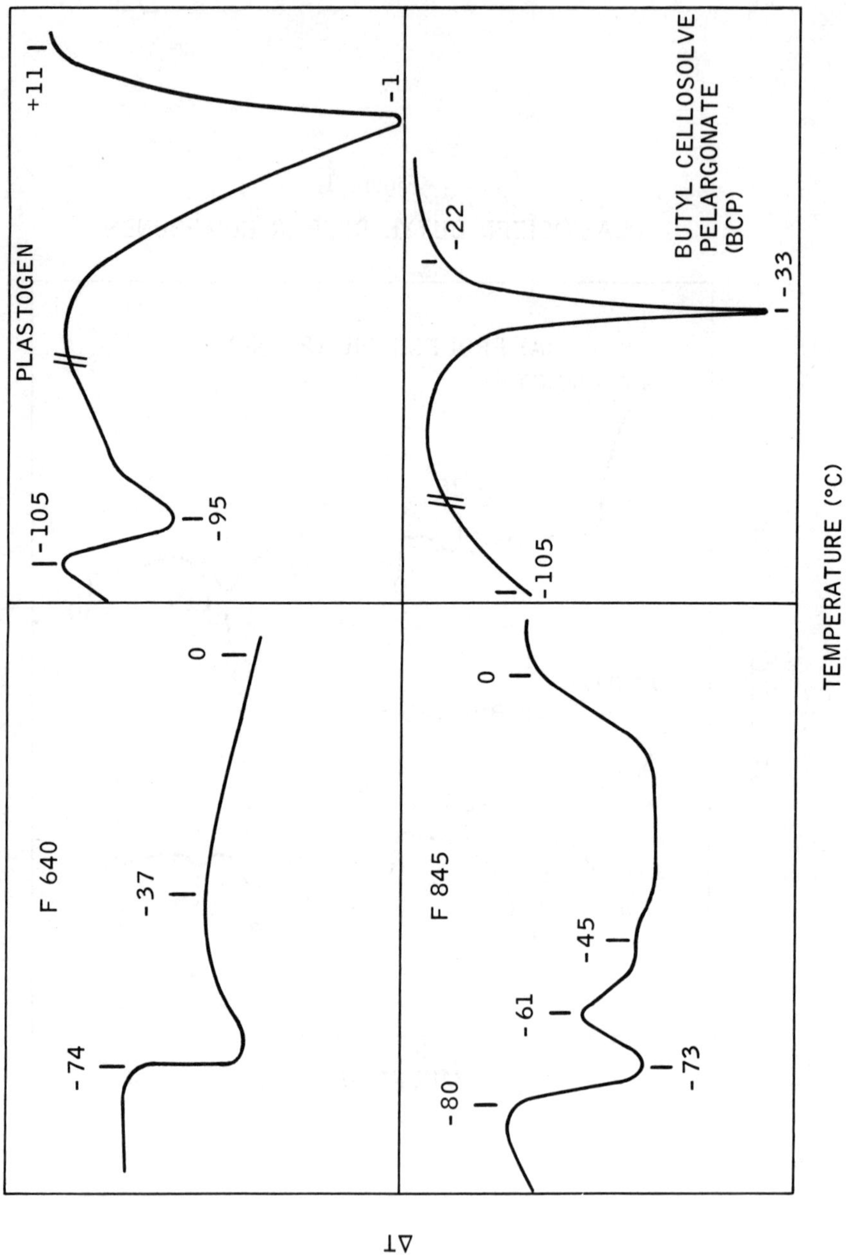

PLASTICIZED ELASTOMER SYSTEMS 65

Relationship Between "Tg" and LTE

In view of the characteristics of F845 and BCP, the low temperature endotherms of Figure 1 suggest that the glass transition regions of these compounds may reflect overlap of polymer and process oil or plasticizer i.e., in addition to (or rather than) a true Tg due to normal plasticizer action. Accordingly, we shall speak of an apparent Tg ("Tg"). Table I presents the observed relationship between "Tg" and LTE. It was encouraging to note that the best (highest) LTE values corresponded to the lowest "Tg" values as anticipated for an ideal system in which only a single transition (Tg) appeared below room temperature. However, the small range of "Tg" was unexpected in view of the magnitude of the corresponding LTE difference. Some possible reasons for this are considered below.

TABLE I

"Tg" OF BUTYL DYNAMICS COMPOUNDS

Plasticizer	% LTE	"Tg" (°C)
Flexon 845 (F845)	50	-69
1:1 mixture of F845 and Plgn	60	-71
Plastogen (Plgn)	65	-74
Butyl cellosolve pelargonate (BCP)	75	-79

Low Temperature Transitions

To provide a basis for examining the significance of the plasticizer endotherms and "Tg" values described in Table I and Figures 1 and 2, a comprehensive analysis was undertaken of the following set of samples: polymer, polymer + oil or plasticizer, accellerated dynamics compound (masterbatch), cured compound (vulcanizate). The low temperature transition behavior of these systems are summarized in Table II.

These data suggest practical applications of DTA for analyzing plasticizer content in these systems and also some thoughts about preferred plasticizer types for obtaining improved LTE. The following points are significant:

- All four plasticizers exhibit apparent melting behavior between "Tg" and 0°C, but to different degrees.

- In the Butyl rubber/plasticizer mixtures, note (a) the relationship between the -7°C LTE reference point and TM values of the plasticizers; (b) that Tg of Butyl rubber is not greatly

TABLE II

"Tg", Tm' and Tm of BUTYL COMPOUNDS (°C)

	Plasticizer			Butyl Rubber + Plasticizer			Accelerated Compound			
	"Tg"	Tm'(1)	Tm(2)	"Tg"	Tm'	Tm	"Tg"	Tm'	Tm	LTE
F845	-80	-10	0	-68	-14	0	-69	-20	0	50
Plgn + F845	-94, -78	-4	+7	-73	-10	+1	-71	-13	0	60
Plgn	-105	-1	+1	-75	-5	0	-74	-10	+5	65
BCP	--	-33	-22	-82	-39	-22	-79	-41	-25	75
Butyl Rubber	--	--	--	-66	--	--	--	--	--	--

(1) endotherm minimum; (2) endotherm termination.

lowered by Plastogen, which has a "Tg" of -105°, but it is
lowered to the greatest extent by BCP in which no transitions
were detected between -100°C and Tm. This difference may indicate (1) that BCP can "solvate" Butyl Rubber to a greater
degree than is possible with Plastogen, (2) that the "Tg" in
BCP systems is due more to pure plasticizer than to polymer +
plasticizer, or (3) that the size and distribution of Plastogen
crystals influence the free volume of the system to a different
degree than do those of BCP.

- The most significant "Tg" value is that of the systems containing BCP since, as noted above, it may reflect true plasticization of butyl rubber.

- There is little or no apparent influence of the carbon black, or of mixing conditions on "Tg".

Practical Significance of DTA Study

The studies described in Table II suggest that, in the Flexon 845, Plastogen, and Flexon 845 + Plastogen systems, some of the plasticizer exists in the crystalline state at the "low" temperature at which LTE is evaluated (-7°C). However, BCP does not appear to exhibit a detectable level of crystallinity at this temperature. This suggests that the efficiency of the former three "low temperature plasticizers" may be significantly reduced as a result of crystallization. The degree of this effect is presently unknown, but it will certainly depend upon several factors, among which are: (a) the degree of polymer-plasticizer interaction, (b) the inherent crystallizability of the plasticizer, (c) the presence of ingredients which can act as nucleating agents for the plasticizer and thus influence its degree of crystallinity, and (d) the rate at which the sample is cooled to the test temperature and the time that it is held at -7°C prior to evaluation of LTE. The observations mentioned above suggest that plasticizers with low Tm' and Tm values would be attractive candidates for improving LTE in these compounds. However, inspection of data for other ester plasticizers[2] and of the reported LTE values for a different compound containing plasticizers whose Tm' and Tm values lie below -7°C shows that these criteria are apparently not sufficient conditions for achieving improved LTE in a dynamics compound[3].

As mentioned above, the degree of polymer-plasticizer interaction is also a factor to be considered when looking for improved plasticizers. Thus, the low "Tg" of the systems containing BCP suggests that this plasticizer may exhibit a higher degree of interaction with the polymer than do the other three plasticizer systems in this study. The high LTE value of this system may thus reflect the fact that BCP has two desirable properties: (1) a relatively

Figure 3
INFLUENCE OF CARBON BLACK LEVEL ON PLASTICIZER ENDOTHERM

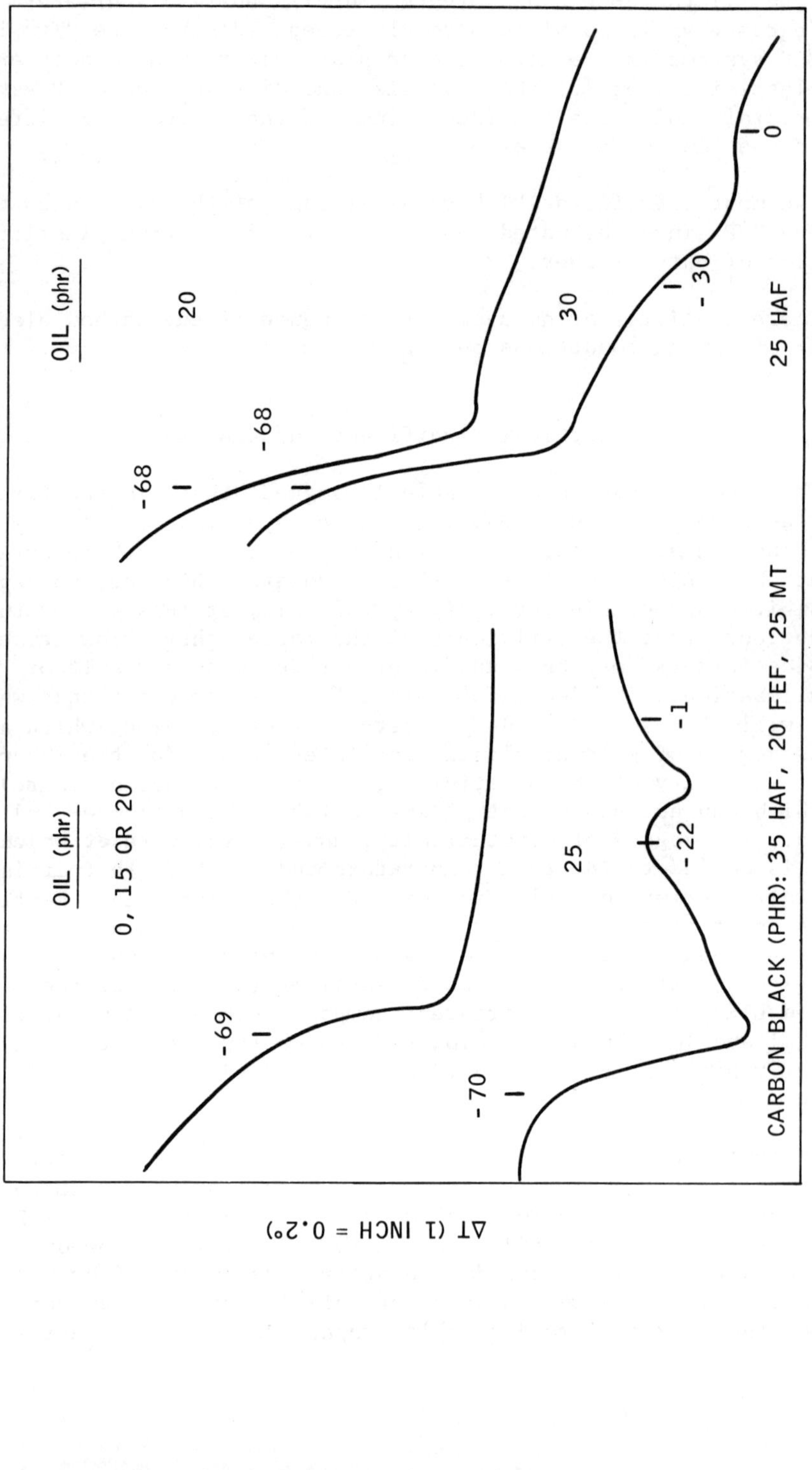

high degree of interaction with butyl, and (2) a Tm value considerably lower than -7°C. Another approach suggested by these observations is that a small amount of BCP, or some other plasticizer <u>with a similar ability to lower "Tg"</u> of Butyl rubber could be used in combination with a larger amount of a less expensive plasticizer.

A survey of the literature uncovered a previous study performed by Henderson and McLeod[4] which demonstrated that as proposed above, plasticizer solubility and Tm do appear to be significant factors in determining the low-temperature properties of plasticized styrene-butadiene copolymers. A simple ASTM test (D-471-49) was used for measuring polymer-solvent interaction in their systems, and may be applicable to other polymers as well. (It should be noted however that misleading results may be obtained for polymers such as some EP types because of low temperature crystallization of the polymer--thus reducing its swelling characteristics from those anticipated for amorphous polymers.

Factors Influencing Plasticizer Endotherms

The detection of the plasticizer endotherms in the compounds cited above suggested that quantitative analyses of the plasticizer content might be feasible. A first test of this idea consisted of examining a series of F845 concentrations in the butyl dynamics compound. As shown on the left side of Figure 3, no endotherm was detected at 20 phr or less; it appeared only at the 25 phr level. This suggested a threshold concentration which might represent the point where "free" or "excess" plasticizer was present (i.e., below this point the plasticizer was associated with solvated polymer and/or filler). A test of this hypothesis is indicated in the right hand side of Figure 3 which represents an F845 concentration study in a different compound containing a lower level of a single carbon black. Here again, there is very little evidence of a plasticizer endotherm at 20 phr, but a pronounced endotherm at 25 phr. The data of Figure 3 suggest that the threshold effect is associated principally with the polymer/F845 ratio. However, the endotherm may be more intense in the more highly loaded system. This may be an effect of the filler on the excess plasticizer, or may indicate some polymer-filler interaction, the degree of which becomes more noticeable at increased polymer/oil ratio.

The final part of the investigation of this subject was a test of the proposal that polymer/oil interaction was the controlling factor in the intensity of the plasticizer endotherm. Comparison of different F845 concentration in three "amorphous" polymers indicated (Figure 4) that this oil endotherm is substantially less evident in butyl rubber. This agrees with the known fact that F845 is a preferred plasticizer for butyl; whereas other types are used

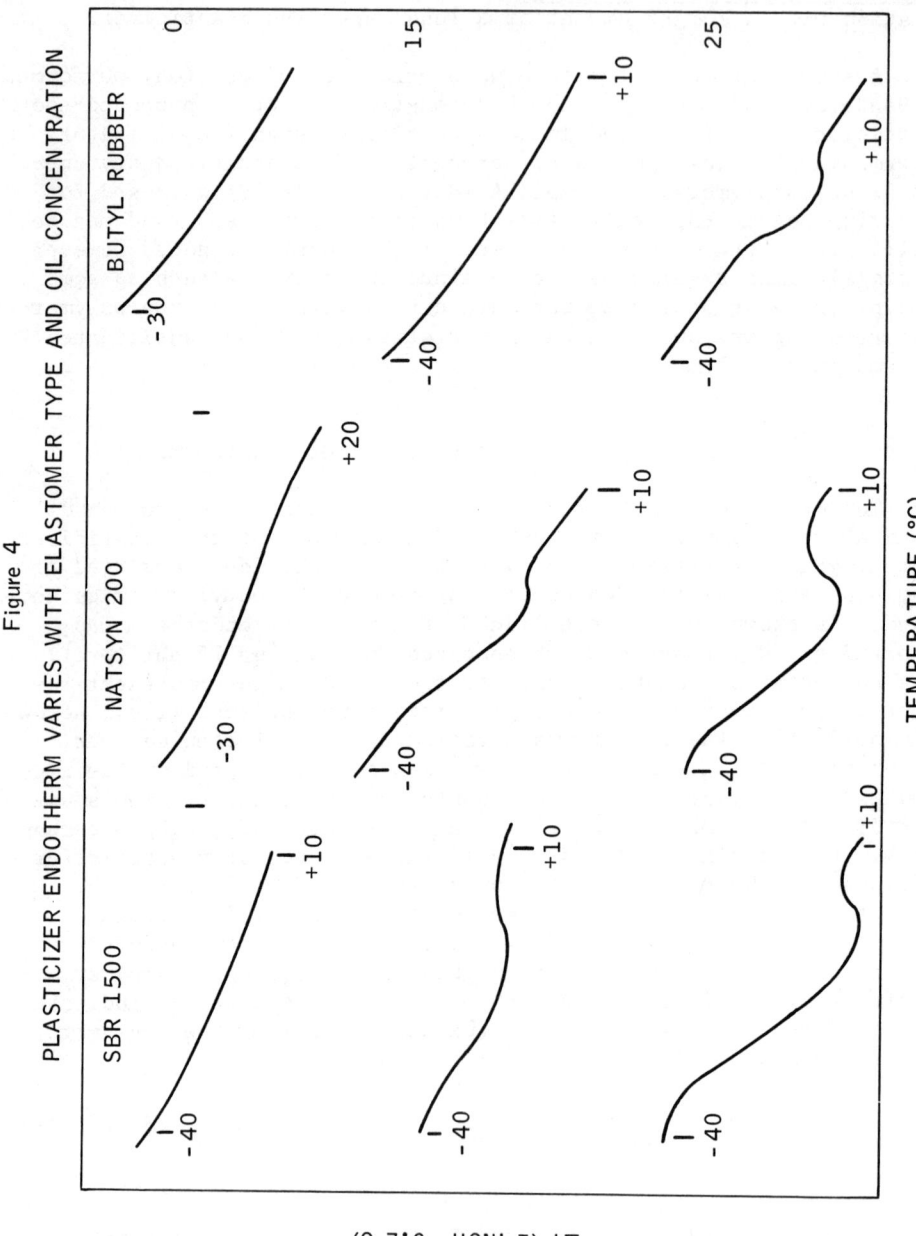

Figure 4
PLASTICIZER ENDOTHERM VARIES WITH ELASTOMER TYPE AND OIL CONCENTRATION

in SBR and Natsyn. This preliminary study suggests a utility for DTA in a variety of polymer/solvent interaction studies.

Consideration of Quantitative Analyses of Plasticizer Endotherms

Several factors which can influence the intensity of these endotherms have been cited above; others are sample uniformity, cooling rate and heating rate. The latter was employed in an analysis of the plasticized dynamics compounds, primarily for the purpose of intensifying the relatively weak F845 endotherm. For a given system, the optimum rate will be a function of the plasticizer type and level, as well as the other components in the formulation. Examples of the thermograms are shown in Figure 5. Area measurements of three different compounds (triplicate DTA determinations of a single sample, in each case) are shown in Table III.

TABLE III

QUANTITATIVE ANALYSIS OF PLASTICIZER ENDOTHERMS

Plasticizer (25 phr)	Area (cm^2/mg) [a]
A (F845)	0.047, 0.046, 0.049
B (Plgn)	0.057, 0.054, 0.057
C (A+B, 1:1)	0.076, 0.074, 0.068

(a) Average of four measurements.

The reproducibility is good for systems A and B, and may be acceptable for system C. In the latter case the cyclic cooling and heating of the sample may have altered the sample (forced some of the plasticizer "out" of the system). Replicates involving fresh samples would be useful to answer this question. The large area of C compared to A and B is also interesting. Possibly one plasticizer displaces the other, thus leading to more crystallization of the latter; DTA may thus be a way to detect and study such events.

Comparison of DTA and DSC

Since in principle Differential Scanning Calorimetry (DSC) should be the preferred technique for quantitative analyses, a comparison was made of DSC and DTA, using an SBR 1500/F845 (25 phr) compound which exhibited a large endotherm (DTA). It was noted

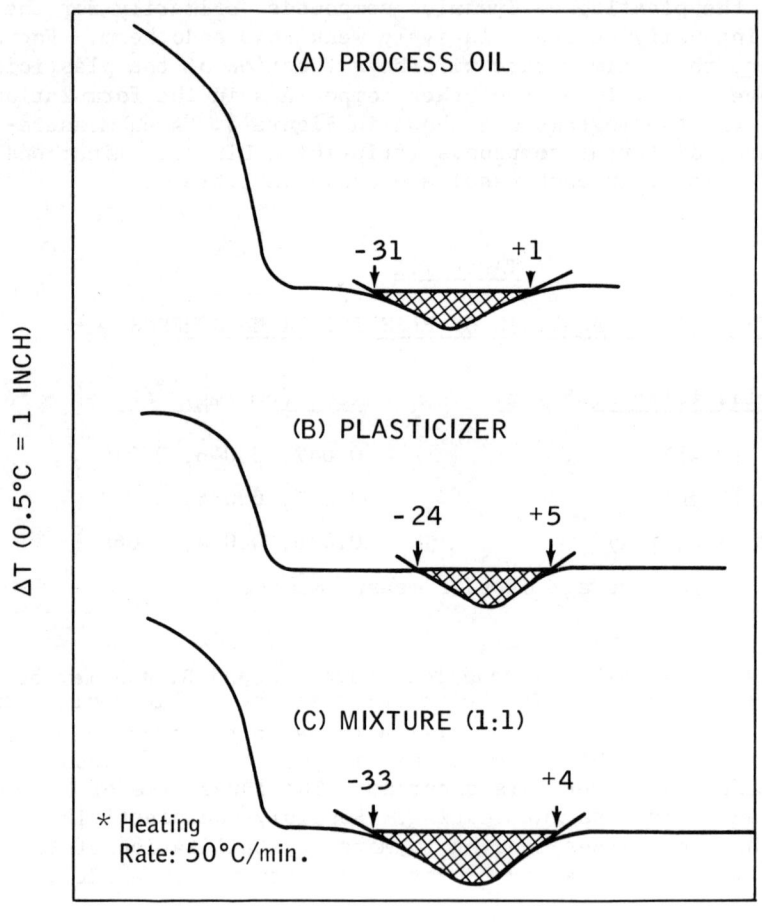

Figure 5
PLASTICIZER ENDOTHERMS IN BUTYL RUBBER VULCANIZATES*

PLASTICIZED ELASTOMER SYSTEMS 73

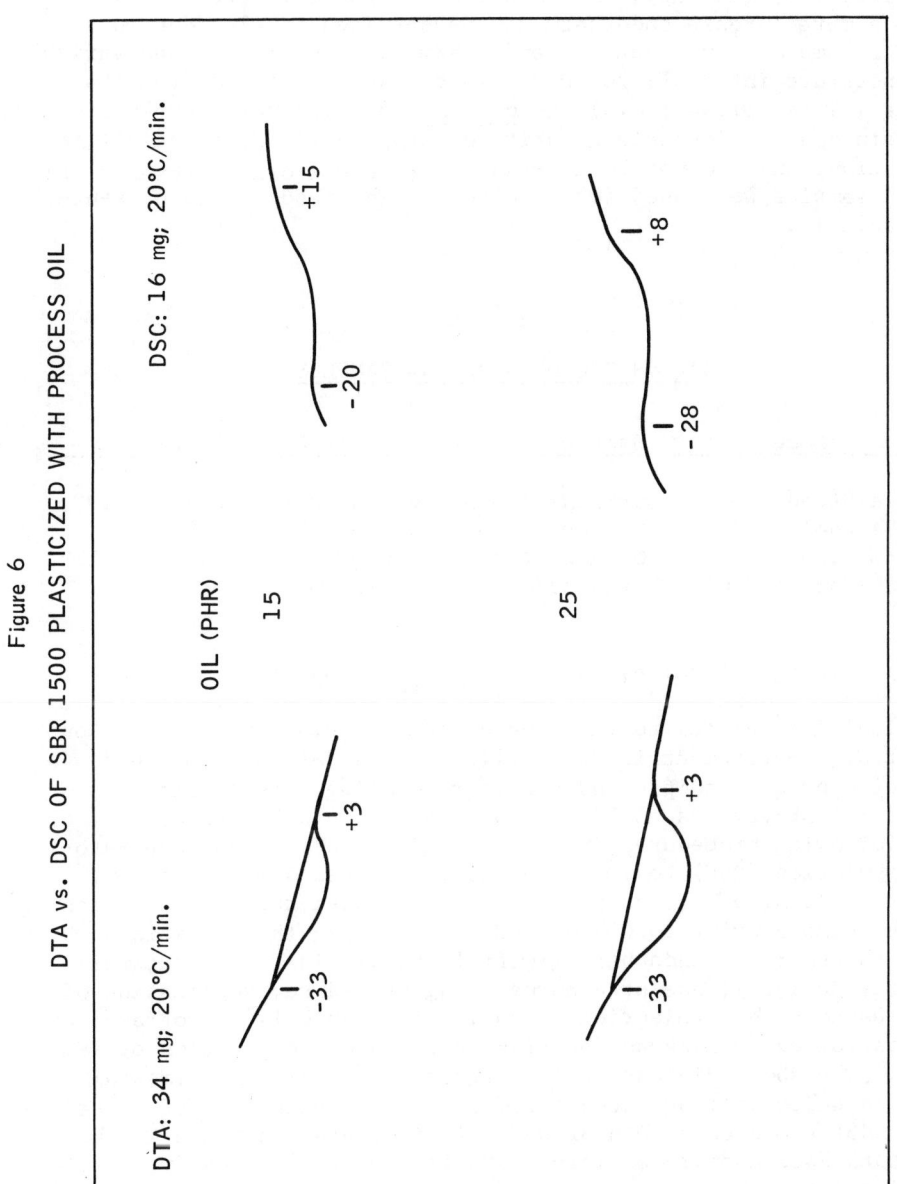

Figure 6
DTA vs. DSC OF SBR 1500 PLASTICIZED WITH PROCESS OIL

that a small sample size was required in DSC to avoid a large, positive shift in the temperature region where the endotherm occurred (thermal lag). A comparison of the two techniques shows that a more intense endotherm can be obtained in DTA (due to a larger sample); and that the endotherm is more diffuse and variable (temperature interval) for DSC. This system differed from that of Table III as follows: (a) the <u>original</u> sample contained no fillers, contained, a considerable quantity of trapped air from the milling operation, and had not been pressed into a uniform thickness; (b) fresh samples were used for each test. The results are presented in Table IV.

TABLE IV

DTA VS DSC OF SBR/F845 SAMPLES

Mode	Sample Type/mg	Areas (cm^2/mg)
DTA-Standard	Original/33.5	0.118, 0.137, 0.113, 0.093
DTA-Cool 10°/min	Pressed/33.5	0.093, 0.107, 0.098
DSC-Standard	Original/15.2	0.098, 0.093, 0.072, 0.119
DSC-Cool 10°/min	Pressed/15.1	0.120, 0.117, 0.120

The range of values for the <u>original</u> samples is much larger than that demonstrated in Table III. This was believed to reflect difficulties related to the type of sample being evaluated. Considerable improvement in the quality of the data was obtained via use of a pressed sample and a uniform cooling rate (Table IV). Further improvements could probably be obtained by one or more of the following procedures (which are under study in our laboratories); encapsulation (DSC) to improve contact between sample and pan; dilution (DTA) of the sample with glass beads; thermal conditioning of the sample prior to DTA or DSC; determining the conditions which give the sharpest endotherm (initial and terminal temperatures clearly defined); use of a normalizing procedure; replication of samples to enable rejection of obviously atypical thermograms; use of controlled cooling and heating rates, and manipulation of sample size. The above discussion illustrates the basic experimental factors which will be encountered in the determination of plasticizer endotherm area by DTA or DSC. As indicated in Table IV the approach does appear feasible, but will apparently require careful development work for each system of interest.

REFERENCES

1. Maurer, J. J., Rubber Chem. and Tech., $\underline{42}$, 110 (1969).

2. "Materials and Compounding Ingredients for Rubber and Plastics", 1965, Publishers Printing Co., Louisville, Kentucky.

3. Enjay Polymer Laboratories, unpublished data.

4. Henderson, D. A., and McLeod, L. A., Rubber Chem. and Tech., $\underline{28}$, 557 (1955).

THERMODYNAMICAL AND STATISTICAL MECHANICAL
IMPLICATIONS ASSOCIATED WITH GRÜNEISEN
RATIOS CALCULATED BY VARIOUS FORMULATIONS

S.R. Urzendowski and A.H. Guenther

Air Force Weapons Laboratory

Kirtland Air Force Base

Albuquerque, New Mexico

INTRODUCTION

The direct relationship between the thermal expansion of a solid material to the motion of the atoms in the interchain potential of the crystal has been the topic of much discussion (1-10). New theoretical approximations were initiated with the theories of Barron (11) and Blackman (12) who used Born-Von Karman lattice dynamics to generalize the Grüneisen theory on thermal expansion. Barron (11) has shown that the calculation of the Grüneisen ratio acts as a useful link between experiment and theory in the consideration of the temperature variation of the frequencies of vibration in solid state dynamics. Childs (13), Collins and White (14) and Vernon and Weintroub (15) have illustrated the relationship of the Grüneisen ratio to the anisotropy of many noncubic materials. Wada, et. al. (10) pointed out that the Grüneisen ratio is a means of determining the interchain specific heat, C_{int}, and the mass, m*, of a segmented unit in the original lattice.

Investigations by Slater (8), Dugdale and MacDonald (16) and others (2, 5, 17) have explained differences in reported Grüneisen ratios by means of the finite strain theory which determines the energy of an atom as a function of lattice parameters.

In this study attempts were made to associate the value of the segmented unit mass with the Debye characteristic temperature, θ_D, calculated from ultrasonic velocity data. High temperature heat capacity and expansivity data were also correlated with proper Grüneisen values for appropriate temperature intervals. Attempts were also made to discuss related thermodynamical and statistical mechanical implications associated with various γ values calculated by the various formulations.

THEORY

The theoretical basis of the equation of state and the Mie-Grüneisen equation of state is well-known (2-10). The discussion of the results of the Grüneisen parameter, γ, are usually given in terms of the thermal expansion by

$$\gamma \cdot C_V/V = \beta/X^T \tag{1}$$

where β is the volume coefficient of expansion, and C_V, V, and X^T are respectively the heat capacity at constant volume, the molar volume and the isothermal compressibility.

From classical thermodynamics it follows that

$$\frac{-\partial^2 F}{\partial V \partial T} = \left(\frac{\partial P}{\partial T}\right)_V = \frac{1}{V}\left(\frac{\partial V}{\partial T}\right)_P \quad \frac{1}{V}\left(\frac{\partial V}{\partial P}\right)_T = \beta/X^T \tag{2}$$

where F is the Helmholtz free energy. In the quasi-harmonic approximation the lattice composed of N atoms is generally represented by a system of 3N loosely-coupled harmonic oscillations. The free energy of the system is

$$F(V,T) = U(V) + \sum_{i=1}^{3N} \frac{1}{2} h\nu_i (V) + kT \sum_{i=1}^{3N} \ln(1-e^{-h\nu_i/kT}) \tag{3}$$

where the first, second and third terms to the right of the equation refer to the lattice, zero point and thermal energy, respectively. To express the effect of change in volume (or lattice constant) on the frequencies ν_i, it is convenient to define ν_i for each normal mode by

$$\gamma_i = -\frac{V}{\nu_i}\frac{d\nu_i}{dV} = -\frac{d \ln \nu_i}{d \ln V} \tag{4}$$

and γ_i is positive if the lattice expands upon heating.

If one uses equation (3) for the free energy and assumes that the ν_i are functions of volume only, and that for a given change in volume the fractional change in ν_i is the same, i.e., all ν_i are equal to each other and have a value, γ, which leads to equation (1) which is the value defined by Grüneisen (6).

Barron (11), Blackman (12) and others (8, 16, 17), have pointed out that 3N values of γ_i are not all equal and γ is therefore a weighed average of γ_i values. If this is true, the force energy term (Equation (3)), defined only the free energy which arises from lattice vibrations and additional terms as electronic, F_e, and magnetic, F_m, should be added. Equation (1) should now be represented as

$$\beta/X^T = (1/V)(\gamma_L C_L + \gamma_e C_e + \gamma_m C_m) \quad (5)$$

where C_L, C_e, and C_m represent the contributions to the specific heat made by the lattice, the free electrons, and the magnetic interactions, respectively. Since no magnetic contributions are expected for aluminum, gold and the polymeric materials studied, the last term of Equation (5) may be eliminated so that only lattice and electronic contributions to the specific heat and the Grüneisen ratio are important.

It can be shown that Equation (1) is related to the heat capacity at constant pressure, C_p, by $\beta \cdot V/X^S = \gamma \cdot C_p$ where X^S now represents the adiabatic compressibility.

The γ of Equation (1) may be defined as the low temperature, γ_D, value when $T \ll \theta$ and the specific heat obeys the T^3 law. For this temperature region the lattice may be treated as an elastic continuum with a Debye characteristic temperature, $\theta_i = h\nu_i/k$, assigned to each mode. Equation (1) may now be defined by

$$\gamma_D = - \frac{d \ln \nu_{max}}{d \ln V} \quad (6)$$

where ν_{max} is Debye's limiting frequency. The ν_{max} and θ values may be obtained from ultrasonic velocity data by

$$\nu_D = \nu_{max} = \left[\{9N/4\pi V\}^{1/3} \{1/c_\ell^3 + 2/c_t^3\}^{-1/3} \right] \quad (7)$$

where C_ℓ and C_t refer to the longitudinal and transverse velocities, respectively.

Slater (8) has shown that for a Debye solid in which Poisson's ratio, σ, is independent of volume, the Grüneisen ratio, γ_S, is defined as

$$\gamma_S = -1/6 + 1/2 \left[\frac{d \ln X^T}{d \ln V}\right]_T = -1/6 - 1/2 \left[\frac{d \ln B^T}{d \ln V}\right] \quad (8)$$

wherein the isothermal compressibility, X^T, and the bulk modulus, B^T, could be obtained ultrasonically and γ_S could be compared to γ_D. The corresponding frequency, ν_S, may be obtained from

$$\nu_S = \{9N/4\pi V\}^{1/3} \{3B^T/\rho\}^{1/2} \qquad (9)$$

where V and ρ refer to the volume and density, respectively.

Since the assumption of a constant Poisson's ratio is hardly justifiable for a real solid under hydrostatic stress, Barker (17), Slater (8) and Sheard (3) have shown that the Debye frequency, ν_D, (Equation (7)) may now be defined by

$$\nu_\sigma = \left[9N/4\pi V\right]^{1/3} \left[3B^T/\rho\right]^{1/2} \cdot f(\sigma) \qquad (10)$$

where $f(\sigma)$ is given by

$$f(\sigma) = \left[\left(\frac{1+\sigma}{1-\sigma}\right)^{5/2} + 2^{5/2}\left(\frac{1+\sigma}{1-\sigma}\right)^{3/2}\right]^{-1/3} \qquad (11)$$

The Grüneisen ratio for this Debye frequency (Equation (10)) is

$$\gamma_\sigma = -\frac{1}{6} - \frac{1}{2}\left(\frac{d \ln B^T}{d \ln V}\right) - \left(\frac{d \ln f(\sigma)}{d \ln V}\right) \qquad (12)$$

which varies from γ_S as Poisson's ratio varies with volume.

Since it can be shown (18) that longitudinal and transverse velocities may be defined in terms of compressibility data, Equations (12) and (7) are the same.

The frequency spectrum for this three-dimensional continuum (18) is the explicit low frequency form defined by Debye (19, 20) and is said to represent the interchain vibrations of polymeric materials. Interchange vibrations are generally defined by the one-dimensional Debye function which exists in the classical specific heat temperature region when $C_V = 3R$.

If γ varies with temperature, equation (1) may be represented as the thermal Grüneisen ratio, γ_G, which varies with each heat increment. At high temperatures γ (Equation (1)) again approaches a constant value γ_∞ which defines the classical specific heat region.

EXPERIMENTAL TECHNIQUE

Thermal Expansion. A DuPont 490 thermomechanical analyzer was used to measure the linear coefficient of thermal expansion, α, from -100 to 200°C. The apparatus includes a simplified form of a dilatometer which makes use of a linear variable differential transformer and is adaptable to automatic recording of length changes of a specimen versus temperature. Details of the experimental procedure were previously described (18-21, 22). Corrected values of length changes, $\Delta \ell/\ell$, adjusted to the adopted reference temperature 25.0°C were used to calculate the linear coefficient of expansion as a function of temperature. Volume coefficients of expansion, ($\beta = 3\alpha$), were calculated with the assumption that all polymers studied were isotropic. After proper application of chromel-alumel thermocouple corrections, individual temperature determinations agreed to within 0.2°C and to within 2.0 to 4.0°C of values reported in the literature. The total probable error for the expansion measurements calculated as the square root of the summation of all errors associated with each component of the instrument was ± 2.5%.

Heat Capacity. The latent heat of phase transitions and heat capacity data were obtained with an appropriately calibrated differential scanning calorimetric cell (DuPont DSC Cell No. 900600). Thermal data obtained by this technique supplement the qualitative data obtained with the differential thermal analyzer and cover the temperature range -100 to 600°C. The present limit of accuracy of the C_p data reported is ± 3.0%.

Ultrasonic Sound Velocities. Sound velocity measurements of longitudinal and shear waves at approximately 1 and 3 Mc/sec., from 0 to 120°C were made at this laboratory by Asay, et al., (18, 22, 23) and were used to calculate the adiabatic bulk modulus, B^s and Poisson's ratio, σ. Longitudinal and shear velocities were obtained by measuring the transit times through various sample thicknesses and performing a least-squares plot to obtain the slope. For most of the velocity-temperature data a quadratic function was found to best fit the data to an accuracy of approximately 1%.

Materials Studied. The solid polymeric materials studies were polystyrene, density 1.046 g/cm^3 (RFQ 7039-1030-Cadco Corporation), cast nylon 6, density 1.15 g/cm^3 (CN 1100-Cadco Corporation), Teflon #1, density 2.20 g/cm^3 (E.I. DuPont de Nemours Co.), and tapewound silicone phenolic (TWSP), density 1.612 g/cm^3 (AVCO Corp), Air Force Weapons Laboratory, Kirtland, New Mexico).

In addition to the polymeric materials two metals, aluminum 1060, density 2.703 g/cm^3 (99.6 aluminum) and gold, density 18.96 g/cm^3 (99.0% gold) were included for comparative purposes.

RESULTS AND DISCUSSION

Linear, α, and volume, β, coefficient of expansion data for the polymeric materials are presented in Tables 1 and 2. The volume expansivity data for Al 1060 and gold, respectively are represented by the following linear least square equations which define $\alpha = a + bT$ from 0° to 200°C by:

Al 1060, $\alpha = 2.09 \times 10^{-5} + 3.00 \times 10^{-8}$ T, °C

Gold, $\alpha = 1.35 \times 10^{-5} + 0.56 \times 10^{-9}$ T, °C

Volume variations were calculated from the expansivity data by $V_T = V_o (1 + \beta T)$ where V_o is the volume at 273°K.

Examination of the expansivity graphs (Figures 1 and 2) indicates well-defined transitional regions at 90°, 48°, and 20°C, respectively, for polystyrene, cast nylon 6 and Teflon #1. Although the cast nylon and Teflon samples are said to have significant property advantages over conventional resins, the mode of preparation did not seem to shift the typical transitional temperature regions.

For the TWSP sample linear thermal expansion measurements were made in two directions, horizontal and vertical to the weave pattern (Figure 2). Both the vertical and horizontal expansivities show a gradual change in slope at approximately 60 to 80°C but at temperatures below this region, the expansion is virtually linear. Since the α values were low, an optical interference method was used to verify the data. The specimen was set between two flat (1/5 wavelength) quartz plates and normal illumination by monochromatic light produced interference fringes which moved past a fiducial point as the specimen expanded. The two methods agreed to within 2.0%.

Heat capacity data in the form of linear equations are given in Tables 3 and 4. For the polymeric materials the equations define all transitional regions and negative values are the result of expansion of the data about 0°C. The lower heat capacity data obtained for the tapewound silicone phenolic resin is associated with the binding power resulting from the fiber weave pattern. Although no definite transitions were noted for TWSP, a marked change in slope was observed on the DTA thermogram at approximately 75°C which may be attributed to the phenol melt within the polymer. The typical transitions were again evident but to a lesser degree than was observed by the more sensitive expansivity measurements.

The isothermal bulk modulus data, B^T, tabulated in Table 5 was obtained from data reported by Asay et al., (22, 23). The

Table I Volume Coefficient of Expansion Versus Temperature Data for the Various Materials Studied ($\beta \times 10^4$, in/in °C)

T, °C	Polystyrene	Nylon 6	Teflon #1
-100	1.49	1.73	2.09
-90	1.73	1.82	2.33
-80	1.82	1.84	2.52
-70	1.84	2.03	2.43
-60	1.82	1.98	2.46
-50	1.81	1.98	2.66
-40	1.82	2.08	2.56
-30	1.91	2.13	2.65
-20	2.03	2.15	2.73
-10	2.28	2.21	3.11
0	2.36	2.42	3.12
10	2.14	2.50	3.35
20	2.11	2.49	6.21
30	2.22	2.68	2.57
40	2.27	3.17	2.39
50	2.26	3.31	2.40
60	2.27	2.61	3.39
70	2.36	2.43	3.97
80	2.50	1.19	6.27
90	2.71	1.36	6.82
100	3.67	1.22	7.27
110	8.69	1.76	7.75

Table II Linear Coefficient of Expansion Data Parallel and Perpendicular to the Weave for Tapewound Silicone Phenolic ($\alpha = a + b T$) in/in °C

Temp. Range, °C	a $\times 10^6$	b $\times 10^8$	Avg. Error $\times 10^7$
Parallel to the Weave:			
-100 to 40	7.16	0.28	11.13
40 to 200	12.03	-10.02	8.59
Perpendicular to the Weave:			
-100 to 50	14.80	2.38	7.29
60 to 200	24.99	-15.50	7.22

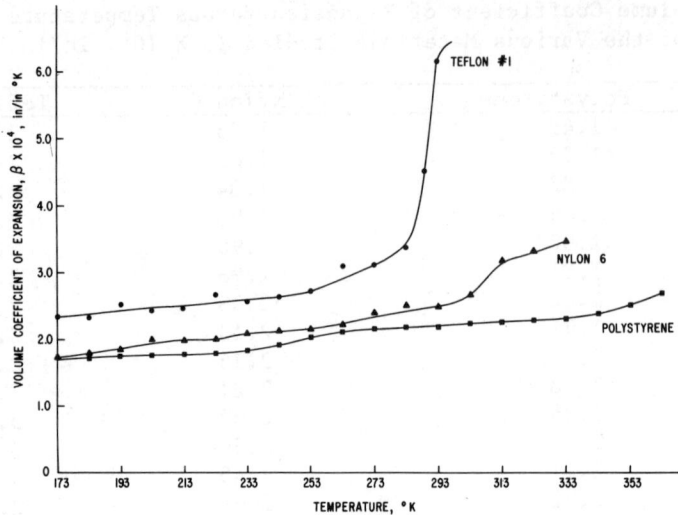

Figure 1. Volume Coefficient of Expansion Versus Temperature for Nylon 6, Teflon #1 and Polystyrene.

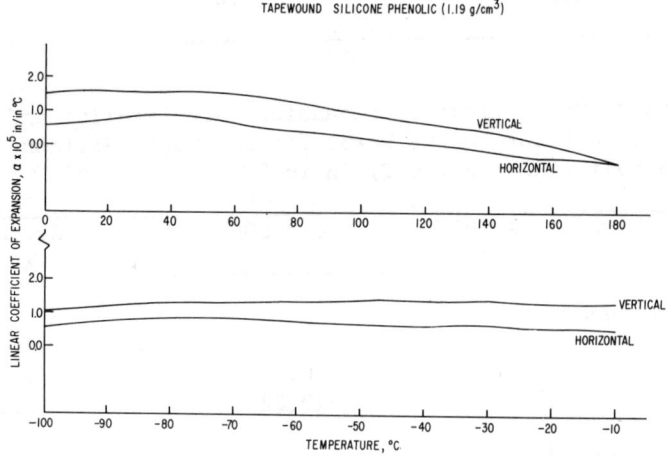

Figure 2. Linear Coefficient of Expansion Versus Temperature for Tapewound Silicone Phenolic (Parallel and Perpendicular to the Weave).

adiabatic moduli were calculated from ultrasonically determined longitudinal and transverse velocities, c_ℓ and c_t by

$$B^S = \rho(c_\ell^2 - 4/3\, c_t^2) \tag{13}$$

and the corresponding isothermal conversion was effected by

$$B^T = B^S/(1+ \beta\gamma_G T) \tag{14}$$

where γ_G is the thermal Grüneisen equation defined by Equation (1).

The experimentally determined C_p data, Tables 3 and 4 was converted to the corresponding C_V data by

$$C_V = \left[C_p - \beta^2 \cdot B^T \cdot T/\rho\right] \tag{15}$$

where ρ is the density and the other terms have their usual significance. Tables 6 and 7 illustrate the differences in the Gruneisen ratio as calculated from the thermal equation (Equation (1)) and from the ultrasonic formulations defined in the previous section. For these calculations the basic thermal and ultrasonic data was linearly extrapolated through the pronounced transitional regions observed in the polymers. The corresponding frequencies and characteristic temperatures are also presented. The ν_S and ν_D values were determined by the proper substitution of ultrasonic and thermal data into Equations (9) and (7). Once the lattice frequencies were obtained, the corresponding θ values were calculated from the relationship $\theta = h\nu_m/k$ where ν_m refers to the appropriately defined frequency. The variation of ν_D with temperature is illustrated in Figure 3. Polystyrene, Teflon #1 and nylon showed the greatest temperature dependence.

The specific longitudinal and transverse ultrasonically determined characteristic temperatures, θ_ℓ and θ_t, respectively as well as θ_{BV} (Tables 6 and 7) were obtained from the modification of Equation (7) based on Born-von Karman (24) lattice dynamics. These calculations gave the maximum frequencies based on a common minimum wavelength, λ_{min}, defined as

$$\lambda_{min} = \left\{4\pi V/3N\right\}^{1/3} = 2r \tag{16}$$

where r defines the nearest neighbor distance. The corresponding longitudinal and transverse frequencies are given by $\nu_\ell = c_\ell (3N/4\pi V)^{1/3}$ and $\nu_t = c_t (3N/4\pi V)^{1/3}$. The Grüneisen ratio defined for the Born-von Karman theory was calculated from the relationship

$$\gamma_{BV} = 1/3\gamma_\ell + 2/3\, \gamma_t \tag{17}$$

Table III Polynomial Fit for Heat Capacity for Polystyrene,
Nylon 6 and Tapewound Silicone Phenolic

Temp. Range, °C	a, cal/g°C	b X 10^3, cal/g°C	Avg. Error C_p X 10^2
Polystyrene:			
-100 to + 100	0.2623	1.27	0.45
115 to 250	0.3855	0.70	0.75
Nylon 6 (Cast):			
-100 to 40	0.3230	1.41	0.30
50 to 210	0.3175	2.22	0.59
210 to 232	-5.9123	31.55	9.50
232 to 250	10.9999	-41.27	7.49
260 to 300	1.8793	-4.82	706.67
Tapewound Silicone Phenolic:			
-100 to 30	0.1899	1.02	0.43
30 to 80	0.1176	3.38	0.83
80 to 225	0.4709	0.04	3.41
285 to 400	0.3776	-0.05	1.46

The equations define all transitions present in the polymers.
Negative values are the result of extrapolating about 0°C and do
not imply negative heat capacity data.

Table IV Heat Capacity Data for Teflon #1, Aluminum 1060 and Gold.
(C_p, cal/g°K)

Temp., °K	Teflon #1	Alumunim 1060	Gold
273	0.2597	0.0301	0.2005
283	0.2662	0.0302	0.2020
293	0.2699	0.0303	0.2025
303	0.3044	0.0304	0.2030
313	0.2777	0.0305	0.2034
323	0.2584	0.0306	0.2040
333	0.2559	0.0308	0.2045
343	0.2534	0.0310	0.2050
353	0.2508	0.0312	0.2055
363	0.2483	0.0320	0.2060
373	0.2457	0.0321	0.2070

Table V Isothermal Bulk Moduli, B^T, for the Various Materials Studied ($B^T \times 10^{-10}$, dyne/cm^2)

T,°C	Polystyrene	Nylon 6	Teflon #1	TWSP	Al 1060	Gold
0	3.96	7.05	3.67	8.57	74.17	1.66
10	3.93	6.83	3.51	8.56	73.96	1.66
20	3.99	6.61	3.36	8.55	73.75	1.66
30	3.98	6.39	3.26	8.53	73.54	1.65
40	3.99	6.15	3.03	8.52	73.32	1.65
50	4.02	6.00	2.91	8.52	73.11	1.65
60	4.03	5.90	2.80	8.49	72.90	1.64
70	4.03	5.78	2.71	8.48	72.67	1.64

Table VI Ultrasonically Determined Frequencies, Characteristic Temperatures and Grüneisen Ratio's for Polystyrene, Nylon 6, and Teflon #1 (T = 293°K)

	Polystyrene	Nylon 6	Teflon #1
ν_D (10^{-12} Hz)	1.32	1.46	1.06
ν_S (10^{-12} Hz)	5.62	6.82	5.84
ν_σ (10^{-12} Hz)	1.29	1.41	0.98
ν_ℓ (10^{-12} Hz)	3.78	4.48	3.76
ν_t (10^{-12} Hz)	1.69	1.86	1.26
θ_D, (°K)	63.26	70.18	47.84
θ_S, (°K)	269.49	326.65	280.20
θ_σ, (°K)	61.83	67.58	46.81
θ_ℓ, (°K)	181.45	215.03	176.30
θ_t, (°K)	80.85	89.47	60.68
θ_{BV}, (°K)	114.39	131.31	99.22
γ_D	3.20	3.88	5.12
γ_S	2.02	3.12	4.00
γ_σ	3.35	3.49	5.32
γ_ℓ	4.01	3.30	6.41
γ_t	3.01	4.24	4.91
γ_{BV}	3.35	3.91	5.39
γ_G	0.70	1.03	0.48

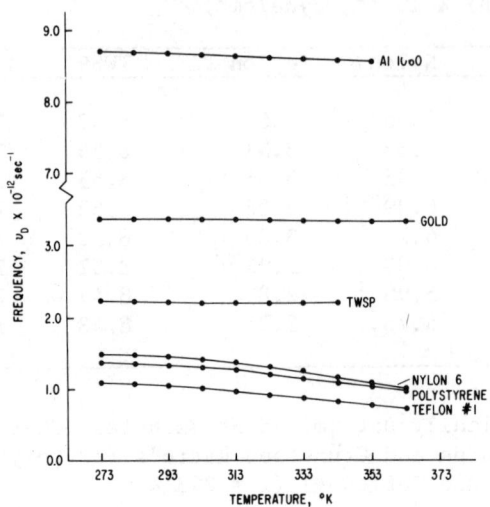

Figure 3. Temperature Variation of the Debye Frequency for the Materials Studied.

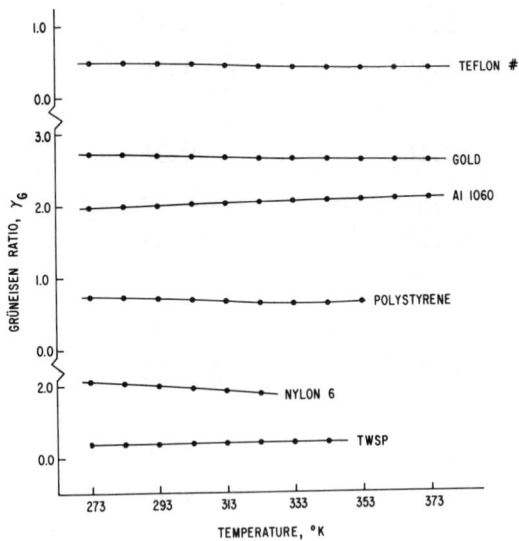

Figure 4. Variation of Grüneisen Ratio, γ_G, with Temperature.

Table VII Ultrasonically Determined Frequencies, Characteristic Temperatures and Grüneisen Ratio's for Tapewound Silicone Phenolic, Aluminum 1060 and Gold (T = 293°K)

	TWSP	Al 1060	Gold
ν_D (10^{-12} Hz)	2.19	8.66	3.36
ν_S (10^{-12} Hz)	6.80	31.84	17.83
ν_σ (10^{-12} Hz)	2.19	8.65	3.36
ν_ℓ (10^{-12} Hz)	5.12	22.43	11.42
ν_t (10^{-12} Hz)	2.84	11.13	4.28
θ_D, (°K)	105.16	415.49	161.45
θ_S, (°K)	326.48	1527.89	855.68
θ_σ, (°K)	105.07	414.93	161.27
θ_ℓ, (°K)	245.46	1076.23	547.38
θ_t, (°K)	136.17	533.93	205.18
θ_{BV}, (°K)	172.97	714.70	319.42
γ_D	4.43	3.98	3.51
γ_S	4.18	2.00	2.09
γ_σ	4.52	5.27	4.71
γ_ℓ	2.54	2.67	2.36
γ_t	4.92	4.06	3.54
γ_{BV}	3.92	3.60	3.15
γ_G	0.18	2.01	2.72

where the respective γ_ℓ and γ_t values were obtained from $\gamma_a = -d \ln \nu_a/d \ln V$ where the subscript a refers to the longitudinal or transverse modes for the room temperature region and $\theta_{BV} = 1/3\, \theta_\ell + 2/3\, \theta_t$.

Tables 6 and 7 indicate that for all the materials studied the longitudinal frequency, ν_ℓ, seems to be larger than ν_t for the transverse modes. This and the other values obtained in the tables may be explained by using the Debye characteristic temperature, θ_D, as a guide (24). Since the θ_D, ν_D and γ_D values for Al 1060 and gold agree well with previously reported values (2, 6, 8, 16), it is assumed that the ultrasonic method is accurate. Since $T \approx \theta_D$, in this temperature range the solid behaves classically so that $\gamma \approx \gamma_\infty$. However, since it was observed that temperature variations do exist for this region, this may be due to the fact that all modes have not been completely excited, or, as observed for polystyrene, anharmonic effects are setting in.

Differences in γ_D and γ_G may be associated with the optical and acoustical branches of the spectra. At low temperatures acoustical modes predominate so that γ_o is determined from elastic constants. This γ_o should be considerably accurate around room temperature where the Grüneisen ratio depends solely on the change of elastic constants with volume. At high temperatures the optical branches also contribute so that γ_{opt} could be estimated by

$$\gamma_{opt} = 2\gamma_\infty - \gamma_o \qquad (18)$$

where $\gamma_\infty \simeq \gamma_G$ and $\gamma_o \simeq \gamma_D$. Since γ_∞ is the average over the whole spectrum, this should approach our γ_G value. The temperature dependence of γ_G is illustrated in Figure 4.

The interchain contributions to the specific heat, C_{int}, and the mass, m^*, of the polymeric units are listed in Table 8. Wada (10) defines the m^* value as the mass of the segment of the polymeric chains inside which the thermal motion is predominantly determined by the intrachain potential. The value of C_{int} was estimated from the equation defined by Wada, et al., (10) where

$$\beta = \gamma_n \cdot X^T \cdot C_{int} \qquad (19)$$

and γ_n refers to the Grüneisen ratios previously defined. No heat capacity data were available for extremely low temperatures, however, it is assumed that β and X^T vary only slightly at low temperatures and that the best value for C_{int} was obtained from the γ_D values.

Applying these assumptions, a C_{int} of 4.3×10^{-2} cal/g deg and m^* of 140.1 g/mole was obtained for a crystalline sample of

Table VIII Estimation of Vibrational Heat Capacity and Segment Size from the Grüneisen Ratio and Thermal Expansion Coefficient

	Polystyrene	Nylon 6	Teflon #1	TWSP
$C_{int}\ \gamma_G$, cal/g deg	0.3250	0.3112	0.2342	0.2129
$m^*\ \gamma_G$, g mole^{-1}	18.34	19.16	25.45	28.03
$C_{int}\ \gamma_\ell$, cal/g deg	0.0480	0.0969	0.0175	0.0156
$m^*\ \gamma_\ell$, g mole^{-1}	124.23	61.50	343.70	397.43
$C_{int}\ \gamma_t$, cal/g deg	0.0639	0.0648	0.0131	0.0098
$m^*\ \gamma_t$, g mole^{-1}	93.25	98.97	459.40	721.68
$C_{int}\ \gamma_D$, cal/g deg	0.0601	0.0655	0.0223	0.0101
$m^*\ \gamma_D$, g mole	99.14	97.59	269.17	695.55
$C_{int}\ \gamma_S$, cal/g deg	0.0952	0.1025	0.0280	0.0402
$m^*\ \gamma_S$, g mole^{-1}	62.58	58.17	214.48	172.29
$C_{int}\ \gamma_{BV}$, cal/g deg	0.0574	0.0584	0.0208	0.0111
$m^*\ \gamma_{BV}$, g mole^{-1}	103.78	102.00	289.01	613.17
$C_{int}\ \gamma_\sigma$, cal/g deg	0.0574	0.0620	0.0210	0.0097
$m^*\ \gamma_\sigma$, g mole^{-1}	103.78	103.25	285.25	716.35
Actual mass of repeat unit	104.15	113.16	50.01	137.09

polyethylene ($\rho = 0.967$ g/cm^3). These values were in close agreement with values reported by Wada (10) in which low temperature specific heat data were used to calculate C_{int}. Wada's values were: $C_{int} = 4.2 \times 10^{-2}$ cal/g deg and $m^* = 148$ g/mole.

The $m^* \gamma_D$ for polystyrene and the cast nylon 6 sample seems to approach the actual mass of 104 and 113 g/mole expected for the monomeric units. For Teflon and TWSP, no definite correlations could be made. The full significance of the m^* value is not understood, however, as postulated by Wada (10), the C_{int} value combined with ultrasonic data is said to account for the fact that $\gamma_G < \gamma_D$.

CONCLUSION

The relation of the Grüneisen ratio to various properties of polymers has been discussed. The proper application of ultrasonic and thermal derivations suggest that the Grüneisen ratio may be a good tool to assist in the understanding of internal interactions in polymeric and other molecular solids. The derived formulae established a relation between the temperature dependence of the lattice vibrational frequencies and thermodynamic properties such as the thermal expansion, heat capacity, and the Grüneisen ratio. The ultrasonic room temperature derivations may be a means to estimate low temperature heat capacity data and to derive Debye temperatures for the polymers. Future measurements using low-temperature heat capacity and thermal expansivity data should provide useful information in predicting relations between harmonic and anharmonic moduli.

ACKNOWLEDGEMENTS

The authors wish to express their appreciation to Leonie D. Boehmer for the data computations.

REFERENCES

(1) G.D. Anderson, D.C. Doran, and A.L. Fabrenbruch, "Equation of State of Solids; Aluminum and Teflon," AFWL-TR-65-147, Kirtland Air Force Base, New Mexico, December, 1965.

(2) J.R. Partington, An Advanced Treatise on Physical Chemistry, The Properties of Solids, Vol III, Longmans, Green and Co., New York, N.Y., 1952.

(3) F.W. Sheard, Phil. Mag., 3, 1381 (1958).

(4) F.G. Fumi and M.P. Tosi, "On the Mie-Grüneisen and Hildebrand Approximations to the Equation of State of Cubic Solids", J. Phys. Chem. Solids, 23, 395 (1962).

(5) P.G. Grodzka, "Grüneisen Parameter Study," LMSC/HREC A 784868, HREC-7087-1, Lockheed Missiles and Space Co., Huntsville, Alabama, October 1967.

(6) E. Grüneisen, Handbuch der Physik, 10, 22 (1926); Ann. Physik, 39, 258 (1912).

(7) T. Rice, R. McQueen and J.M. Walsh, "Solid State Physics," F. Seitz and D. Turnbull, Ed., Academic Press, Inc., New York, N.Y., 6, 1958.

(8) J.C. Slater, Introduction to Chemical Physics, New York, N.Y. 1939.

(9) E. Mie. Ann Phys, II, 657 (1903).

(10) Y. Wada, A. Itani, T. Nishi, and S. Nagai, J. Poly. Sci., 2, 201 (1969); Y. Wada, Conferences on Relaxation Phenomena in Polymeric Systems (Sponsored by the Institute of Physical and Chemical Research of Tokyo), Tokyo, Japan, September, 1966.

(11) T.H.K. Barron, Phil. Mag., 46, 720 (1955); Ann Phys., 1, 77 (1957).

(12) M. Blackman, Proc. Roy. Soc., A 148, 365 (1934); Phil. Mag., 3, 831 (1958); Proc. Phys. Soc., 74, 17 (1959).

(13) B.G. Childs, Mod. Phys., 25, 665 (1953).

(14) J.G. Collins and G.K. White, Prog. in Low Temp. Phys., 4, 450 (1964).

(15) E.V. Vernon and S. Weintroub, Proc. Phys. Soc., B, 66, 887 (1953).

(16) J.S. Dugdale and D.K.C. MacDonald, Phys. Rev., 98, 1751 (1955).

(17) R.E. Barker, J. Appl. Phys., 34, 107 (1963).

(18) S.R. Urzendowski, A.H. Guenther, and J.R. Asay, Analytical Calorimetry, Edited by R.S. Porter and J.F. Johnson, Plenum Press, New York, N.Y., 1968, p. 119.

(19) E.A. Moelwyn-Hughes, Physical Chemistry, Pergamon Press, New York, N.Y., 1957, p. 327.

(20) P. Debye, Ann. Physik, 39, 789 (1912).

(21) S.R. Urzendowski and A.H. Guenther, Thermal Analysis, Vol I, Edited by R.F. Schwenker, Jr., and P.D. Garn, Academic Press, New York, N.Y., 1969.

(22) J.R. Asay, S.R. Urzendowski, and A.H. Guenther, "Ultrasonic and Thermal Studies of Selected Plastics, Laminated Materials and Metals," AFWL-TR 67-91, Air Force Weapons Laboratory, Kirtland Air Force Base, New Mexico, 1966, p. 493.

(23) N.D. Arnold and A.H. Guenther, J. Appl. Poly. Sci., 10, 731 (1966).

(24) M. Born and T. von Karman, Phys. Z., 13, 297 (1912); 14, 15 (1914).

MOLECULAR WEIGHT DETERMINATION OF CARBOXYL-TERMINATED POLYSTYRENE
BY THERMOMETRIC TITRATION

Robert Mermelstein, John Short, and Linda Flannery

Xerox Corporation, Rochester, New York   14603

INTRODUCTION

Thermometric enthalpy titration (TET) is one of the newer methods in the field of analytical calorimetry. It was introduced to this symposium last year by J. Jordan.[1] TET utilizes the enthalpy change of the reaction involved to define the titration endpoint. Theoretically, it can be applied to any reaction involving a significant enthalpy change. This factor gives TET wider applicability than the more conventional titration techniques like potentiometric, conductometric, and amperometric, because the latter depend on free energy changes. This report describes a novel application of the TET technique to the determination of number average molecular weight of carboxyl-terminated polystyrene. While the results presented here deal solely with carboxyl termination, TET should be applicable to polymers terminated with any of the functional groups shown in Table I.[2]

EXPERIMENTAL

A 500 ml flask, equipped with stirrer, gas inlet tube, thermometer and condenser, was charged with 50 g (0.48 mole) of styrene (low inhibitor grade) and 150 g of anhydrous benzene. Argon was passed through the solution followed by the addition of 0.125 g (0.00045 mole) of 4,4'-azobis(4-cyanopentanoic acid) (Aldrich). The mixture was heated to 80 ± 2°C for 24 hours. Complete solution of the initiator was noted after 5 hours. The reaction mixture was cooled to room temperature and the polymer isolated by precipitation with methanol in a Waring blender. After filtration and additional washing of the filtrate with methanol,

TABLE I

FUNCTIONAL GROUP ANALYSIS BY THERMOCHEMICAL METHODS

Functional Group	Reagent	Reactions	Precision and Accuracy Attainable
-COOH	Strong base	Acid-base neutralization	1%
$-COO^-$ or $-NH_2$	Strong acid	Acid-base neutralization	1%
-C=O or -CHO	$NH_2OH \cdot HCl$	Oximation	0.6%
-CHO	$H_2SO_4 - Na_2SO_3$	Bisulfite addition	0.6%
$Ar-NH_2$	$NaNO_2$	Diazotization	0.2%
$Ar-SO_2NH_2$	NaOCl	Oxidation	3 - 5%

32 g (64% yield) of polymer was collected. The fine powder was air dried overnight followed by 12 hours in vacuo at room temperature. Infrared analysis confirmed the presence of trace amounts of carboxyl absorption.

Gel permeation chromatograms were obtained on a Waters Associates Model 100 GPC utilizing a four-column series in 10^6, 10^5, 10^4, and 10^3 Å porosity. Samples were run using 0.5% concentration of polymer in Eastman Kodak reagent grade tetrahydrofuran at 25°C. Light scattering measurements were made on a Brice-Phoenix Light Scattering Photometer at 5460Å. The tetrahydrofuran solutions of the polymer were clarified by pressure filtration through a Gelman 0.2μ filter to remove dust particles. Molecular weight values were calculated from absolute turbidity data. Solution viscosity measurements in tetrahydrofuran were carried out in a 50N317 Cannon-Ubbelohde Viscometer at 25°C.

The Titra-Thermo-Mat, manufactured by the American Instrument Company, is the only commercially available thermometric titrator and was the equipment employed in this work. It incorporates a constant-speed, motor driven buret; an adiabatic titration tower; a Styrofoam plastic housing for thermal insulation of the titration beaker, a thermistor bridge circuit for temperature measurement; a recording potentiometer; a stirrer; and a small heater useful for calibration in calorimetric applications.

In all titrations 0.1N NaOH in 2-propanol was used, which was standardized against N. B. S. benzoic acid. The titrant was delivered at a rate of 500µℓ/minute. Polymer samples ranging from 100 - 300 mg were weighed into a 30 ml beaker and dissolved in 5 ml of chloroform. Fifteen ml of acetone was added and the solution titrated. Data reported are the average of two determinations. The number average molecular weight was calculated assuming two carboxyl groups per molecule. A typical titration curve is shown in Figure I.

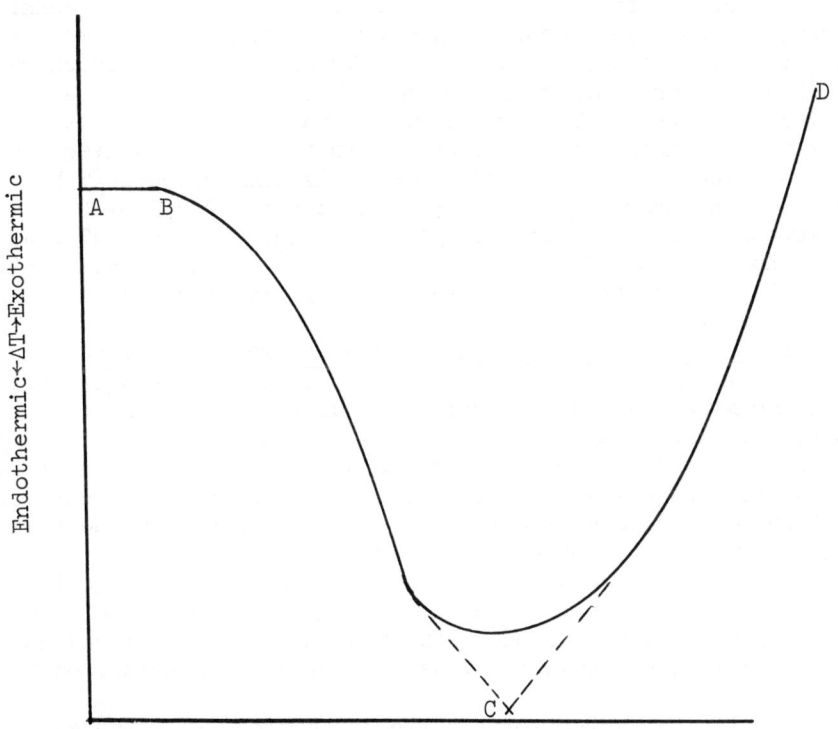

Time or Volume of Titrant

FIGURE I. Typical Enthalpy Titration Curve

Section AB is baseline before titrant addition, BC is an endotherm representing the heat of mixing of the three solvent system, point C is taken as the endpoint and exotherm CD is the formation of diacetone alcohol.

RESULTS AND DISCUSSION

Bamford and Jenkins[3] have demonstrated that the chain termination in the polymerization of styrene using 4,4'-azobis(4-cyanopentanoic acid) as initiator is substantially by combination. Under suitable polymerization conditions, chain transfer to monomer, solvent, and polymer are minimal and, on the average, each radical will carry one functional group at its inactive end, and will result in a chain carrying two terminal functional groups. When polymerization of styrene is initiated by 4,4'azobis(4-cyanopentanoic acid), yielding radicals of $HOOC(CH_2)_2C(CH_3)CN$ structure, the resultant product is substantially terminated by carboxyl groups. One would anticipate chain transfer to solvent to be lower in the present case,[4] benzene solution 80°C, compared to dichlorobenzene at 100°C, the experimental conditions used by Bamford and Jenkins. However, the polymers reported here are all higher in molecular weight than in the above study, and a lower initiator concentration, hence fewer growing chains, would favor increased chain transfer to solvent. The two opposing and roughly equivalent forces thus balance each other and chain transfer to solvent is not expected to be a serious complication. As summarized in Table II a series of polymers ranging in intrinsic viscosity from 0.19 - 0.35 were prepared by minor variations of the experimental conditions. As anticipated, an increase in the initiator concentration results in the isolation of polymer with lower viscosity.

While conventional acid-base titration was considered as a method of determining the concentration of carboxyl end-groups, the thermometric method was preferred due to higher sensitivity resulting in lower sample requirements. Vaugham and Swithenbank[5] developed a nonaqueous method for the determination of weak acids using acetone as a thermometric indicator. The rate of neutralization and diffusion of the titrant is more rapid than the base catalyzed exothermic conversion of acetone into diacetone alcohol. The end point of the neutralization is evident by a change in slope of the titration curve. Since the polystyrene samples were not adequately soluble in acetone, 25% by volume chloroform was added to attain sufficient solubility as well as a satisfactory endpoint.

The molecular weight of a series of carboxyl-terminated polystyrene samples, prepared above were measured by TET, GPC, and light scattering techniques, and the data are summarized in Table III. An indirect comparison for number average molecular weight determinations was necessary, as considerable experimental difficulties were experienced during attempted vapor pressure and membrane osmometry determinations. Osmometry data are not reported, because diffusion of low molecular weight solute through the S-S08 membrane led to erroneously high number average molecular weights. The molecular weights of the polymers were beyond the practical detection limits of the vapor pressure osmometer. As GPC is not

TABLE II

SYNTHESIS OF DICARBOXYL-TERMINATED POLYSTYRENE

Sample	Mole Ratio Monomer/Solvent	Mole Ratio Initiator/Monomer	% Yield	$[\eta]^{25}_{THF}$ (dl/g)
1	0.249	0.00929	64	0.193
2	0.249	0.00929	64	0.192
3	0.149	0.00559	47	0.225
4	0.374	0.00390	56	0.312
5	0.374	0.00279	51	0.347
6	0.374	0.00267	49	0.349

TABLE III

MOLECULAR WEIGHT OF DICARBOXYL-TERMINATED POLYSTYRENE

Sample	Number Average Molecular Wt.		Weight Average Molecular Wt.	
	Thermometric	GPC	GPC	Light Scattering
1	14,650	14,050	27,800	32,300
2	14,650	14,100	29,350	
3	19,500	16,300	34,800	
4	24,750	20,700	58,400	55,900
5	31,200	28,450	66,600	
6	34,300	32,400	70,900	68,300

an absolute method for molecular weight determinations, the weight average molecular weight of three of the above samples was measured by light scattering techniques. Light scattering and GPC measurements were carried out on a series of standard polystyrene samples purchased from ArRo Laboratories, ranging in molecular weight from 3,600 to 867,000. These measurements indicated that a Q factor of 41 correlates molecular size and molecular weight. Similar measurements on three of the above samples confirmed that the above Q factor is valid for carboxyl-terminated polymers. At the lower range of molecular weights examined, there is an excellent correlation between the number average molecular weight measured by TET and GPC. Near $\overline{M}_n$ = 15,000, the results obtained from the two methods agree within 4%. Based upon the six samples shown in Table III, the ratio between the number average molecular weight as measured by thermometric and GPC methods varies from 1.04 - 1.20 with an average value of 1.10 ± .06. A replicate of eight individual titrations of sample 1 gave an average of 14,650 ± 200, or a precision of 1.5%. No detectable endothermic temperature change was observed, on either the titration of a 3:1 acetone:chloroform mixture or titration of NBS 20K polystyrene sample. Therefore, the "blank" was determined to be negligible.

In the present technique, all base consuming compounds are titrated. Accordingly, the presence of even trace amounts of low molecular weight acidic compounds such as unchanged initiator or a decomposition product thereof, would result in large errors in the resulting $\overline{M}_n$. A 3.0 g. quantity of sample 1 was washed an additional three times with 250 ml. portions of methanol in a Waring blender and dried. The $\overline{M}_n$ of this sample, 14,500, is within experimental error, identical to that of starting sample 1. Therefore, contamination by initiator or other low molecular weight acidic impurities is not a significant factor. A substantial portion of the differences in number average molecular weight between the two methods are due to experimental error. However, we cannot exclude the possibility of a small portion (less than 10%) of the polymer chains not being dicarboxyl terminated.

CONCLUSION

A novel application of the thermometric titration technique has been demonstrated in the determination of number average molecular weight of carboxyl-terminated polystyrene. The convenient and rapid method permitted small sample size and afforded excellent precision. A good correlation was obtained between molecular weights measured by thermometric enthalpy titration and gel permeation chromatography. Application of this method to other polymer systems is underway.

ACKNOWLEDGEMENT

We are indebted to Dr. C. B. Murphy and Dr. R. A. Braun for stimulating discussions. We appreciate the assistance of D. Alliet and H. Walser in performing gel permeation chromatography measurements.

BIBLOOGRAPHY

1. Porter, R. S. and Johnson, J. F., Analytical Calorimetry, Plenum Press, New York, 1968, pp. 203 - 209.

2. Ibid., p. 207.

3. Bamford, C. H. and Jennings, A. A., Nature, 176, 78, 1955.

4. Young, L. J., Brandrup. G., and Brandrup, J., Polymer Handbook, Interscience, 1966, II, p. 91.

5. Vaugham, G. A. and Swithenbank, J. J., Analyst, 90, Oct. 1965, pp. 594 - 599.

DTA OF POLYSULFONE

H. T. Lee

Polymer Research Branch, Picatinny Arsenal, Dover, N. J.

INTRODUCTION

Polysulfone, a specific poly(aryl ether), is composed of phenylene units linked by sulfone and isopropylidene groups in addition to the ether groups. The repeating unit is

$$\left[\phenyl - \underset{\underset{CH_3}{|}}{\overset{\overset{CH_3}{|}}{C}} - \phenyl - O - \phenyl - \underset{\underset{O}{\|}}{\overset{\overset{O}{\|}}{S}} - \phenyl - O \right]_n$$

The electron-withdrawing sulfone group has a stabilizing influence upon oxidative degradation in this polymer system. A mixed ether and sulfone linked material is more thermally stable than a material linked with a sulfone group alone.[1] The thermal and oxidative stability of this polymer can be useful in potential Military high temperature materials application.

The purpose of the present study was to evaluate the kinetics of thermal degradation of polysulfone by using differential thermal analysis (DTA).

EXPERIMENTAL

The material used was Union Carbide Bakelite polysulfone, Grade P-1700, a transparent light amber thermoplastic resin with no additives. Its structure was nearly in a complete amorphous state. Its number average molecular weight determined by membrane osmometry was 25,000. The sample was cut into pieces of about 2 mg. each by a metal cutter, and vacuum dried at 70-80°C. for a total of 20 hours within 3 consecutive days.

A duPont 900 DTA was used in this study. The types of DTA cells tested were the standard cell, the calorimeter cell, and the differential scanning calorimeter (DSC) cell. With open sample cups or tubes, all these cells produced unstable baselines and peak temperatures as well as greatly distorted peak shapes with very poor reproducibilities. The non-hermatically crimped sample pan with either an aligned or an inverted cover in the DSC cell gave reproducible peak temperatures and less distorted peak shapes. However, the baselines were not stable above 350°.

Satisfactory results were achieved by using the hermetically sealed sample and reference cups in the DSC cell programmed at nominal heating rates ranging from 2° to 40°/min. Unsatisfactory results obtained from open tubes or cups and crimped pans could be due to the deposition of pyrolysis products, especially the corrosive SO_2 on the thermocouples.

A single piece of sample weighing 2.0 $\pm$ 0.1 mg. was sealed in the passivated aluminum cup and lid on an encapsulating press. The reference cup contained no inert material except atmospheric air and was sealed in the same manner as was the sample cup. The actual heating rates were $\pm$ 5% of the nominal rates at temperatures over 300°.

Most of the hermetically sealed cups withstood the pressure produced during pyrolysis. The shape or contour and the weight of these sealed cups after pyrolysis remained unchanged. However, some of the sealed sample cups expanded or changed in shape, and some showed a loss in weight. In these cases, the thermograms were unsatisfactory and discarded.

RESULTS AND DISCUSSIONS

Typical primary DTA thermograms above 500° at 6 heating rates are shown in Fig. 1. Fig. 2 shows the plot of log (Φ/T_p^2) versus $1/T_p$. Here Φ is the heating rate in °C/min., and T_p the peak temperature. From this linear plot, activation energy was determined from its slope by using Kissinger's equation[2]

$$d(\log \Phi/T_p^2)/d(1/T_p) = -E/2.303R$$

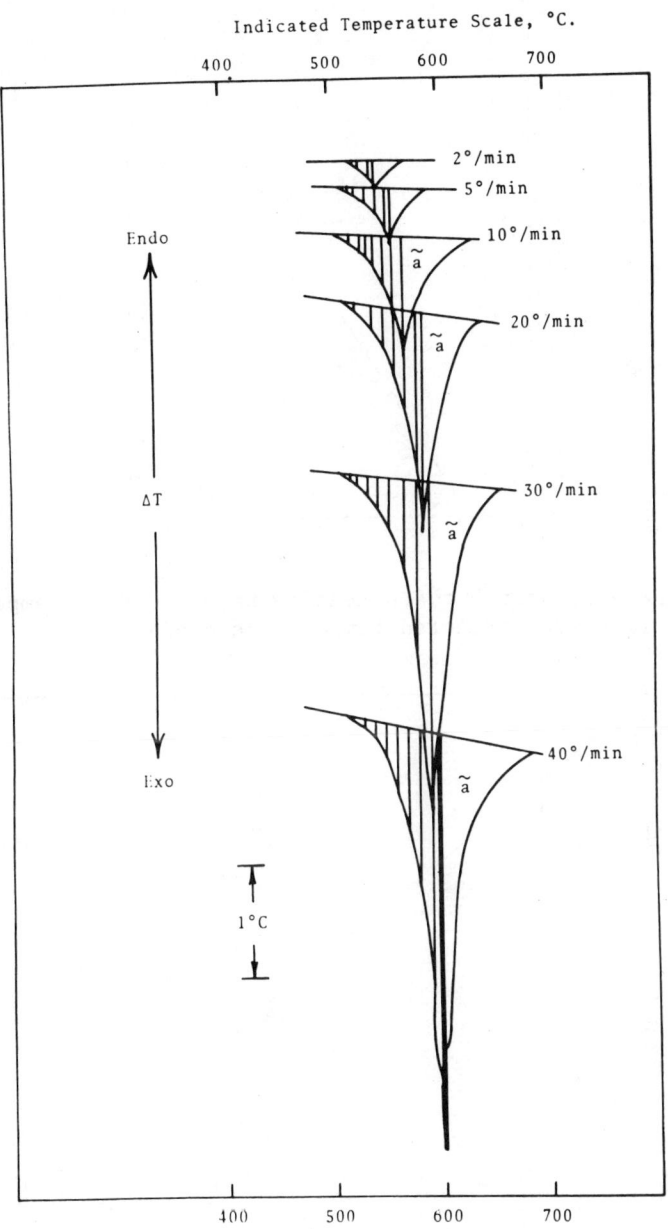

Fig. 1. Typical DTA Thermograms

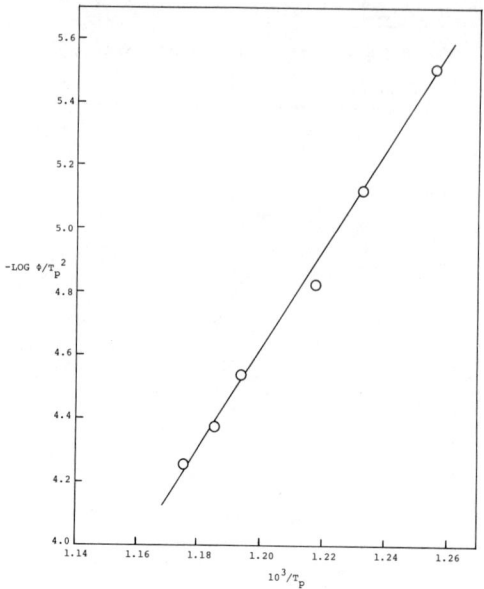

Fig. 2. Relationship between Heating Rate and Peak Temperature (T_p's are corrected temperature scale)

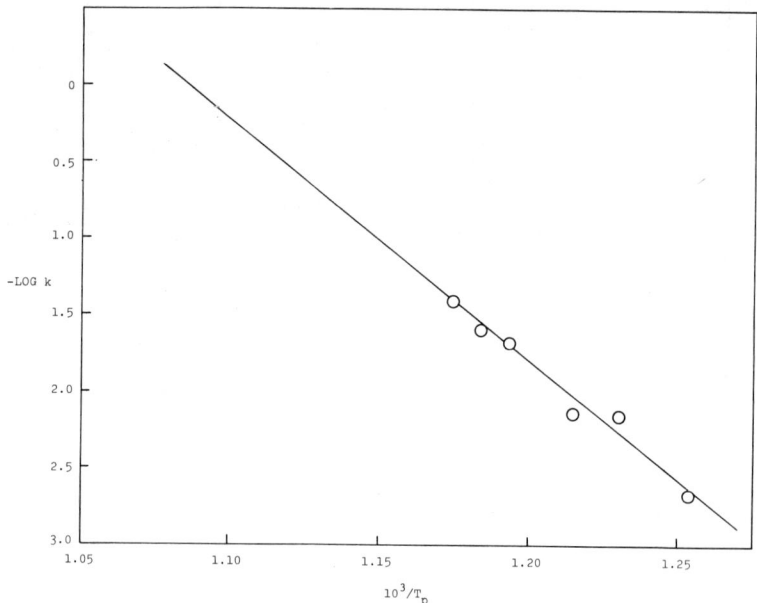

Fig. 3. Reaction Rate Constants Vs Reciprocal of Peak Temperature (T_p's are corrected temperature scale)

The E value obtained was 70 Kcal/mole. There were several reports (3,4) which criticized Kissinger's method. However, this E value was in good agreement with results obtained from two other methods which will be discussed later. In the author's opinion, Kissinger's method was applicable to the present work.

The area of the exothermic peaks was measured by a planimeter. The portion of the area between the peak temperature and the return of the curve to the baseline, $\tilde{a}$ (as shown in Fig. 1), and the peak height, ΔT are listed in Table 2. The constants of the reaction rate at the 6 peak temperatures were calculated from the equation used by Barrett[5]

$$k = \Delta T / \tilde{a}$$

These k's are plotted in Fig. 3. The value of activation energy obtained from the classical Arrhenius plot, assuming that a first order reaction took place[1], was 76 Kcal/mole. The frequency factor was $10^{16.5}$.

The distances between the baseline and the curve at several temperatures below the peak temperature are listed in Table 3. Fig. 4 shows the logarithms of these distances plotted versus $1/T_p$. From these linear plots, the value of activation energy was determined by the equation according to Rogers and Morris[6]

$$-E = R \frac{\ln d_1 - \ln d_2}{1/T_1 - 1/T_2} = 4.58 \frac{\log (d_1/d_2)}{1/T_1 - 1/T_2}$$

The E value was 73 ± 5 Kcal/mole.

Both the work of Barrett and of Rogers and Morris used Perkin-Elmer's DSC. Barrett applied the differential enthalpic analysis because this DSC measured energy changes directly. Du Pont's DSC produced data similar to the conventional DTA which measured temperature difference. However, the area under the output curve was directly proportional to the total amount of energy transferred in or out of the sample. The peak area could be used for measuring enthalpic changes.[7]

When sealed cups were used for both sample and reference, the heat flow from the sample to reference, or vice versa, was reduced. The thermocouple was positioned underneath the sample platform. There was no direct contact between the sample and thermocouple. This temperature sensing device enabled the elimination of some disadvantages of the conventional DTA.

TABLE 1

Heating Rates and Peak Temperatures

Heating Rate, Φ (°C/min)	T_p (°C) (Ind. scale)	T_p (°K) (Corr. scale)	Φ/T_p^2 (°C/min)/(°K)^2
2	542	797	3.149×10^{-6}
5	557	811.5	7.593×10^{-6}
10	568	822	1.480×10^{-5}
20	585	836	2.862×10^{-5}
30	592	844	4.212×10^{-5}
40	599	851	5.523×10^{-5}

TABLE 2

Peak Heights and Areas

ΔT (inch)	Total Peak Area (in^2)	$\tilde{a}$ (in^2)
0.22	0.063	0.034
0.55	0.145	0.059
1.08	0.539	0.218
2.05	0.790	0.317
2.96	1.334	0.589
4.04	1.432	0.677

TABLE 3

Distances from Baseline, in Units of 0.05°C

°C (Ind. Scale)	°K (Corr. Scale)	2°/min	5°/min	10°/min	20°/min	30°/min	40°/min
520	777	1.0	1.1	1.3	1.0	1.4	1.0
525	781		1.6		1.5		
530	786	1.9		2.6		2.5	1.8
535	791		2.8	3.0	3.0		
540	795	3.2		4.5		4.3	3.1
545	800		4.5		4.5		
550	805			8.0		7.1	6.4
555	810		8.1	11.0	8.0		
560	814			15.0		12.0	11.3
565	819						
570	824				18.0	20.0	18.3
575	828						
580	833				28.5	35.0	28.0
585	838						
590	842						41.5
595	847						64.3

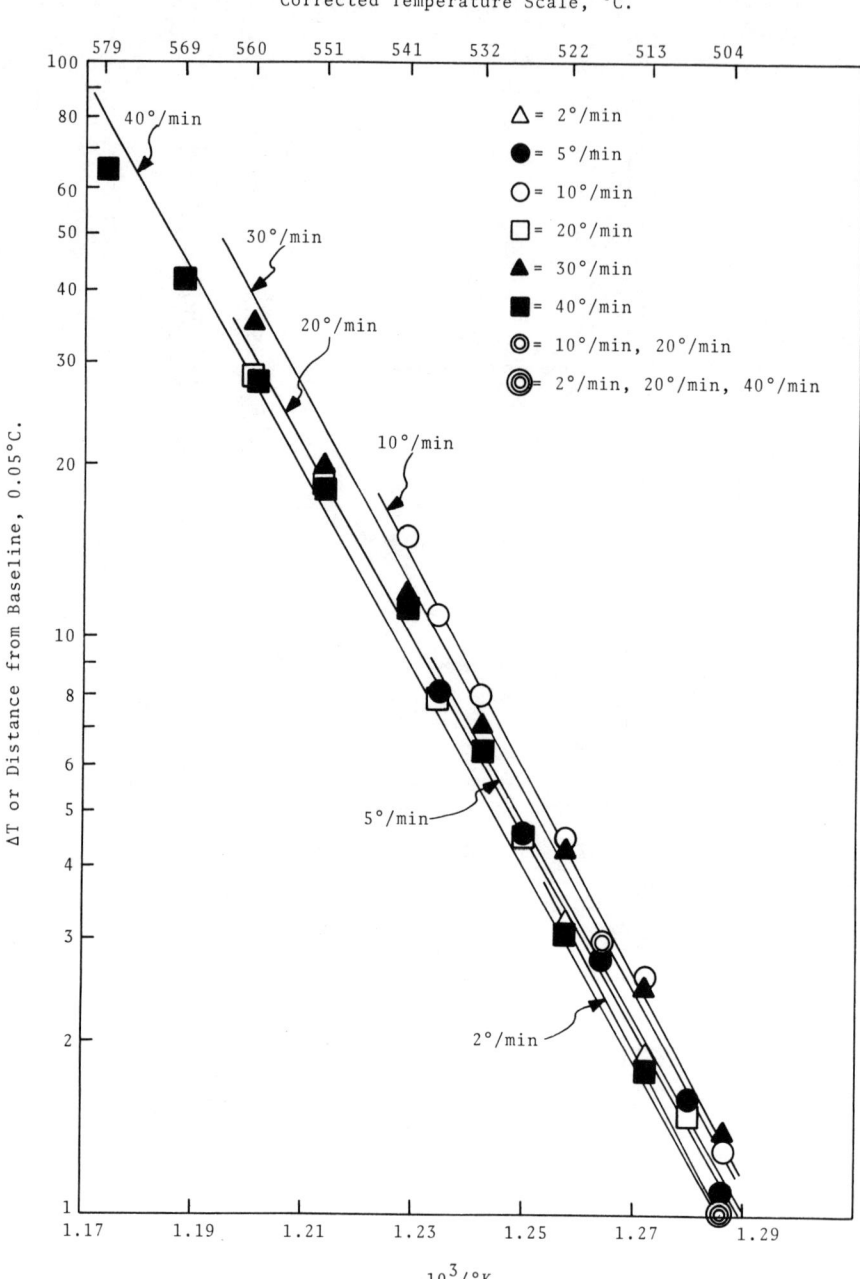

Fig. 4. Distances from Baseline Vs Reciprocal Temperatures

The effects of the sample holder design on the peak shape of DTA curves have been investigated by Wilburn et al[4] using the electrical analogs of some mathematical models. They pointed out that to minimize the distortion of peak shapes, the holder must have a high heat capacity compared with the sample and reference materials; also a high diffusivity was necessary in order to pass heat rapidly to sample and reference. It must also effectively isolate the reference and sample materials.

In the present work, the sealed sample and reference could be considered as effectively isolated. But the aluminum cup had a lower heat capacity than the sample.[8] However, the heat conductivity of aluminum was higher than the sample[8,9]. The heat passed rapidly from the cup to the sample. In addition, the heat loss from the sample was at a minimum due to the fact that gaseous products were unable to escape from the holder. This reduced the temperature gradients within the sample, thus stabilizing the peak shape.

There was a baseline shift observed on all the thermograms including those using open cups and crimped pans. The initial departure of this shift from the baseline was between 190 and 200°. This baseline shift was believed to be the glass transition of polysulfone.

The single exothermic peak that occurred after 500° possibly was the result of net heat changes due to oxidation and decomposition or other reactions and transitions. The gaseous products of thermal decomposition of polysulfone from mass spectrometry study by Hale, et al[1] indicated that there were complex fragmentation processes at temperatures over 500°. In the present study, these gaseous products were confined within the sealed cup. This could probably affect the actual physical transitions and chemical reactions and further complicate the net heat changes before the peak temperatures. Therefore it does not seem feasible to make a particular assignment for this exothermic peak.

In a future study, the sample and reference cups should be sealed under oxygen and under inert gas atmosphere. If the thermograms are significantly different from the results of the present study, a comparison of the kinetic data under various atmospheres would be possible.

REFERENCES

1. W.F. Hale, A.G. Farnham, R.N. Johnson, and R.A. Clendinning, J. Polymer Sci. A-1, $\underline{5}$, 2399 (1967).

2. H.E. Kissinger, Anal. Chem., $\underline{29}$, 1702 (1957).

3. R.L. Reed, L. Weber, and B.S. Gottfried, Ind. Eng. Chem. Fundamentals, $\underline{4}$, 38 (1965).

4. F.W. Wilburn, J.R. Hesford, and J.R. Flower, Anal. Chem. $\underline{40}$, 777 (1968).

5. K.E.J. Barrett, J. Appl. Polymer Sci. $\underline{11}$, 1617 (1967).

6. R.N. Rogers and E.D. Morris, Anal. Chem. $\underline{38}$, 412 (1966).

7. R.R. Currell, Proceedings 2nd International Conf. on Thermal Analysis, Aug. 1968, Editors, R.F. Schwenker and P.D. Garn, Academic Press, p. 1185 (1969).

8. E. Spingler, Plastica, $\underline{19(7)}$, 269 (1966).

9. R.C. Weast, Handbook of Chem. & Phys., Chem. Rubber Co., 48th ed. p. E-10 (1967).

SCANNING CALORIMETRY OF MESOPHASE TRANSITIONS: MARKER'S ACID

William R. Young,[1] Edward M. Barrall II,[2] and Arieh Aviram[1]

[1]International Business Machines Corporation
Thomas J. Watson Research Center
Yorktown Heights, New York 10598
[2]International Business Machines Corporation
Research Laboratory
San Jose, California 95114

Marker's acid (MA), first reported by Marker, Oakwood and Crookes in 1936 (1), contains a 3β-carboxylic acid group in the place of a 3β-hydroxyl found in cholesterol. The structure of MA has been rigorously demonstrated by both Roberts, Shoppee and Stephenson (2) and Corey and Sneen (3) to be cholest-5-ene-3β-carboxylic acid (alternatively named 3β-carboxy-Δ^5-cholestene). This configuration makes MA an isomeric form of cholesteryl formate. As such, it might be expected to exhibit a cholesteric mesophase at some temperature.

Marker's Acid   Cholesteryl Formate

Earlier studies of cholesteryl esters, carbonates and thio analogues (4,5,6,7) have indicated that the chain length of the side group determines the entropy of transition for both the solid and mesophases. MA was synthesized to evaluate this proposition with a material in which only the order of the atoms in the side chain was altered. The carbonates previously studied (7) are not adequate for this test since it is well known that bond angles and distances about the carbonate group are somewhat different from those about a carboxylate group. The polarization of the sulfur group in the thio analogues also introduces some dubious effects when direct comparison is attempted between thio cholesterol and cholesterol esters (6). If chain length is the controlling factor in transition entropy, then MA should exhibit a single mesophase (cholesteric) and the following transition entropies:

solid → mesophase ∼14 cal/mole/°K
mesophase → isotropic liquid ∼0.25 cal/mole/°K

This is by analogy with the previously reported cholesteryl formate (4,5).

EXPERIMENTAL

Synthesis

Marker's acid (cholest-5-ene-3β-carboxylic acid) was prepared by the method of Roberts, Shoppee and Stephenson (2). The last traces of non-acidic impurities were removed by treatment of the product with hot alcoholic potassium hydroxide solution. The insoluble impurities were removed by filtration. The cool filtrate was acidified, and the liberated acid was collected and recrystallized from benzene. Earlier studies have excluded the possible formation of the 3α isomer (2).

Microscopy

Hot stage microscopy was carried out on a 1 mm thick layer of MA between coverslips on a Mettler microscopy hot stage. The transitions were viewed between crossed polarizers using a Zeiss Ultra Phot II microscope. The acid exhibits a distinctly spherulitic solid phase up to 225°C at a heating rate of 2°C/min. At 225°C the material transforms sharply to a typical cholesteric "net." This net consists of a set of white mobile filaments against a brilliant blue texture. The cholesteric mesophase abruptly vanishes at 259°C to yield the isotropic liquid (black field between polarizers). Bubbles accumulated between the slide and coverslip at 264°C indicating decomposition of MA. Preliminary experiments indicate the presence of carbon dioxide in the evolved gas when the acid was melted in air. No gas was similarly evolved when melting occurred in a nitrogen atmosphere.

Calorimetry

The heats of transition for the solid → mesophase and mesophase → isotropic liquid were determined in the heating mode. For the transition heat determinations three samples, 2.561 mg, 2.450 mg and 2.550 mg, were encapsulated under nitrogen in volatile sample sealers. To insure good thermal contact between sample pan and sample an additional small disk of aluminum (∼3 mm) was pressed atop the sample prior to sealing. The DSC-1B used in this study was calibrated on the temperature axis with 99.9999% mercury, 99.999% gallium, 99.999% stearic acid, NBS benzoic acid, and 99.9999% indium melting. All temperatures from MA thermograms were corrected using the calibration curve and indium slope as described elsewhere (8). The calorie/unit area factor was obtained from the same standards. For the determination of transition heats a heating rate of 5°C/min at 4 and 2 millicalories sensitivity was used. The thermograms are shown in Figures 1 and 2 and the data tabulated in Table I. Care was exercised to insure precise re-positioning of the samples and standards in the DSC-1B platforms.

The sample purity was determined using the solid to mesophase transition partial areas. The Van't Hoff equation was used in the data reduction as reported by several workers (7,8,9). The areas from onset of melting to the peak vertex were used in the T_s versus 1/F plots. For purity measurement 4.015 mg, 3.586 mg and 4.121 mg samples were scanned at 1.25°C/minute at a 2 millicalorie sensitivity. The average purity on three determinations was 99.158 mole% ±0.042. The extrapolated temperature for the melting point of 100% MA was 225.2°C.

TABLE I

Thermodynamic Data for 99.158% Marker's Acid

T_b °C	T_m °C	T_e °C	ΔH kcal/mole	ΔS cal/mole/°K	Transition
217.3	224.5	226.5	6.73	13.5	solid → mesophase
254.8	258.6	260.3	0.744	1.40	mesophase → isotropic liquid

DISCUSSION

MA exhibits one of the highest temperature cholesteric ranges yet reported in the literature. The solid → mesophase and mesophase → isotropic liquid transitions at 224.5° and 258.6°C, respectively, may be near some sort of upper limit for the cholesteric mesophase. Since decomposition was noted only a few degrees above the clear point, it is possible that this temperature was depressed by traces of decomposition products. Therefore, the upper temperature, 258.6°C, may be even higher in a non-decomposing system. However, 258.6°C is near the upper temperature limit of unconjugated double bonds.

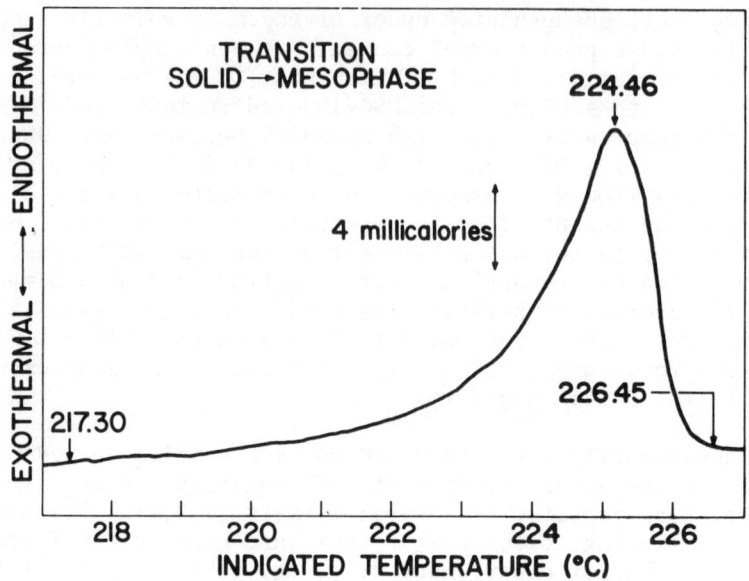

Figure 1. First Heating of Marker's Acid MW = 414.68
2.5612 mg Sample Heated at 5°C/min.
Sens. 4 millicalories/inch
Heating Range: 207° to 226°C

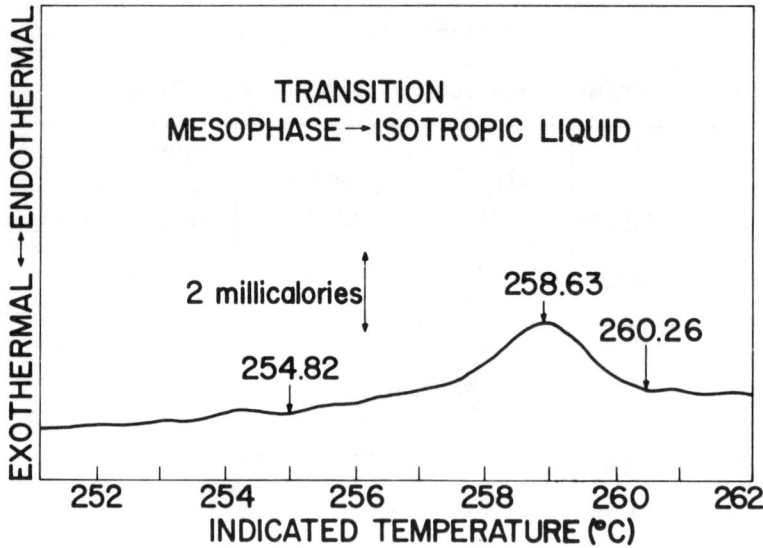

Figure 2. First Heating of Marker's Acid
2.5612 mg Sample Heated at 5°C/min.
Heating Range: 231° to 263°C
Transition
Mesophase → Isotropic Liquid

Cholesteryl-p-anisoate exhibits a higher clear cholesteric mesophase, but this compound also decomposes.(10).

With the exception of the high transition temperatures, Marker's acid appears under the polarizing microscope to be closely akin to the lower esters of cholesterol. Unlike most cholesteryl esters, Marker's acid decomposes in the isotropic liquid range. It decarboxylates at elevated temperatures in the presence of oxygen. After a single DSC heating scan, the solid melted directly to the isotropic liquid over a very broad range. No trace of a mesophase could be observed with the DSC on second heating or cooling.

The total entropy change from solid → isotropic liquid for MA is in excellent agreement with the analogous cholesteryl ester, cholesteryl formate. The entropy change for cholesteryl formate solid → isotropic liquid is 14.2 cal/mole/°K at 97.0°C (4). For the same total transition MA requires 13.5 + 1.40 or 14.9 cal/mole/°K. Summation of ΔS values for MA is necessary since MA is not monotropic where as cholesteryl formate is monotropic with respect to the cholesteric mesophase. The structural similarity is very close between MA and cholesteryl formate insofar as the <u>length</u> of the side group is concerned:

Marker's Acid — ~4Å

Cholesteryl Formate — ~4.2Å

An earlier study of over twenty-five cholesteryl esters has shown that given the cholestene ring structure, the magnitude of the entropy change (order → disorder) is determined predominantly by the length of the side chain. The case of Marker's acid serves to confirm this observation in the presence of large chemical differences.

The mesophase → isotropic liquid transition entropy change of Marker's acid is much larger than that of the same transition in cholesteryl formate, 1.40 cal/mole/°K as compared to 0.248 cal/mole/°K. However, the 1.40 cal/mole/°K is in very good agreement with the mesophase → isotropic liquid transition in dicholesteryl adipate and sebacate. These materials obviously contain two cholesterol rings. The relationship is shown graphically in Figure 3 as a function of the estimated distances between the two cholesterol rings. It is assumed that MA in the dimer form has ~5.5Å between cholesterol rings.

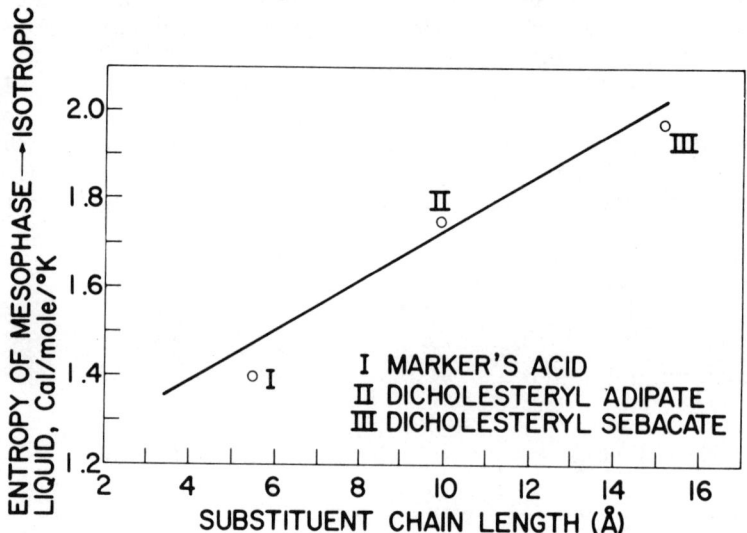

Figure 3. Mesophase Transition Entropy for Materials Having More Than One Cholesterol Ring Per Molecule Contrasted with Marker's Acid

CONCLUSIONS

Marker's acid exhibits entropy changes which are very similar to the analogous cholesteryl esters. The total entropy change for the solid → mesophase → isotropic liquid transition is 14.9 cal/mole/°K for Marker's acid and is 14.5 cal/mole/°K for cholesteryl formate. This is a very satisfactory example of entropy conservation for two molecular types of the same length undergoing the same type of order → disorder rearrangement. Such close agreement verifies the great importance molecular geometry plays in compounds exhibiting the cholesteric mesophase. The higher transition temperatures of the Marker's acid are probably due to polar interactions of the acid hydrogen. Work in progress on the transition temperatures of Marker's acid esters indicates that they are actually lower than the transition temperatures of analogous cholesteryl esters. This observation supports the role of the acid hydrogen in increasing the intermolecular attraction and the melting point.

The large transition entropy of the mesophase → isotropic liquid transition of Marker's acid clearly indicates that the acid is dimerized in the cholesteric mesophase. The dimerization is supported by the thermodynamic data available on the dicholesterol esters. At the present time, the authors are unable to show a model which would permit both ring systems in a dimer MA or dicholesteryl ester to participate in a cholesteric mesophase helix. Nevertheless, thermal evidence indicates that this is the case.

REFERENCES

1. R. E. Marker, T. S. Oakwood, H. M. Crookes, J. Am. Chem. Soc. 58, 481 (1936).
2. G. Roberts, C. W. Shoppee and R. J. Stephenson, J. Chem. Soc., 1954, 2705.
3. E. J. Corey and R. A. Sneen, J. Am. Chem. Soc. 75, 6234 (1953).
4. E. M. Barrall II, Julian F. Johnson and R. S. Porter, Thermal Analysis, R. F. Schwenker, P. D. Garn eds., Academic Press, N. Y., Vol. I (1969) p. 555.
5. E. M. Barrall II, Julian F. Johnson and R. S. Porter, Molecular Crystals, 8, 27 (1969).
6. R. D. Ennulat, Molecular Crystals, 8, 247 (1969).
7. R. D. Ennulat, Analytical Calorimetry, R. S. Porter and J. F. Johnson eds., Plenum Press, N. Y. (1968) p. 219.
8. E. M. Barrall II and J. F. Johnson, Fractional Solidification, M. Zief, ed., Vol. 2, M. Dekker, N. Y. (1969), p. 77.
9. G. L. Driscoll, I. N. Duling and F. Magnotta, Analytical Calorimetry, op. cit., p. 271.
10. E. M. Barrall II, J. F. Johnson and R. S. Porter, Molecular Crystals 8, 27 (1969).

EFFECTS OF SUBSTITUENT CHAINS ON THE MESOMORPHISM OF THE SCHIFF'S BASES OF p-AMINOCINNAMIC ACID ESTERS

Edward M. Barrall II

International Business Machines Corporation
Research Laboratory
San Jose, California 95114

Compounds of the general structure

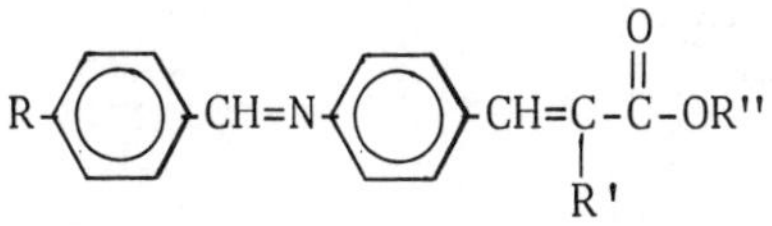

$$R-\bigcirc-CH=N-\bigcirc-CH=\underset{R'}{C}-\overset{O}{\underset{\|}{C}}-OR''$$

are known to form variously smectic and nematic mesophases (1,2). Leclercq, Billard and Jacques (4) have presented thermodynamic data on fifteen derivatives of the above structure. Very little work has been presented on the effect of various substituent groups on the order-disorder magnitude (entropy) of various transitions in these or other nematogenic series. The availability of the data of Leclercq et al. (4), Table I, permits a preliminary consideration of the following questions for the lower members of the series:

1. Which end of the molecule controls predominantly the order-disorder of the solid → smectic, smectic → nematic and nematic → isotropic liquid changes?
2. To what extent is pure geometric size a controlling factor in the above transitions?
3. To what extent does the polarity of the substituent groups control transition entropies and types?
4. What effect does an increase in molecular thickness have on transition entropy and mesophase type?

DISCUSSION

The previously stated questions will now be addressed insofar as the data of Leclercq et al. (4) and supporting evidence permit.

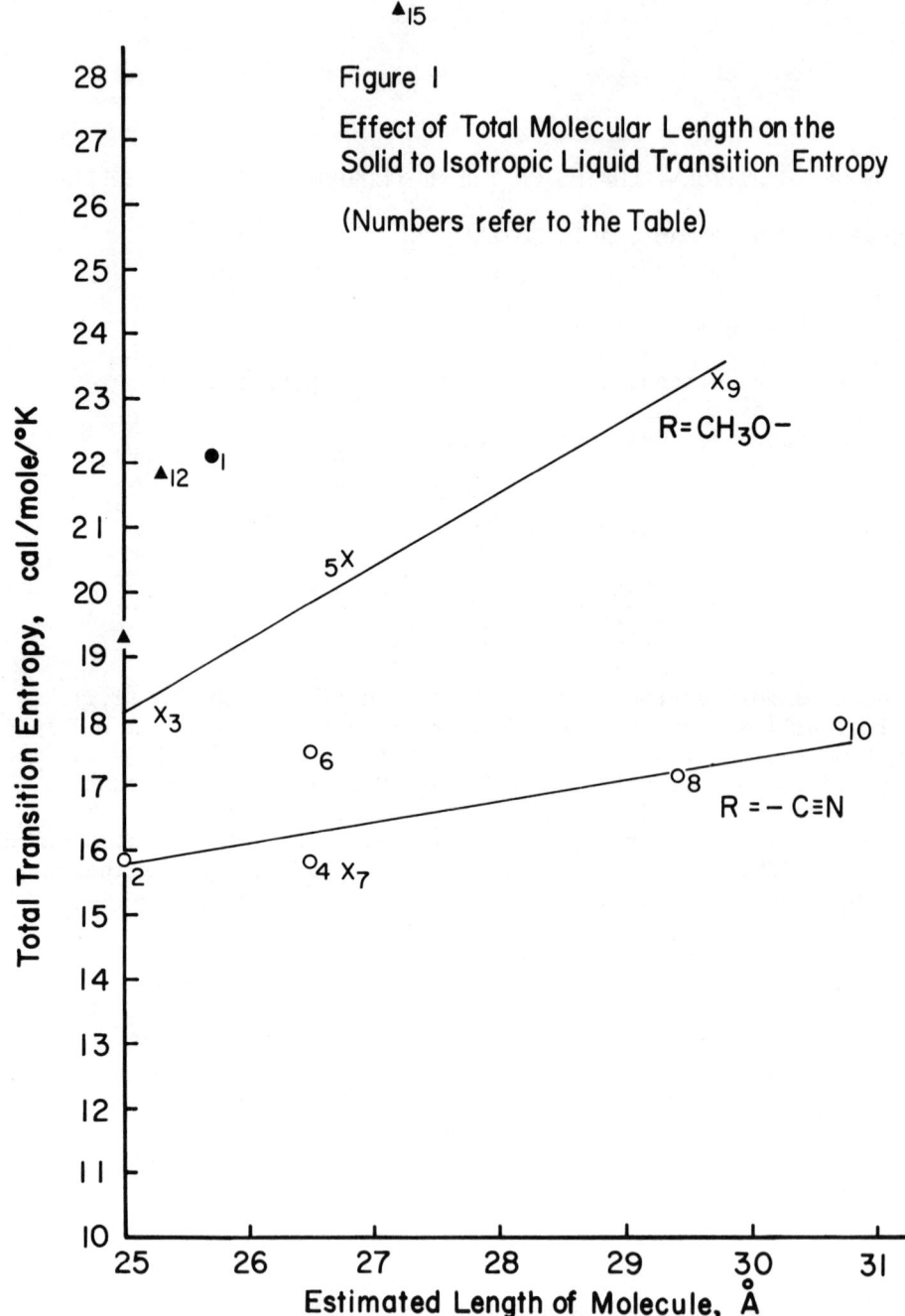

Figure 1

Effect of Total Molecular Length on the Solid to Isotropic Liquid Transition Entropy

(Numbers refer to the Table)

1. Molecular end effect. Referring to Table I and Figure 1 it is obvious that changing either the R, R', or R" by one carbon atom produces unique and highly different effects on the total order (entropy) of the system. Given R = -C ≡ N and R' = H, an increase of one carbon atom at R" causes an increase in the total entropy of fusion. But, the role of R" chain cannot be very great, since in going from R" = propyl to R" = octyl (compounds 2, 4, 8 and 10, Table I) an entropy increase of only 2 cal/mole/°K is noted. In the cholesteryl ester series the same change in chain length results in an 11 cal/mole/°K increase in total entropy (5). However, if the nitrile group is changed to the longer methoxy group and R" is held at C_4 (see compounds 4 and 5), a much greater increase in entropy is noted, 5.2 cal/mole/°K. Also, compound 1, while having a shorter R" tail is one carbon longer at R than compound 3. The addition of this carbon atom results in a 4 cal/mole/°K entropy increase. Thus, it appears that the R end of the molecule more profoundly affects molecular crystal order than the R" end. Although the evidence is not extensive, the R" chains could be out of the main molecular plane and have a smaller role in determining molecular order.

The data plotted in Figure 1 indicate that with the addition of a carbon at R a new series of molecular order develops at R". The methoxy materials follow a much steeper entropy trend than the nitrile at R materials. Compound 1 indicates that yet another series is formed by the ethoxy at R materials.

The data in Table I indicate that the formation of the smectic mesophase is controlled by the substitution at R. Two forms of the smectic can be exhibited. This could be due to axial tilt of the smectic layers (6). None of the nitrile materials yet reported (2,4) develop a smectic phase.

From Figure 2 it is apparent that substitution at R, R', or R" has little affect on the entropy of the nematic → isotropic liquid conversion. This indicates that the nematic order is regulated almost completely by the interior parts of the molecule. Compound 10 suggests that above R" = C_8 some affect may occur outside the central portion of the molecule.

2. Effect of absolute geometric size of the molecule. The molecular size within the total group of fifteen materials appears to be of secondary importance compared to other variables. However, within given series (R = nitrile, methoxy and othoxy) a systematic increase in total order is noted with increasing geometric size. One group of materials (compounds 4, 5, 6, 7) when contrasted show the R" chain branching is more important in the total molecular order than sample geometrical size. The data in Figure 2 indicates the mesophase entropy of the nematic → isotropic liquid transition appears to be independent of substitution.

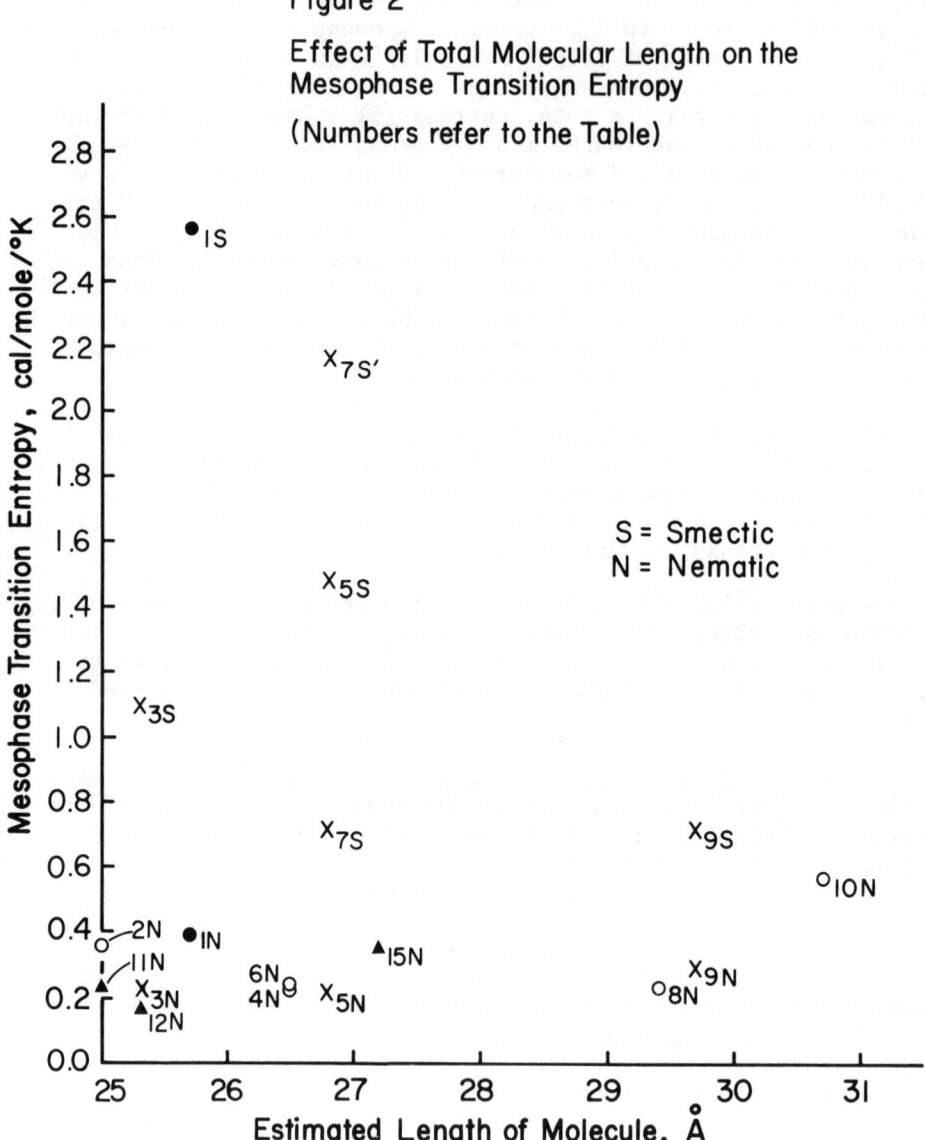

Figure 2

Effect of Total Molecular Length on the Mesophase Transition Entropy

(Numbers refer to the Table)

TABLE I

Structures and Thermodynamic Properties of a
Group of Schiff's Basis of p-Aminocinnamic Acid Esters (4)

Code	R	R'	R"	Smectic Transition			Nematic Transition			Liquid Transition			ΔS Total
				T,°C	ΔH, cal/M	ΔS cal/M/°K	T,°C	ΔH cal/M	ΔS cal/M/°K	T,°C	ΔH cal/M	ΔS cal/M/°K	cal/M/°K
1	C_2H_5O	H-	$-C_2H_5$	82^I 113^{II}	6.45 0.41	18.2 1.1	157	1.10	2.56	160	0.17	0.39	22.2
2	-C≡N	H-	$-C_3H_7$				102.5	5.8	15.5	163	0.16	0.37	15.9
3	CH_3O-	H-	$-C_3H_7$	$[71]^I$ 79^{II}	0.19 5.90	0.55 16.8	101	0.41	1.10	134	0.10	0.25	18.2
4	-C≡N	H-	$-C_4H_9$				87	5.6	15.6	133	0.090	0.22	15.8
5	CH_3O-	H-	$-C_4H_9$	66^I $[70]^{II}$	6.4 0.34	18.9 0.99	91	0.54	1.48	109	0.080	0.21	20.6
6	-C≡N	H-	$i-C_5H_{11}$				100	6.45	17.3	109	0.090	0.24	17.5
7	CH_3O-	H-	$i-C_5H_{11}$	47^I 76^{II}	4.10 0.25	12.8 0.72				97	0.80	2.2	15.7
8	-C≡N	H-	$-C_6H_{13}$				59	5.60	16.9	119	0.090	0.23	17.1
9	CH_3O-	H-	$-C_6H_{13}$	58.5	7.40	22.3	83.5	0.26	0.73	101	0.11	0.29	23.3
10	-C≡N	H-	$-C_8H_{17}$				62	5.50	16.4	111	0.22	0.57	18.0
11	-C≡N	$-CH_3$	$-C_3H_7$				120	7.50	19.1	122	0.090	0.23	19.3
12	CH_3O-	$-CH_3$	$-C_3H_7$				67.5	7.40	21.7	83.5	0.080	0.17	21.9
13	-C≡N	$-C_2H_5$	$-C_3H_7$							81.5	?		?
14	CH_3O	$-C_2H_5$	$-C_3H_7$							62	?		?
15	C_2H_5O-	$-C_2H_5$	$-C_3H_7$				[63]	0.12	0.36	99	10.8	29.0	29.4

Code refers to Figures 1 and 2 and the text.
Temperatures in [ ] indicate monotropic transitions.
I and II refer to two forms of the smectic mesophase.

3. Polarity effects appear to be less important than location of the substituent group in total molecular order. The lack of a smectic phase in the R = nitrile series may be due to polarity. However, the geometrical configuration of the nitrile group (coplanar with the benzene ring) may be the controlling factor. The methoxy and ethoxy groups are aplanar. Polarity also does not appear to control the transition temperature or range.

4. Molecular thickness appears to be a very important function at both R and R' as well as at R". An increase in the molecular thickness at R' greatly increases the molecular order, see Figure 1, compounds 11, 12, 15. The melting point or mesophase transition temperature is lowered with increasing size of R'. Thickness does not appear to contribute to the total order of the mesophase in the nematic form, see Figure 2. The present data suggest that increased thickness at R' inhibits the formation of the layered smectic structure. However, information is scant. Increasing molecular thickness at R" by branching, compounds 6 and 7, lowers the mesophase transition temperature when R ≠ nitrile. At present insufficient data are at hand to be more precise.

CONCLUSIONS

In terms of molecular order based on considerations of entropy change the following conclusions appear warranted for the arylidene amino cinnamates:

1. Substitution at R gives rise to a unique group of compounds as R" is extended. R substitution appears to dictate the primary statement of molecular order.
2. The total order in the nematic mesophase is a function of the basic molecular internal structure and not a function of side chains at R, R' or R".
3. Substitution of ethoxy or methoxy at R introduces a smectic order as well as a nematic order.
4. Substitution at R' favors nematic mesophase formation and greatly increases the total order of the system.
5. Substitution at R' reduced the solid to nematic transition temperature.

REFERENCES

1. W. Kast, Landolt Tabellen II 2a, Springerverlag, Berlin (1960) p. 304.
2. H. Arnold and P. Roediger, Z. phys. Chem. 231, 407 (1966).
3. M. Leclercq, J. Billard and J. Jacques, Molecular Cryst. 8, 367 (1969).
4. M. Leclercq, J. Billard and J. Jacques, submitted to Molecular Cryst., March 1970.
5. Edward M. Barrall II and Julian F. Johnson, Molecular Cryst. 8, 27 (1969).
6. A. Saupe, Molecular Cryst. 7, 59 (1969).

ACKNOWLEDGEMENTS

The author is greatly indebted to Dr. Jean Jacques and his colleagues, M. Leclercq and J. Billard, for their permission to use their extensive data on the Schiff's bases of p-aminocinnamic acid esters in the preparation of this paper.

SYNTHESIS AND CALORIMETRY OF SOME DERIVATIVES OF DIBENZAZEPINE

Edward Gipstein, Edward M. Barrall II and Karen E. Bredfeldt
International Business Machines Corporation
Research Laboratory
San Jose, California 95114

INTRODUCTION

Dibenzazepine consists of two aromatic rings connected by a single double bond bridge. The aromatic rings are prevented from assuming a planar configuration by an imide nitrogen bridge. This is shown in the drawing of a model of propionyl dibenzazepine in Figure 1. Construction of such models indicates that substitution on the nitrogen may give rise to compounds in which free rotation in the derivative is seriously hindered. Indeed, chemical evidence of hindrance was found in several dibenzazepine derivatives synthesized as monomers for polymerization studies (1). The N.M.R. spectrum for each derivative revealed rapid inversion of the non-planar dibenzazepine ring and restricted rotation about the C-N and C-C bonds.

These preliminary observations suggested that dibenzazepine derivatives may have unique thermodynamic and electronic properties. Therefore, a group of eleven derivatives were synthesized from purified dibenzazepine to test the effect of various N-side chains on the thermodynamic properties of the ring system. The effect of the side chain on the N.M.R. spectra was also of interest, and the results of that investigation will be reported in a future communication.

EXPERIMENTAL

Synthesis of Dibenzazepine Derivatives

5-H-Dibenz(b,f)azepine (compound 1, see Table I). "Iminostilbene," Aldrich Chemical Co., m. 193-196°, was recrystallized twice

Figure 1

Perspective Drawing of 5-Propionyl-5H-dibenz(b,f)azepine

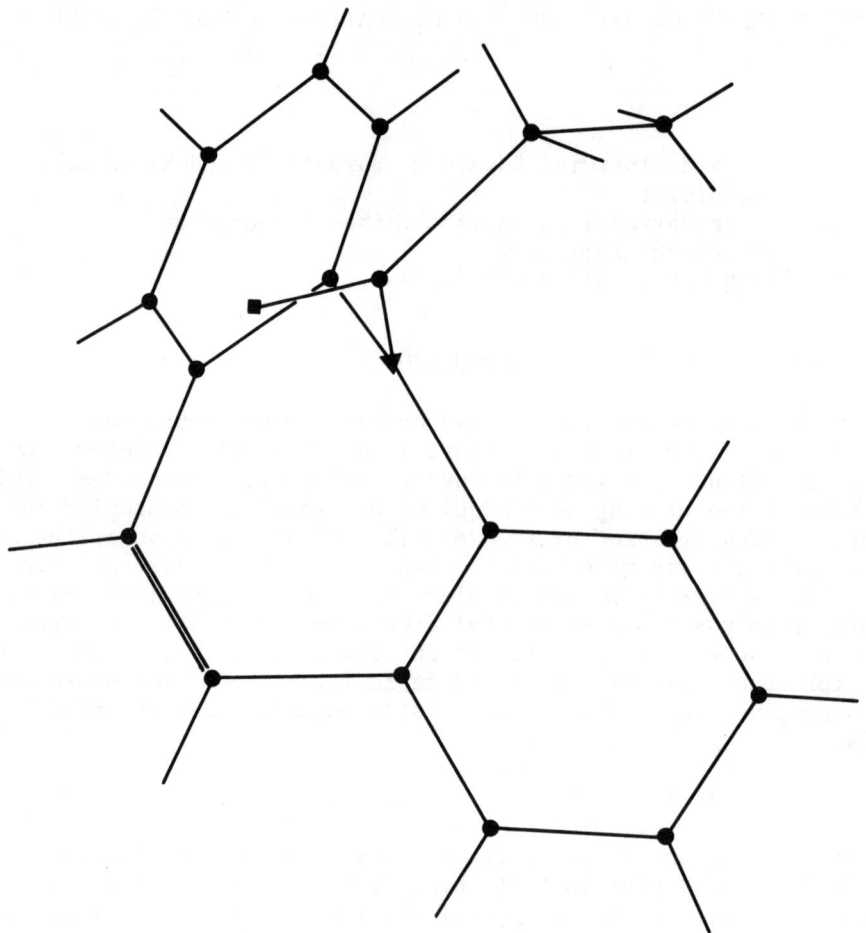

● = carbon, ▲ = nitrogen, ■ = oxygen

from heptane with decolorizing charcoal to give yellow platelets, m. 197.5-198.0° (lit. 196-198° (2)).

Dibenzazepine was condensed with a number of acid halides in benzene or toluene solvents to give N-substituted dibenzazepines. A N-formyl derivative was prepared by the use of excess formic acid. N-Acetyl derivatives were prepared using acetic anhydride and trifluoroacetic anhydride with zinc chloride. N-Acrylyl and N-methacrylyl derivatives (compounds 11 and 12) reported in a previous publication (1) were prepared by dehydrohalogenation of the chloropropionyl and bromoisobutylryl dibenzazepine derivatives (compounds 6 and 9), respectively. The syntheses of those derivatives not previously given (1,2) are outlined below.

5-Formyl-5H-dibenz(b,f)azepine (compound 2). A mixture of 19.3 g (0.1 mole) dibenzazepine and 230.2 g (5.0 moles) of formic acid (97%) was refluxed 2.5 hr, cooled and poured into ice water to give a white solid. Two recrystallizations from heptane with decolorizing charcoal gave 18.3 g (78%) of pure product as fine white needles. Anal. Calcd. for $C_{15}H_{11}NO$: C, 81.43; H, 5.01; N, 6.33. Found: C, 81.27; H, 5.02; N, 6.15.

5-Acetyl-5H-dibenz(b,f)azepine (compound 3). A mixture of 5.2 g (0.027 mole) dibenzazepine, 8.2 g (0.081 mole) acetic anhydride and 3.7 g (0.027 mole) zinc chloride was refluxed 1 hr in 200 ml dry benzene. The mixture was cooled, washed with water in a separatory funnel and the organic layer was separated and dried over Na_2SO_4. The solvent was evaporated under vacuum to give a yellow oil which crystallized on standing overnight. The crude solid was recrystallized twice from heptane with decolorizing charcoal to give 4.2 g (65.6%) of pure product as fine white needles. Anal. Calcd. for $C_{16}H_{13}NO$: C, 81.68; H, 5.57; N, 5.95. Found: C, 81.67; H, 5.58; N, 5.80.

5-Trifluoroacetyl-5H-dibenz(b,f)azepine (compound 10). The trifluoroacetyl derivative was prepared similarly using 17.0 g (0.081 mole) trifluoroacetic anhydride. The crude solid obtained was recrystallized three times from heptane with decolorizing charcoal to give 3.5 g (45.2%) of white crystals. Anal. Calcd. for $C_{16}H_8F_3NO$: C, 66.91; H, 2.81; F, 19.84; N, 4.87. Found: C, 66.52; H, 3.09; F, 19.93; N, 4.84.

5-Propionyl-5H-dibenz(b,f)azepine (compound 4). A mixture of 17.4 g (0.09 mole) dibenzazepine, 8.4 g (0.09 mole) propionylchloride and 9.2 g (0.09 mole) triethylamine was refluxed 18 hours in 300 ml dry benzene. The mixture was cooled, washed with water in a separatory funnel and the organic layer was collected and dried over Na_2SO_4. Evaporation of the benzene solvent under vacuum gave a yellow oil which crystallized to a solid mass by washing repeatedly with a solvent mixture of benzene/petroleum ether. The crude solid was recrystallized twice from heptane/petroleum ether (60-90°) (80:20) with decolorizing charcoal to give 8.2 g (36.1%) of white crystals. Anal. Calcd. for $C_{17}H_{15}NO$: C, 81.90; H, 6.06; N, 5.62. Found: C, 81.94; H, 5.97; N, 5.63.

5-(Isobutyryl)-5H-dibenz(b,f)azepine (compound 5). A mixture of 19.3 g (0.10 mole) dibenzazepine and 21.2 g (0.2 mole) isobutyrylchloride in 450 ml dry toluene was refluxed 18 hours. The mixture was cooled and concentrated under vacuum to give a yellow colored solid. The crude product was recrystallized twice from a 50/50 benzene/petroleum ether (60-90°) mixture with decolorizing charcoal to give 9.4 g (35.7%) of a white crystalline solid. Anal. Calcd. for $C_{18}H_{17}NO$: C, 82.10; H, 6.51; N, 5.32. Found: C, 82.09; H, 6.44; N, 5.21.

5-Chlorobutyryl-5H-dibenz(b,f)azepine (compound 7). A mixture of 5.2 g (0.027 mole) dibenzazepine, 7.61 g (0.054 mole) chlorobutyrylchloride in 200 ml dry toluene was refluxed 18 hours. The mixture was cooled and then rotary evaporated to give an orange colored oil. The oil after washing with petroleum ether (30-60°) and cooling to 0° for 12 hours solidified to an orange colored solid. Four recrystallizations from a 4:1 heptane/benzene solvent mixture with decolorizing charcoal gave, after cooling at 0° for 24 hours, 1.1 g (12.5%) of a white crystalline solid. Anal. Calcd. for $C_{18}H_{16}ClNO$: C, 72.60; H, 5.42; N, 4.70; Cl, 11.91 Found: C, 72.20; H, 5.42; N, 4.10; Cl, 11.80.

5-Bromoacetyl-5H-dibenz(b,f)azepine (compound 8). A mixture of 5.2 g (0.027 mole) dibenzazepine, 10.9 g (0.054 mole) bromoacetylbromide in 200 ml dry toluene was refluxed 18 hours. The mixture was cooled and then rotary evaporated to give a yellow oil. The oil solidified to a crystalline light yellow solid after repeatedly washing with absolute ether. Three crystallizations from a 5:1 heptane/benzene solvent mixture with decolorizing charcoal gave 4.3 g (50.5%) of fine white needles. Anal. Calcd. for $C_{16}H_{12}BrNO$: C, 61.16; H, 3.85; N, 4.46; Br, 25.44. Found: C, 60.88; H, 3.85; N, 4.26; Br, 25.48.

SCANNING CALORIMETRY

Differential scanning calorimetry (DSC) was carried out in a Perkin-Elmer DSC-1B scanning calorimeter at a heating rate of 1.25°C/min and a sensitivity of 2 millicalories per inch. The samples were encapsulated under nitrogen in aluminum volatile sample sealers. These sealers used an additional small insert to obtain good thermal contact. The samples were positioned in the center of the cells and covered with aluminum domes. The technique is described in detail elsewhere (3). The calorimeter temperature axis was calibrated using 99.9999 mole % gallium, benzoic acid, indium, tin and lead. The thermal response was also calibrated in the same way. The purity analysis was made using the van't Hoff equation described previously

(3-7). The data are given in Table I. All temperatures represent the vertex of the endothermal minimum and are corrected for thermal resistance. The data are all obtained on first heating of the solvent recrystallized solids. The solids formed oils on melting which super-cooled without recrystallization. IR and UV spectra did not indicate decomposition.

DISCUSSION

As was recognized by Pirsch (8) and restated by Bond (9) the topological conformation of each molecule is extremely important in analyzing the entropy change experienced when a molecular aggregation undergoes an order-disorder transition. For the purposes of discussion molecules are classified into a) rod-like, b) sheet or planar, c) globular or spherical and d) oblate spheroid. In each of the above groups the entropy of transition changes in a unique fashion as the shape of various members in an homologous series is altered.

By far the greater portion of the thermodynamic literature has been concerned with homologous series that are more or less rod-like. A fairly complete understanding of the transition entropy of normal hydrocarbons, alcohols, aldehydes, acids and esters has been obtained (9). These series usually show a regular systematic change in thermodynamic properties from member to member or as small changes are made on the backbone structure (9). In particular, the entropy of melting is usually some simple function of carbon chain length or arrangement given a constant functionality. The esters of such complex molecules as cholesterol show a regular entropy of melting increase with increasing ester chain length (10). This is almost independent of the group causing the chain length increase (aromatic, aliphatic, halogen). The total molar entropy of transition from ordered solid (inclusive of all solid → solid transitions below melting) to isotropic liquid, when plotted against molecular length, is generally a simple function. Any departure from such a sample relationship is generally due to the introduction of a new crystal order or the introduction of an unsuspected solid → solid transition.

Spherical or globular molecules exhibit a steady increase in entropy of fusion with increasing diameter (9). Wong and Westrum have shown bicyclo[2.2.2]octane and octene to have total solid → liquid transition entropies differing by only 0.27 cal/mole/°K presumably because of similar spherical molecular radii (11). In addition, Pirsch has shown for a large number of spherical and oblate molecules that entropy generally increases with increasing melting point (8). However, the general linear dimensions or membership in an homologous series contributed in only a minor way.

As shown in Table I, the derivatives of dibenzazepine do not show a regular variation in entropy of fusion as the N-alkyl side chain is increased in length. If this were the case, the entropy of fusion should increase as $-CH_2-$ groups are added to compounds 2, 3 and 4. A careful examination of side chain length, bulk and polarity does not reveal any systematic entropy effect for the derivatives of dibenzazepine shown in Table I. However, if the assumption that these molecules are oblate spheroids is made, and the entropy data are plotted as a function of melting point, Figure 2, a very systematic relationship appears. The materials shown in Table I fall predominantly along two straight lines. The series showing the largest entropy change with increasing melting point contains the formyl (compound 2), isobutyryl (compound 5), bromoacetyl (compound 8) and methacrylyl (compound 12) derivatives. The majority of the derivatives lie on a second line in which there is less dependence of entropy on melting point. The parent dibenzazepine (compound 1) lies well above both lines on Figure 2. Thus, it appears that the dibenzazepine derivatives form two sets of uniquely oblate spheroid molecules. The two sets appear to converge at a melting point of 144°C and an entropy of fusion of 18.6 cal/mole/°K. Although no long chain members of the series have been investigated, it is possible to speculate that the intersection may mark the point at which the side chain no longer forms an oblate molecular cluster with the dibenzazepine core.

The interaction of the side chain with the core must be very complex. No obvious explanation arises from the study of models. It was postulated that perhaps our data had missed a solid → solid transition. The DSC records were searched down to boiling liquid nitrogen and no phase transitions or heat capacity anomalies were noted. Although such a search does not rule out a meta-stable state, this would be very unusual for an entire series of 12 compounds.

Hydrogen bonding is responsible for the high melting point of dibenzazepine. On the basis of the present work dibenzazepine should melt nearer to 130°C rather than at 198°C in the absence of such bonding. Hydrogen bonding also lowers the entropy of fusion due to association in the melt (9).

The acrylyl derivative (compound 11) probably belongs to the same series as the formyl and bromoacetyl derivatives. The melting peak probably consisted of two transitions. Some spontaneous polymerization also could have occurred. This would lower the melting point.

CONCLUSIONS

On the basis of the calorimetry of eleven derivatives of dibenzazepine the following statements can be made:
1. The actual length of the derivative side chain has little affect on the heat or entropy of transition.
2. The molecules appear to be oblate spheroids for which the entropy of transition and the melting point form the only basis of correlation.

DERIVATIVES OF DIBENZAZEPINE

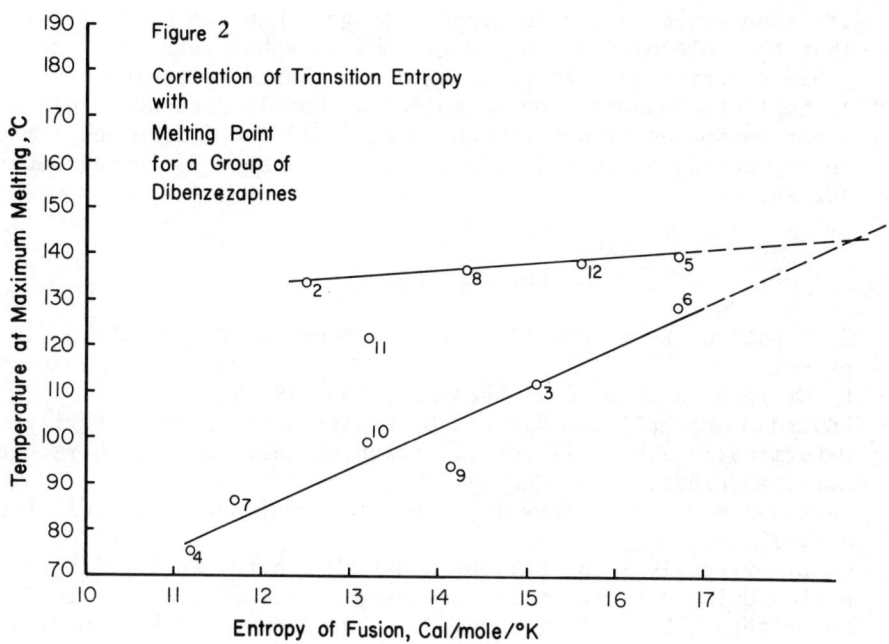

Figure 2
Correlation of Transition Entropy with Melting Point for a Group of Dibenzezapines

TABLE I

Thermodynamic Properties of Some Dibenzazepines

Graphical Code	Compound	Tm°C	ΔH cal/mole	ΔS cal/mole/°K	Purity*	Remarks
1	Dibenzazepine	198.0	6840	14.5	99.6	
2	Formyl Dibenzazepine	133.5	5092	12.5	99.4	
3	Acetyl Dibenzazepine	112.1	5847	15.1	97.4	
4	Propionyl Dibenzazepine	75.7	3904	11.2	----	Poorly defined peak, perhaps two.
5	Isobutyryl Dibenzazepine	140.4	6890	16.7	>99.99	
6	Chloropropionyl Dibenzazepine	129.3	6707	16.7	98.6	
7	Chlorobutyryl Dibenzazepine	86.3	4188	11.7	~96.3	
8	Bromoacetyl Dibenzazepine	136.8	5856	14.3	~97.8	
9	Bromoisobutyryl Dibenzazepine	86.3 94.1	2220 2930 Σ=5150	6.2 7.98 Σ=14.18	----	Two peaks on first melt.
10	Trifluoroacetyl Dibenzazepine	00.2	4010	13.2	99.67	
11	Acrylyl Dibenzazepine	121.9	5202	13.2	98.9	
12	Methacrylyl Dibenzazepine	138.3	6435	15.6	98.95	

* Calculated from Differential Scanning Calorimetry

3. Since the materials fall into two distinct groups, on the basis of item 2 there appears to be two fundamental groups of molecular shape.
4. It is possible, at chain lengths longer than the C_4 limit studied, that the molecules are no longer oblate spheroids. This could yield a series more comparable to rod shaped molecules.
5. A simplistic examination of molecular models does not yield a clear reason as to why compounds 2, 8, 12 and 5 form one series and compounds 4, 7, 10, 9, 3 and 6 form a second clearly separate series.

REFERENCES

1. E. Gipstein, W. A. Hewett and O. U. Need, J. Polymer Sci., in press.
2. P. Craig et al., J. Org. Chem. 26, 135 (1961).
3. Edward M. Barrall and Richard D. Diller, Precision of Purity Determinations by Differential Scanning Calorimetry, Thermochemica Acta, in press.
4. Perkin-Elmer Corp., Norwalk, Conn., Thermal Analysis Newsletter, Nos. 1-6.
5. G. L. Driscoll, I. N. Dulling and F. Magnotta, Purity Determination Using a Differential Scanning Calorimeter, Analytical Calorimetry, R. S. Porter and J. F. Johnson eds., Plenum Press, N. Y. (1968) pp. 271-278.
6. N. J. DeAngelis and G. J. Papariello, J. Pharm. Sci. 57, 1868 (1968).
7. C. Plato and A. R. Glasgow Jr., Anal. Chem., 41, 330 (1969).
8. J. Pirsch, Chem. Ber. 70, 12 (1937).
9. A. Bondi, Chem. Rev. 67, 565 (1967).
10. R. S. Porter, E. M. Barrall and J. F. Johnson, Accounts of Chem. Res. 2, 53 (1969).
11. W. Wong and E. F. Westrum Jr., J. Phys. Chem. 74, 1303 (1970).

WATER BINDING INDEX OF PROTEINS AS DETERMINED BY DIFFERENTIAL MICROCALORIMETRY

Endel Karmas and G. Robert DiMarco

Department of Food Science

Rutgers University, New Brunswick, New Jersey

INTRODUCTION

The functional and protective role of water in biological materials is critical and supposedly has an important relationship with its state and binding. The water binding aspects of muscle tissue have been reviewed by Hamm (1960). He states that the water binding of muscle proteins is dependent on the pH-value, the minimum being at the isoelectric point. Giese (1962) reports that an average molecule of living muscle protein is associated with some 17,000 water molecules.

Any method devised to determine water binding depends on the definition of "bound" water. This term is not easily defined. Most workers define "bound" water as the water which remains in an unchanged form when the system is subjected to a particular treatment, e.g., water which does not freeze at low temperatures or does not combine with a chemical desiccant or otherwise behaves differently from "free" water. On the other hand, Kuprianoff (1958) states that all water in biological systems is bound and cannot be separated into "free" and "bound" water.

Methods for the determination of water binding are divided into three major groups:

(1) Methods based on water abstraction -- such as freezing, heat dehydration, chemical dehydration, sedimentation, centrifugation, filtration, pressing, desorption. The general procedure of these methods is that free water content has to be defined on the basis of the method of determination and subsequently subtracted from the amount of total water in order to determine the amount of bound water. There, however, does not seem to be a clear borderline. Sometimes one is unable to separate any juice from tissues at even very high gravitational forces of centrifugation.

(2) Methods based on water uptake -- such as adsorption, osmotic pressure, hydration. The starting materials have to be "bone-dry". It is known, however, that the removal of initial bound water leads to irreversible changes in biological systems which subsequently lead to erroneous experimental results regarding bound water.

(3) Methods based on various other principles -- such as vapor pressure lowering, dielectric constant, nuclear magnetic resonance, dilatometry.

The literature reveals that the agreement of the results obtained by different methods is frequently poor. Wide deviation is found between results on the same kind of material not only when using different methods, but also when the same method is used. It appears that with the freezing method (Riedel, 1961) the amount of non-freezable water can be determined with reasonable accuracy (Kuprianoff, 1958).

To measure the water binding of proteins means to measure the energy input that is necessary to liberate the water from proteins. Differential microcalorimetry (Watson et al., 1964) seems to be especially suited to measure such energy directly. In addition, there are strong indications that a relationship exists between the dehydration energy and water binding of proteins. Therefore, the aim of this study was to establish a quantitative method to compare the protein dehydration with the evaporation of distilled water which may be expressed as a "water binding index".

EXPERIMENTAL

Beef muscle tissue and egg albumin as well as soy and milk protein isolates and gelatin were used as sample materials. In addition to these, salt-soluble fraction of beef muscle tissue was extracted (Karmas, 1968), the final moisture content of which was 81.2%.

Protein isolates at 25% concentrations of the total weight were mixed with water in a Waring blender provided with Variac control. In addition, to some samples 2% sodium chloride was added.

Gelatin gels were prepared as described by Karmas (1968).

Samples, calculated to contain approximately 2 mg of water, were weighed out in open sample pans by means of Cahn Electro-balance. Pure water was used as reference material. Each sample was dehydrated isothermally at a desired temperature in Differential Scanning Calorimeter, Model DSC-1B (The Perkin-Elmer Corporation, Norwalk, Connecticut), at 0.032 cal/sec for a full-scale deflection. The areas underneath the obtained curves were measured by a planimeter and the corresponding dehydration energies determined. Subsequently, the following empirical ratio was established and calculated for each sample:

$$\text{Ratio} = \frac{\text{Energy required to evaporate 1 mg water from protein}}{\text{Energy required to evaporate 1 mg distilled water}}$$

A preliminary study was performed at four temperatures -- $45°$, $68°$, $82°$, and $100°C$ -- to determine the variation in the above-mentioned ratio as a function of temperature. As a result of this study, $105°C$ was selected as the appropriate temperature for dehydration of samples. The ratio determined at $105°C$ was called the Water Binding Index (WBI).

A standardization curve was constructed, since it would have been extremely difficult to weigh each sample to contain exactly 2 mg of water.

RESULTS AND DISCUSSION

In a preliminary experiment, the variation of the above mentioned ratio was studied as a function of temperature. Four temperatures -- $45°$, $68°$, $82°$, and $100°C$ -- were selected. The sample materials were beef muscle tissue, salt-soluble beef muscle protein, and egg albumin, both raw and cooked. The results were plotted in Fig. 1. The values for the ratio of "energy required to evaporate 1 mg water from protein" to "energy required to evaporate 1 mg distilled water" for these proteins at $45°C$ and $100°C$ were particularly distinct, although they were of almost opposite ranking order at these temperatures. At the two middle temperatures, the ratios were not significantly different from the reference ratio to permit any comparison.

The low values of the ratios obtained at $45°C$ were probably due to a radiation difference of the reference and sample pans placed in the calorimeter. While the sample pan was partially shielded with an isolating proteinaceous layer, thus diminishing the radiation, the entire surface of the reference pan was radiating freely. This heat radiation effect was compensated for by the differential control loop of the instrument, and partially neutralized the energy required to dehydrate proteins. The apparent ratio values were thus lower. This effect increased markedly with larger sample size and, hence, longer dehydration time. Due to this fact, time economy involved, and a meaningful numerical value, it was more advantageous to obtain the ratios at $100°C$. However, instead of $100°C$, $105°C$ seemed to be more appropriate. This temperature, slightly higher than the boiling point of water, has been widely accepted as a standard temperature for dehydration of materials for moisture determination. Therefore, all WBI values were determined at $105°C$.

The preceding discussion pointed out that variation in sample size caused radiation differences. This fact emphasized the importance of uniform sample size. Since it was extremely difficult to weigh out each sample to contain exactly 2 mg of water, a calibration curve was devised to diminish the error in WBI determination. The curve in Fig. 2 shows that the smaller the sample weight, especially below 2 mg, the more sensitive the results were in terms of energy units but the larger also the standard deviation or error. Also the error due to preliminary

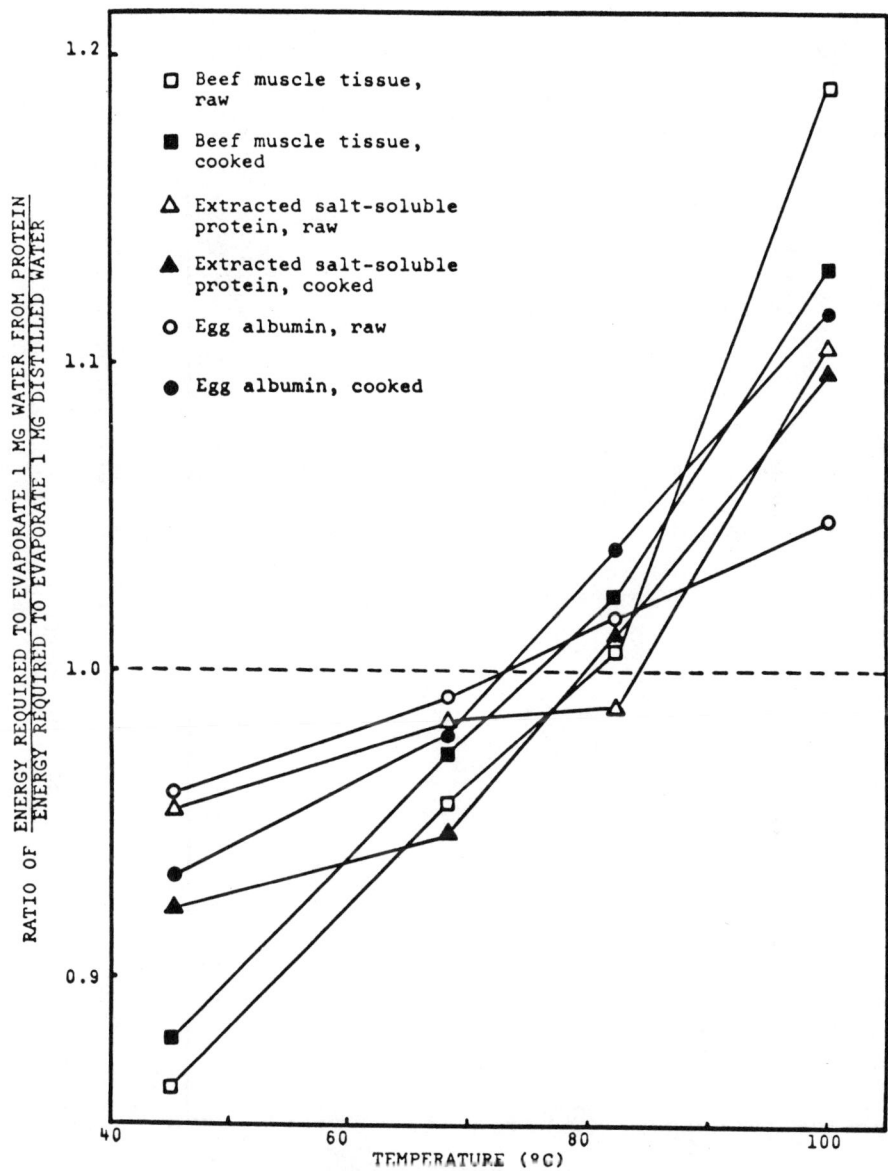

Fig. 1. Change in the ratio of "energy required to evaporate 1 mg water from protein" to "energy required to evaporate 1 mg distilled water" as a function of temperature.

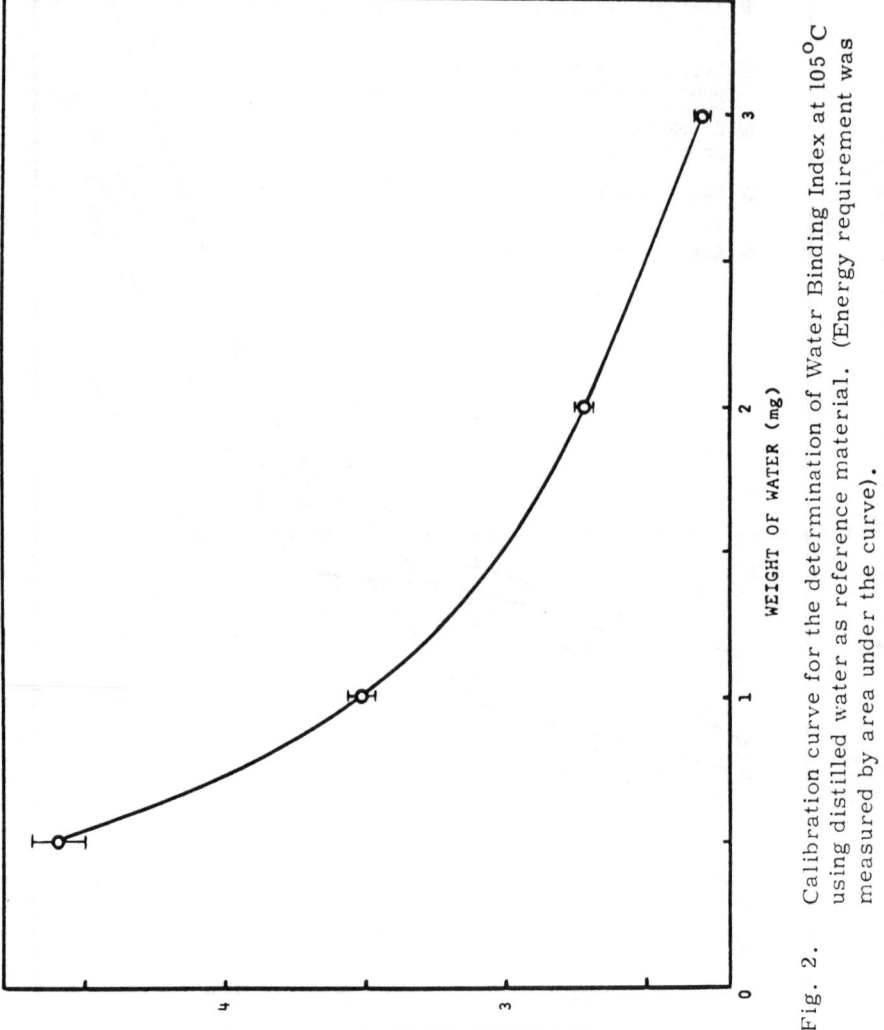

Fig. 2. Calibration curve for the determination of Water Binding Index at 105°C using distilled water as reference material. (Energy requirement was measured by area under the curve).

water evaporation during the sample preparation and weighing was relatively more serious for smaller samples. According to the calibration curve, an appropriate sample weight seemed to be between 2.0-2.5 mg of water. The instrument had its limit for 0.032 cal/sec sensitivity range at about 5 mg sample weight.

The results of a preliminary experiment are given in Table 1.

Table 1. Water Binding Index of various proteins

Material	WBI
Sodium chloride, 2% aq. solution	1.01 ± 0.00
Isolated soy protein, 25% in water, pH=4.6	1.09 ± 0.02
Beef muscle tissue, raw (73.8% average moisture content)	1.16 ± 0.03
Isolated soy protein (sodium form), 25% in water, pH=7.0	1.33 ± 0.02
Sodium caseinate, 25% in water	1.45 ± 0.02
Sodium caseinate, 25%, + sodium chloride, 2%, in water	1.49 ± 0.03

A WBI was obtained when only 2% sodium chloride was added to water. This effect was increased with the presence of proteins. Soy protein isolate at the isoelectric point, pH=4.6, had considerably lower WBI than the isolate at pH=7.0. Hence, the WBI seems to confirm higher water binding at pH values other than the isoelectric point. The WBI for sodium caseinate was higher than that for the sodium form of soy protein isolate; however, both proteins possessed a much higher WBI than beef muscle tissue. Since the moisture content of the proteins was similar (73-75%) and the rest of the samples was mostly protein, the sensitivity of the WBI from one protein to another may be considered significant.

Table 2 tabulates the results, together with standard deviations, for various proteins.

Table 2. Water Binding Index of raw and cooked animal proteins.

Protein	WBI
Raw:	
Beef muscle tissue (73.8% moisture)	1.175 ± 0.009
Extracted salt-soluble beef muscle protein (81.2% moisture)	1.095 ± 0.005
Egg albumin (87.7% moisture)	1.056 ± 0.011
Cooked:	
Beef muscle tissue	1.124 ± 0.005
Extracted salt-soluble beef muscle protein	1.090 ± 0.007
Egg albumin	1.105 ± 0.013

It is interesting to note that the WBI values are rather uniform for all cooked samples pointing out similarities in denatured proteins. The WBI value for raw beef muscle tissue was very agreeable with that in Table 1. This fact indicates good precision of the method. WBI for raw egg albumin was surprisingly low. This may be due to the higher moisture and, hence, the lower protein content. Since the weight calibration curve was used for this experiment, the standard deviation averaged less than 1%.

In a further experiment, the WBI of gelatin gels, up to 8% gelatin content (Bloom value = 260), were determined. The results with standard deviations are plotted in Fig. 3. As the concentration of gelatin increased, the WBI, seemed to increase sigmoidally. The curve should contain more data points to permit exact interpolation of the curve. If the curve presented in Fig. 3 is correct, the WBI values for gelatin, and probably for other proteins, seem to approach eventually a certain

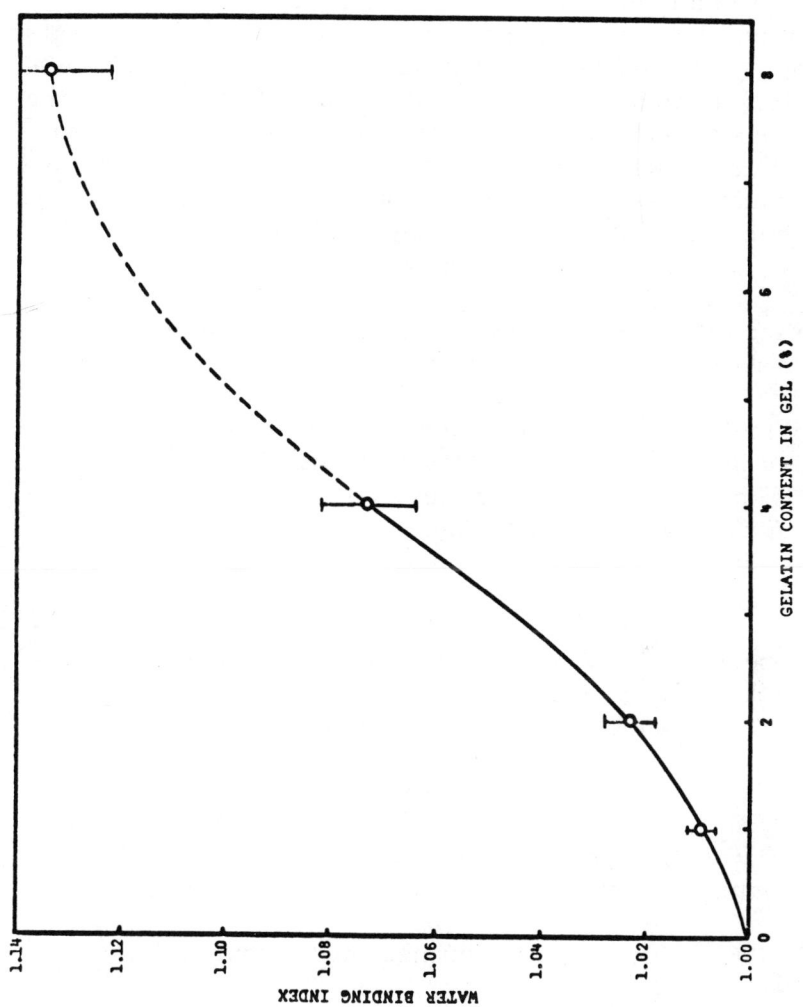

Fig. 3. Water Binding Index of gelatin gels.

limiting asymptotic value. This appears to be a very important aspect and might be a distinguishing characteristic for every protein and, therefore, merits further investigation. The average standard deviation in this experiment was again considerably less than 1%.

The WBI is an empirical method for the determination of water binding. It belongs to the group of methods based on water abstraction. The total energy required to remove water from a water-protein system, to transport it through the network of the protein matrix to the surface, and to evaporate it is measured isothermally in a microcalorimeter. The methods referred to in the introductory section are based on the measurement of various other properties. For this reason, no comparison of this method with other methods was attempted.

Since the temperature of $105^\circ C$ applied to a water-protein system for a relatively short time is not sufficient to remove the "tightly bound" or chemical hydration water, the WBI apparently measures the abstraction of water which has been immobilized by a protein matrix (such as muscle cells or gelatin gels) or in a protein solution (such as egg albumin). The empirical nature of the index is further emphasized by the fact that beside the species also the amount of the protein has to be considered as parameter.

The concept of the WBI may be extended to other substances that interact with water (such as carbohydrates) and combinations of such substances with proteins in aqueous systems.

REFERENCES

Giese, A. C. 1962. "Cell Physiology", 2nd Ed.; p. 35. W. B. Saunders Co., New York, N. Y.

Hamm, R. 1960. Biochemistry of meat hydration. Adv. Food Res. 10, 355.

Karmas, E. 1968. Interactions of water with amino acids and proteins as determined by differential microcalorimetry. Ph.D. thesis. Rutgers University, New Brunswick, N. J.

Kuprianoff, J. 1958. "Bound water" in foods. In "Fundamental Aspects of the Dehydration of Foodstuffs," p. 14. Society of Chemical Industry, London; ed. The Macmillan Co., New York, N. Y.

Riedel, L. 1961. The problem of bound water in meat. Kaltetechnick 13, 122.

Watson, E. S., O'Neill, M. J., Justin, J., and Brenner, N. 1964. A differential scanning calorimeter for quantitative differential thermal analysis. Anal. Chem. 36, 1233.

THE MOLECULAR WEIGHT DEPENDENCE OF THE TRANSITION ENTHALPY OF

POLY-γ-BENZYL-L-GLUTAMATE

A. Kagemoto* and F. E. Karasz

Polymer Science and Engineering, University of Massachusetts, Amherst, Massachusetts   01002

INTRODUCTION

It is now well known that the poly-γ-benzyl glutamate (PBG) molecule adopts an α-helical conformation in helicogenic solvents such as 1,2-dichloroethane (DCE) and a random-coil conformation in strong hydrogen bonding solvents such as dichloroacetic acid (DCA) (1). The significant thermodynamic parameters of this prototypical helix-coil transition have been measured by a variety of techniques. Thus the overall van't Hoff heat of transition (ΔH_{VH}) is most conveniently obtained from optical rotatory measurements; the transition enthalpy per peptide residue (ΔH_o) has been found from measurements of heat capacities (2), heats of solution (3) or dilution (4), or indirectly from studies of transition temperatures as a function of solvent composition (5). Such measurements have provided considerable insight, also, into the effects of variables such as transition temperature and solute concentration on the primary parameters ΔH_o and ΔH_{VH}, and their ratio $\sigma^{1/2}$, where σ is the Zimm-Bragg cooperativity parameter (6).

Because of the cooperative nature of this transition, it is also of interest to consider the molecular weight dependence of these thermodynamic parameters. From the terms of the Zimm-Bragg theory, we can estimate that at the midpoint of the transition, the mean number of residues in either a random-coil or helical sequence will be of the order of $\sigma^{-1/2}$ or about 10^2, corresponding to a PBG molecular weight of $\approx 2.5 \times 10^4$. Thus it may be anticipated that size effects could persist to molecular weights perhaps

*Permanent Address: Department of Chemistry, Osaka Institute of
                     Technology, Osaka, Japan

a few times larger than this. (This qualitative point remains valid regardless of which of the many theoretical models of the cooperative transition is used to analyze the data.) Optical rotation studies of PBG from which ΔH_{VH} can be obtained have recently been published for a range of molecular weights (7). Earlier studies of ΔH_{VH} were actually used, in effect, to arrive at the original estimate for σ (8,9). However, no systematic direct studies of ΔH_o as a function of molecular weight, permitting an unequivocal estimate of σ, have been previously reported. Miller, et al. (10), have been able to study these parameters in the water soluble polypeptide, poly-L-glutamic acid, using a titration curve analysis.

As a first objective of the present paper, therefore, we examine ΔH_o for PBG samples with M_v ranging from 35,000 to 550,000. By combining the results with existing ΔH_{VH} data we can also discuss the corresponding dependency in σ. A second objective of this work was to develop a technique for measuring ΔH_o using a commercially-available device, the Differential Scanning Calorimeter (DSC) (Model 1-B, Perkin-Elmer Corporation). Previous investigations had shown that ΔH_o could be obtained from C_p measurements using conventional adiabatic single chamber or twin calorimeters (2). Such measurements have certain advantages over heat of solution or dilution techniques: it is relatively easy to make measurements over a wide range of transition temperatures, that is, in effect, solvent compositions, and it is possible to estimate ΔC_p (the difference in heat capacities of the ordered and disordered species) directly. However this technique suffers to a greater or lesser degree from experimental complexity, the requirement of large samples (several hundred milligrams of dry polypeptide), and relatively slow measurement speeds. The DSC, although not providing the same precision in C_p measurements as classical calorimeters, or the same flexibility in working at low heating rates, is inherently an apparatus which uses samples of the order of several milligrams and measurements are comparatively rapid. As has been recently pointed out, the study of transition phenomena in many polypeptides, proteins and nucleic acids would be greatly facilitated by the development of small-scale techniques (11).

EXPERIMENTAL

Apparatus

The heats of transition in this study were measured by using a modified Perkin-Elmer DSC-1B. Two modifications were adopted to provide sufficient sensitivity for the present purpose.

(a) A D.C. voltage amplifier (Keithley Model 150B) was interposed between the normal output of the DSC and the strip-chart recorder (Leeds and Northrup Model G; 1 mv. span) to provide additional gain by a factor of from three to ten. Electromagnetic shielding and line-voltage filtering were found helpful in reducing noise at maximum amplification.

(b) The sample containers provided by the manufacturers, normally used for measurements involving volatile liquids, were replaced by special sample cells of approximately three times greater capacity (70 µℓ). These were punched from aluminum sheets and sealed with aluminum lids using a crimper of design similar to that originally provided, but appropriately enlarged. Complete sealing was aided by carefully coating the circumference of the cell and lid with Eastman 910 adhesive.

Measurements were made at heating rates of 10°C per minute with the reference cell containing the same solvent mixture as the sample cell. The apparatus was calibrated by measuring the heats of fusion of zone-refined samples of appropriate low molecular weight organic compounds.

Materials

Seven samples of PBG used in this study were purchased from Pilot Chemical and Miles-Yeda Companies. The reported viscosity average molecular weights were 3.5×10^4, 4.5×10^4, 9.9×10^4, 2.9×10^5, 3.35×10^5, 4.35×10^5 and 5.5×10^5. Because of the method of synthesis, it is expected their polydispersity were relatively small. The solvents used, DCA and DCE, were of reagent grade.

For each measurement, 3 wt. % PBG solution was prepared in a mixed solvent containing 75 vol. % DCA and 25 vol. % DCE. This solvent composition located the transition at about 25°C.

RESULTS

It was established that with the apparatus modified as described above, the transition enthalpy of a typical sample, corresponding in this case to an average of about 6 millicalories, could be measured with a precision of about ± 1 millicalorie. This relative precision is comparable to that attained in earlier adiabatic calorimeter experiments (2a), though is inferior by a factor of perhaps 5 to 10 to that obtained with a more appropriately designed twin solution calorimeter (2b). The latter, however, also usually require relatively large samples. It is felt that in many studies of polypeptides, the lack of high precision is offset by the reduction in sample required.

The transition enthalpy per mole of residue for PBG is recorded in Table I. Included are two earlier results using solutions of the same solvent composition and solute concentration obtained respectively with an adiabatic calorimeter and with a DTA apparatus. These are in good agreement with the present data.

TABLE I

TRANSITION ENTHALPY OF THE HELIX-COIL TRANSITION
OF PBG 3 WT. % SOLUTION IN
75 VOL. % DCA/25 VOL. % DCE MIXTURE

$10^4 M_V$	ΔH_o (Cal./Residue Mole)
3.5	300 $\pm$ 80
4.5	380 $\pm$ 80
7.7	330 $\pm$ 100*
9.9	490 $\pm$ 50
25.0	525 $\pm$ 80**
29.0	615 $\pm$ 80
33.5	590 $\pm$ 80
43.5	600 $\pm$ 50
55.0	515 $\pm$ 60

*A. Kagemoto, K. Tada and Y. Baba, unpublished result.
**F. E. Karasz and J. M. O'Reilly, Biopolymers 4, 1015 (1969).

The plot of ΔH_o vs. log M_V, Figure 1, demonstrates that the transition enthalpy is considerably dependent on the molecular weight in the lower part of the range studied. The asymptotic value of 550 $\pm$ 60 cals/mole residue (for a 3% solution of the stated solvent composition) appears to be reached around a molecular weight of about 10^5. The finding is in close accord with what can be deduced from the optical rotation results of Teramoto, et al. The latter authors have demonstrated by a self-consistent examination of the Moffitt parameter b_o that PBG does not form a completely helical conformation at high temperatures even in pure DCE, until the molecular weight exceeds about 1×10^5. It is assumed that at shorter chain lengths the helical sections are, to a greater or lesser degree, interrupted by one-dimensional lattice imperfections which permit rotational motion of the stiff helical segments. The numbers and locations of these faults are presumably a dynamic equilibrium in each chain.

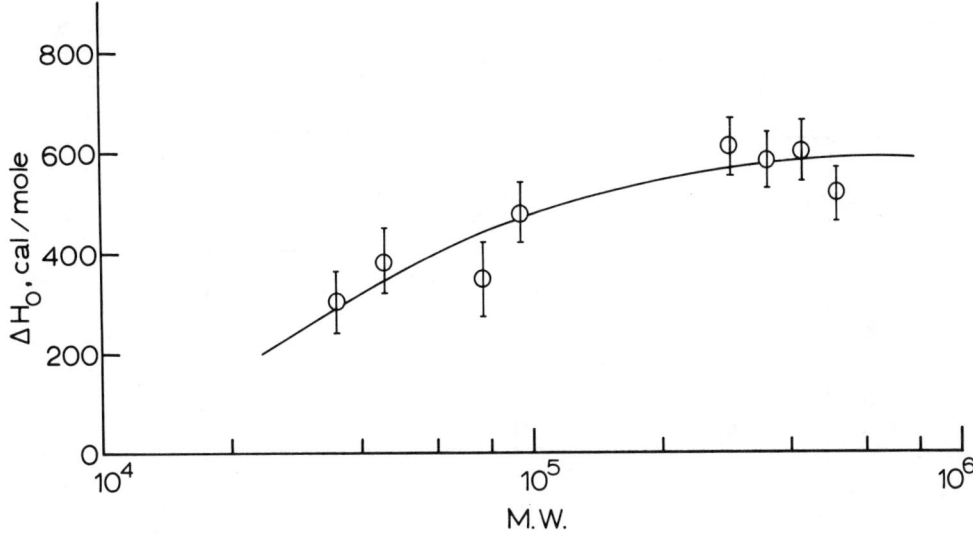

Figure 1. Calorimetric heat of transition per residue mole as function of molecular weight.

The lower values of ΔH_o found in the present study are therefore attributed to the failure for the conversion to proceed completely and not to an inherent molecular weight dependent effect in ΔH_o. The latter is a function of difference between the two intramolecular hydrogen bonding enthalpies in the polypeptide and the dimerized acid and the intermolecular peptide-acid bond, and it would be unlikely for ΔH_o to be influenced by comparatively long range conformational effects. We exclude also end effects occurring at helix-coil interfaces, because even at the shortest chain lengths considered here the number of such interfaces at high temperature is small.

This view is substantiated when we examine the cooperativity parameter σ. The failure to attain complete helicity is reflected also in a change in the calculated ΔH_{VH}. We have taken the ORD results of Teramoto, et al., and calculated ΔH_{VH} by standard techniques (Table II). Included in this tabulation are three other measurements taken from earlier work by Yang and by Karasz and O'Reilly. These results together show a comparable decrease in ΔH_{VH} with decreasing molecular weight.

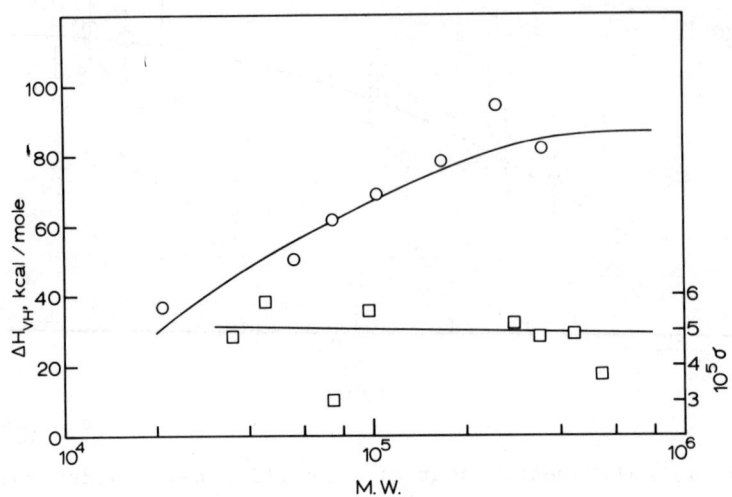

Figure 2. Van't Hoff heat of transition and Zimm-Bragg cooperativity parameter as function of molecular weight.

TABLE II

VAN'T HOFF HEATS OF TRANSITION
FOR PBG FROM ORD DATA
(FROM REF. 7 UNLESS OTHERWISE INDICATED)

$10^4 M_V$	ΔH_{VH} (Kcal/Mole)
2.0	31*
5.6	40
7.6	66
10.8	67
17.6	68
35.0	99*

*J. T. Yang, Tetrahedron 13, 143 (1961).

The plot of σ against molecular weight, Figure 2, indicates, in contrast, that the cooperativity parameter is, within the limits of experimental error, independent of molecular weights as low as 5×10^4. We would expect, of course, that at still lower molecular weights σ would increase, inasmuch as $\sigma^{-1/2} = \bar{n}$ in the Zimm-Bragg theory, where $\bar{n}$ is the average number of residues in a helical or a random-coil sequence at the midpoint of the transition. This increase of σ found in PGA by Miller, et al. (10), seems to occur at a chain length of about 70 residues or approximately $5\sigma^{-1/2}$. At PBG molecular weights of 50,000 we are already at about $2\sigma^{-1/2}$ and an upswing in σ is to be expected. Unfortunately, the decrease in ΔH_o at the comparatively low molecular weights increases the difficulty of making an accurate determination of this parameter in the present apparatus, and the quantities of PBG of still lower molecular weights available to us were not sufficient to use an adiabatic calorimeter.

CONCLUSION

It has been found experimentally that the thermodynamic parameter ΔH_o, corresponding to the heat of transition per residue mole, although apparently decreasing for molecular weights below about 10^5, is actually independent of this parameter when account is taken of the less than complete conversion of the coil to helix conformation.

The experiment has also demonstrated that the signal/noise ratio of the DSC-1B is low enough for modifications to be made so that it becomes capable of measuring transitions involving an enthalpy of at least 5 millicalories per sample, provided that the half-width of the transition is less than about 10°.

ACKNOWLEDGEMENT

This work was supported by NSF Grant GB-8080.

REFERENCES

1. P. Doty and J. T. Yang, J. Am. Chem. Soc. 78, 498 (1956).
2. (a) F. E. Karasz and J. M. O'Reilly, Biopolymers 4, 1015 (1966).
   (b) T. Ackermann and E. Neumann, Biopolymers 5, 649 (1967).
3. G. Giacometti, A. Turolla and R. Boni, Biopolymers 6, 441 (1968).
4. A. Kagemoto and R. Fujishiro, Makromol. Chem. 14, 139 (1968).
5. F. E. Karasz and J. M. O'Reilly, Biopolymers 5, 27 (1967).

6. B. H. Zimm and J. K. Bragg, J. Chem. Phys. 31, 526 (1959).
7. A. Teramoto, K. Nakagawa and H. Fujita, J. Chem. Phys. 46, 4197 (1967).
8. B. H. Zimm, P. Doty and K. Iso, Proc. Natl. Acad. Sci. U.S., 45, 1601 (1959).
9. Y. Hayashi, A. Teramoto, K. Kawahara and H. Fujita, Biopolymers 8, 409 (1969).
10. R. L. Snipp, W. G. Miller and R. E. Nylund, J. Am. Chem. Soc. 87, 3547 (1965).
11. T. Ackermann in BIOCHEMICAL MICROCALORIMETRY, H. D. Brown, ed., Academic Press, New York, 1969.

DERIVATIVE THERMOGRAVIMETRIC ANALYSIS OF AN EPOXY ADHESIVE AND A PHENOLIC-SILICA PREPREG

Stanley E. Gordon and Ross C. Caballero

McDonnell-Douglas Astronautics Company

Santa Monica, California 90406

INTRODUCTION

The aerospace industry has a need for an improved method for the analyses of cured epoxies and phenolic-silica prepregs. This analysis is needed for quality control on incoming material, for continuing control on inventory material, for material control during and after manufacture, and for material analysis after failure.

A technique which can be used to perform the required analyses is thermogravimetric analysis (TGA) concurrent with its derivative. The use of the derivative of the TGA curve is a very old technique;[1,2,3,4] however, its use in the analysis of polymers has been neglected or unpublished. This apparent lack of interest is unfortunate since so much more information can be gained from the derivative curve than from the typical or integral TGA curve which usually, in polymer analysis, results in a monotonically decreasing line with slight inflections in it. Alternatively, the derivative exhibits a series of maximum and minimum rates of weight loss areas which will be shown to have much greater value in analysis than the typical TGA curve.

EXPERIMENTAL

The equipment used was an E. I. du Pont Model[a] 950 Thermogravimetric Analyzer[5] to which was attached a Cahn[b] Time Derivative Computer Mark II. The derivative block diagram (Figure 1) shows the interconnection of the analyzer's meter coil, the Derivative

---

a. E. I. du Pont, Wilmington, Delaware 19898.
b. Cahn-Ventron, Paramount, California 90723.

Computer and the auxiliary X-Y recorder with the Model 900 Thermal Analyzer Console. The plus-input of the Derivative Computer is connected between the operational amplifier (A-1) and the meter coil. This connection can be made conveniently on either the meter coil side of R-47 (a 1.5 megohm resistor) or C-2 (an 8-microfarad capacitor). The minus-input must be connected to one of three possible points and the point to be used must be determined experimentally. The three points are (1) between R-6, the span calibration potentiometer and the meter coil, (2) between R-6 and R-27, the balance zero potentiometer, and (3) direct to the instrument ground or Y-minus. The magnitude of the derivative signal is highest at (1) and lowest at (3), but due to the electronics in the TGA system, the linearity of the signal is different at each connection. A baseline needs to be determined with each of the three connections to see which is preferred; also, the baseline should be checked periodically since it slowly changes with the age of the TGA components.

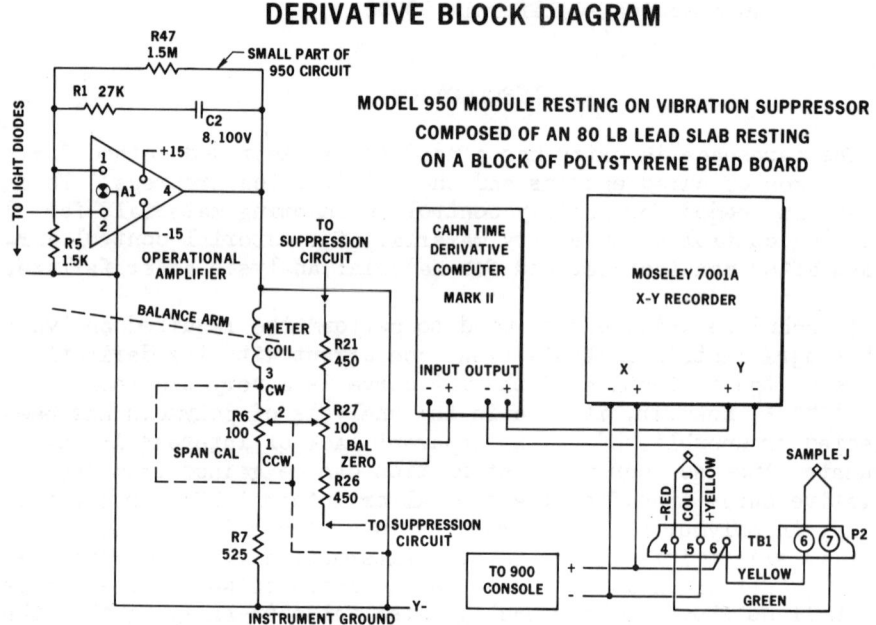

Figure 1. Block diagram showing interconnection of 950 meter coil, Derivative Computer, and X-Y recorder.

The epoxy was Lefkoweld 109, a filled polyamide cured epoxy manufactured by Leffingwell Chemical Company[a]. The cure conditions were 160°F for 24 hours. The epoxy analytical samples were scraped from tensile test blocks and from lap shear specimens as a fine powder. The sample size was 10 - 13 mg. A heating rate of 15°C/min.

---

a. Leffingwell Chemical Company, Brea, California 92621.

was used up to 700°C, in argon flowing at a rate of 700 cc/min. The weight loss was recorded using a scale of 0 - 100%.

The phenolic-silica prepreg was DP 24-2 (Buna N modified) impregnated into C 100-96 silica cloth and B-staged by Western Backing Company.[a] The phenolic analytical sample was a single small rectangle with a weight of 50 - 58 mg. cut from impregnated broad goods. The sample was analyzed by heating in clean air (dry air should not be used as total lack of water vapor delays oxidation) flowing at 700 cc/min. The heating rate was 5°C/min. up to 300°C at which point the heating rate was increased to 45°C/min. The weight loss scale recorded was 90 - 100% up to 300°C, where the scale was changed to 60 - 100%.

RESULTS - DISCUSSION

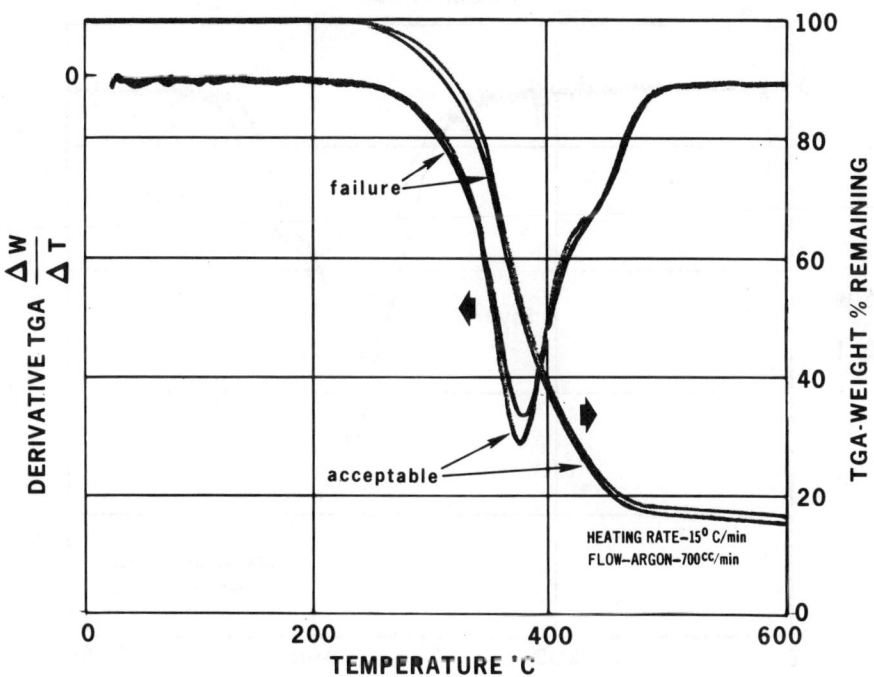

Figure 2. Derivative and integral TGA curves of the epoxy samples from both acceptable and failure test blocks.

---

a. Western Backing Company, Culver City, California 90230.

The investigation of Lefkoweld 109 epoxy adhesive was prompted by a randomly occurring adhesive failure between the adhesive and a sealed anodized aluminum substrate.[a] The failure was observed when tensile test blocks were immersed in liquid nitrogen. All standard methods of failure analysis, material composition analysis, and physical and mechanical analyses were unsuccessful in determining the cause of the random failure. It was obvious, therefore, that the failure was caused by some obscure phenomenon. Based on previous experience with derivative TGA, an investigation of the adhesive was made in an effort to determine the material and/or physical variation which could cause the observed failure.

The initial analyses were performed using powdered epoxy samples taken at random from a failed test block and from an acceptable test block (Figure 2). No significant difference in either the integral or derivative TGA curves from either the failed or acceptable samples was observed.

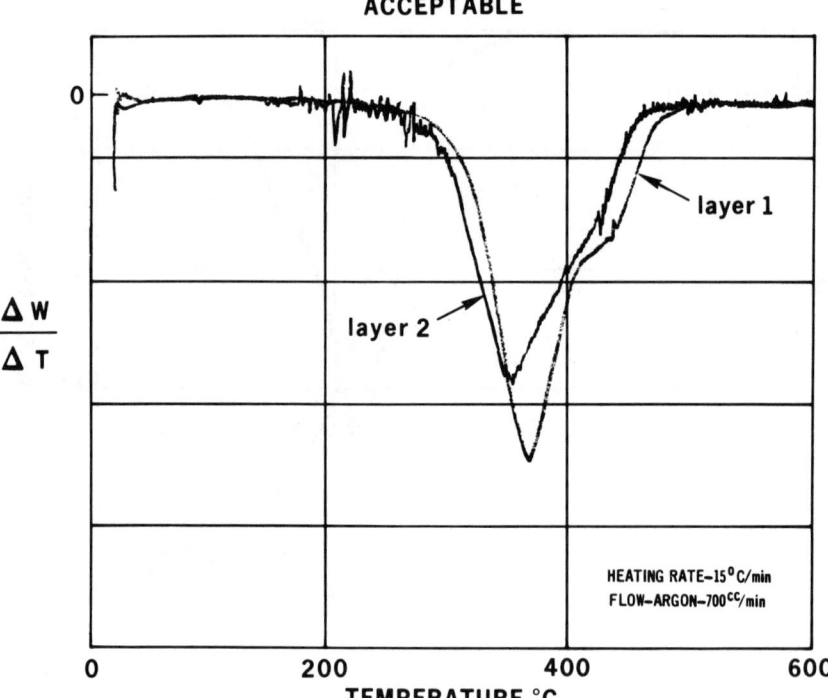

Figure 3. Derivative TGA curves of different layers of epoxy from accetable samples.

---

a. This failure did not occur when unsealed anodized or bare aluminum was used as a substrate.

Then a series of layers of the adhesive on each test block was analyzed. Again, no significant difference between failed or acceptable samples was observed; but one item was observed: the ratio[a] of polyamide to epoxy was not uniform throughout each adhesive (Figures 3 and 4).

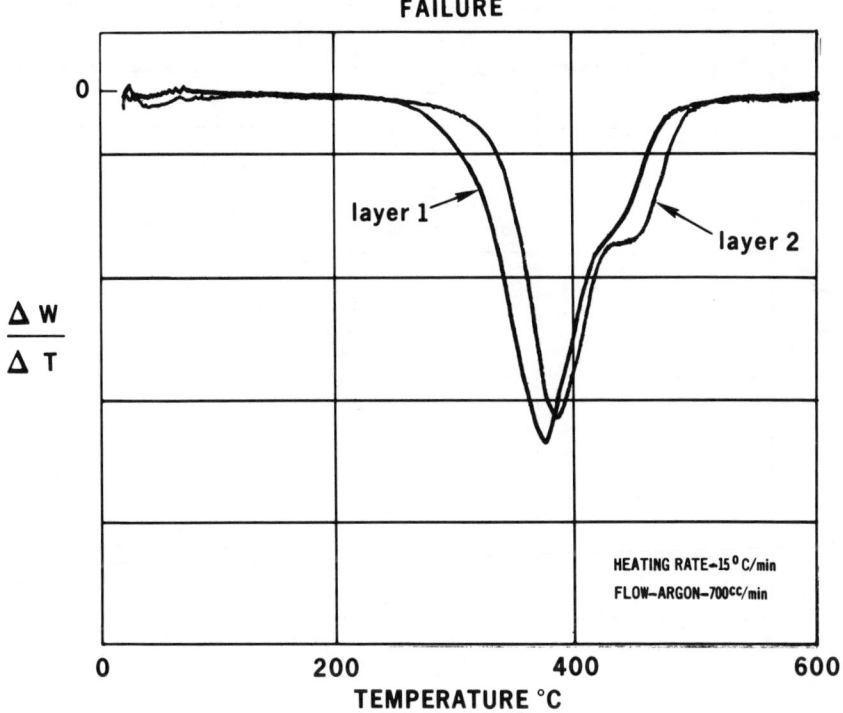

Figure 4. Derivative TGA curves of different layers of epoxy from a failed sample.

With this observation, it was believed that the adhesive was not wholly mixed and that the low temperature adhesion may be a function of the degree of mixing. As a result, the adhesive at the bond line on two batches was analyzed. In every case, when comparing the same batch of adhesive, the bond line failure samples had a larger polyamide-to-epoxy ratio than the sample from the acceptable bond line (Figures 5 and 6). The analysis of a series of single test blocks that contained both bonded and non-bonded (failed) adhesive

---

a. The higher temperature peak or shoulder in the derivative has been found to be due to the decomposition of polyamide while the lower temperature peak is formed from the decomposition of the epoxy.

areas gave further evidence that mixing was the cause of the failures. On analysis[a] of the bond line adhesive in both of these areas, it was found that the adhesive in the area of failure contained the higher ratio of polyamide-to-epoxy (Figures 7 and 8).

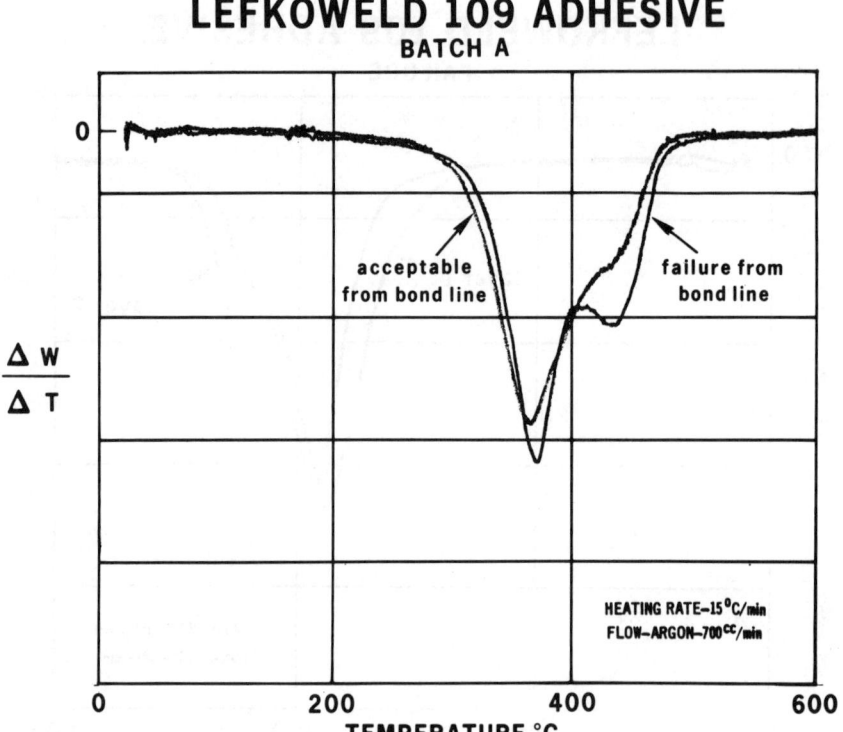

Figure 5. Adhesive samples from the bond line of an acceptable test block and a failed test block containing the same batch of adhesive.

The analysis was performed on two sets of lap shear specimens fabricated from the same batch of Lefkoweld 109 adhesive at different times, but having a 40 percent difference in strength. It was found that the lower strength set adhesive was not mixed while the adhesive on the high strength set was very well mixed (Figures 9 and 10).

---

a. The bond line of the failed parts was easily analyzed since it was exposed. The bond line on the acceptable area was first exposed by slowly milling the aluminum down to a thin sheet and pealing the aluminum off the adhesive.

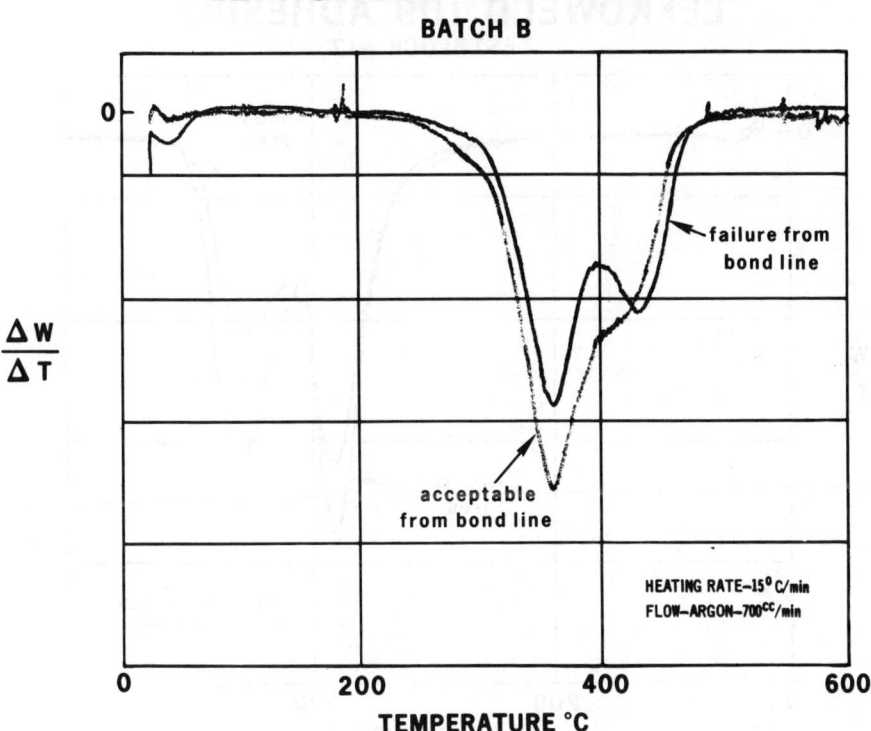

Figure 6. Adhesive samples from the bond line of an acceptable test block and a failed test block containing the same batch of adhesive.

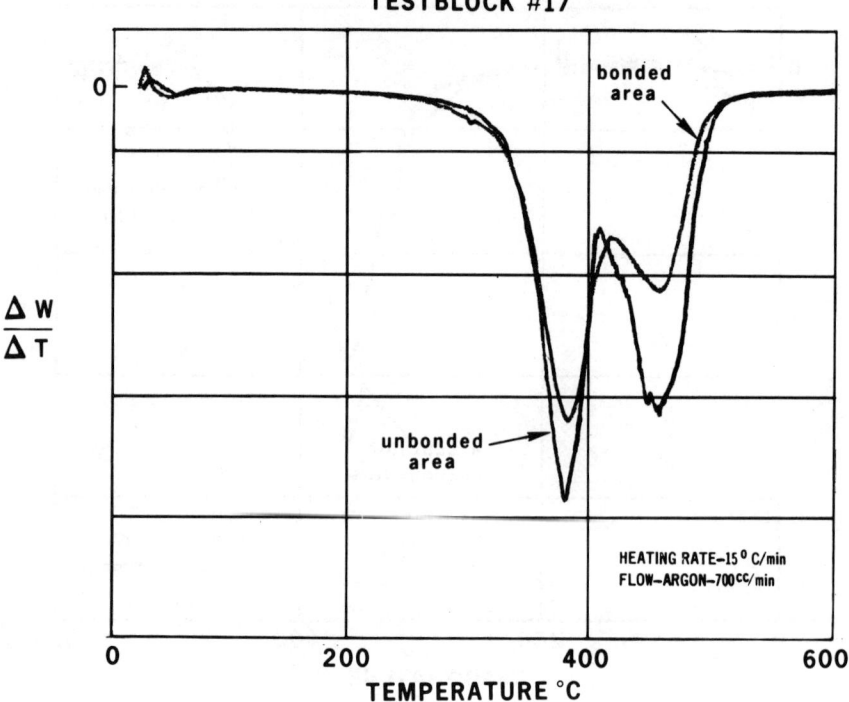

Figure 7. Samples of adhesive from both bonded and unbonded areas found on the same test block.

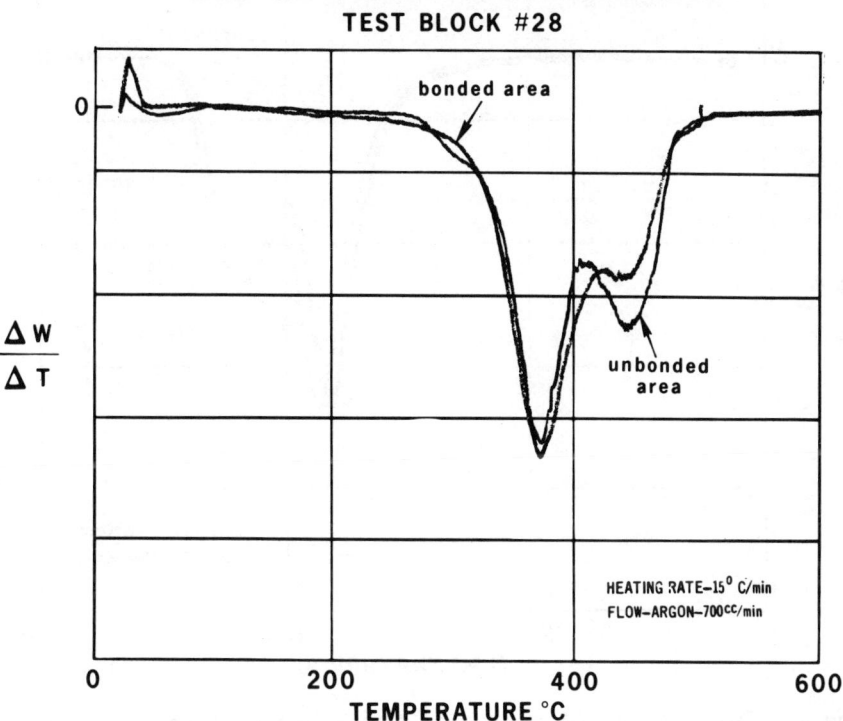

Figure 8. Samples of adhesive from both bonded and unbonded areas found on the same test block.

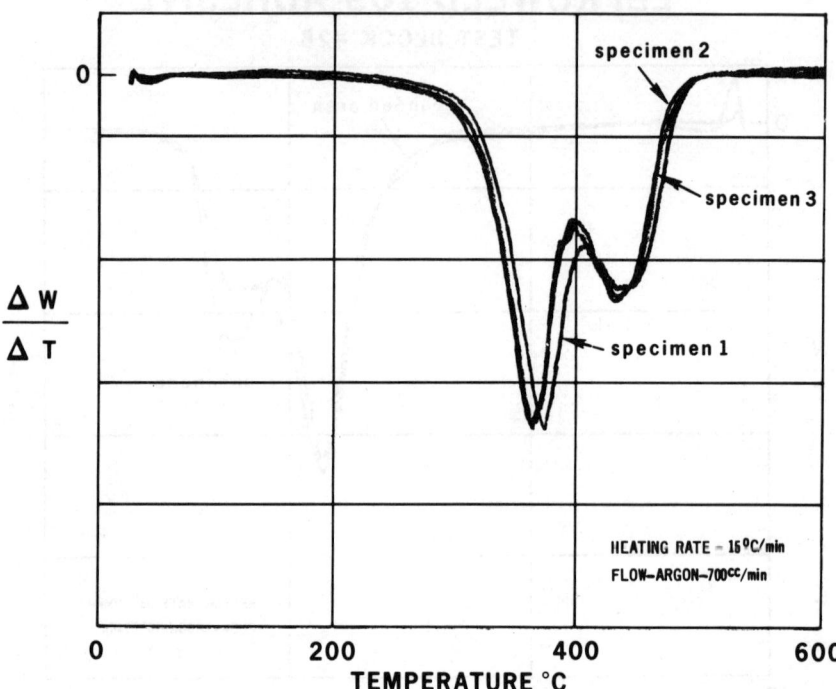

Figure 9. Derivative TGA curves of the adhesive from different specimens of a lap shear test.

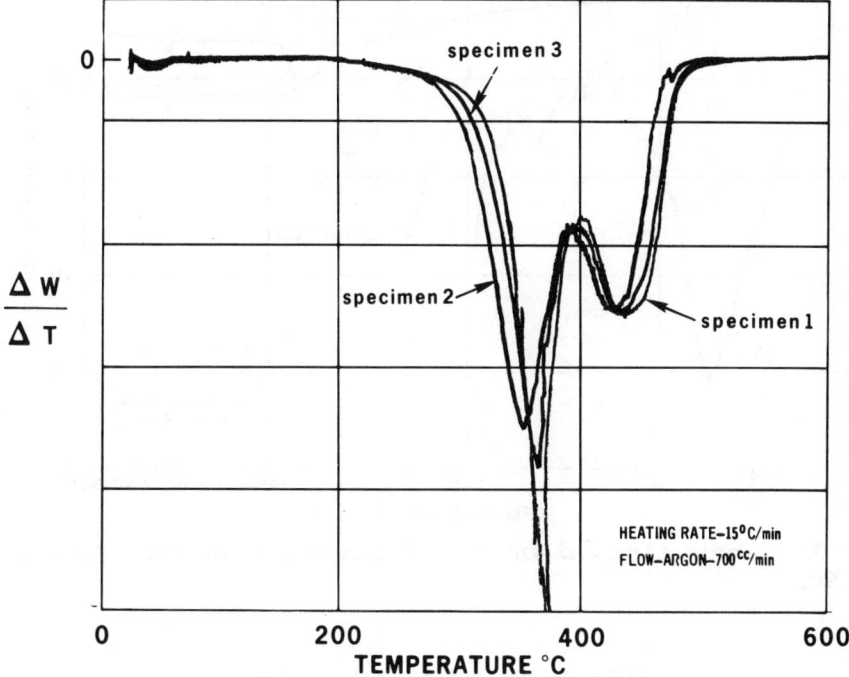

Figure 10. Derivative TGA curves of the adhesive from different specimens of a lap shear test.

The analysis of phenolic-silica prepregs was performed to determine if TGA could replace two of the quality control methods normally used: percent volatiles and resin content. It was believed that TGA would offer the following advantages over the standard methods of analysis: increased accuracy, reproducibility, and speed, especially when only a small number of samples are involved. In addition, a much smaller sample of an expensive material can be analyzed.

Initially, a heating rate of 15°C/ min. in air was used with the weight scale range being 0 - 100% (Figure 11). It can be seen in the figure (11) that the derivative has a minimum rate of weight loss at 300°C. This minimum occurs at a point where the cure is almost completed and oxidation and/or decomposition is still proceeding at a low rate. In light of the above observation, 300°C was chosen as the dividing line between volatile evolution (volatile content) and resin burn-off (resin content). The volatiles were determined at 180°C while the resin content was determined at 700°C and 870°C. These temperatures were used only because they are used in standard tests. Since volatile content was less than ten per-

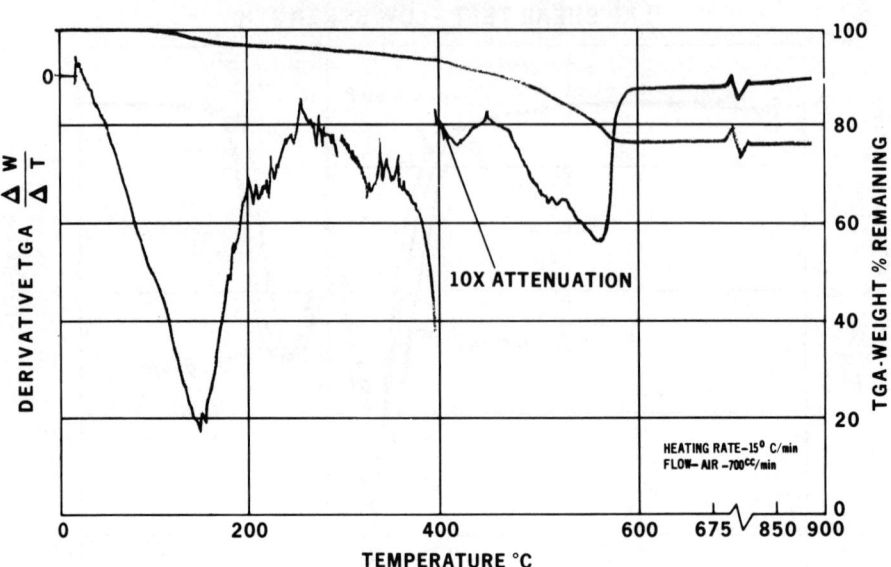

Figure 11. Derivative and integral TGA curves of DP 24-2/Buna N/ C 100-96.

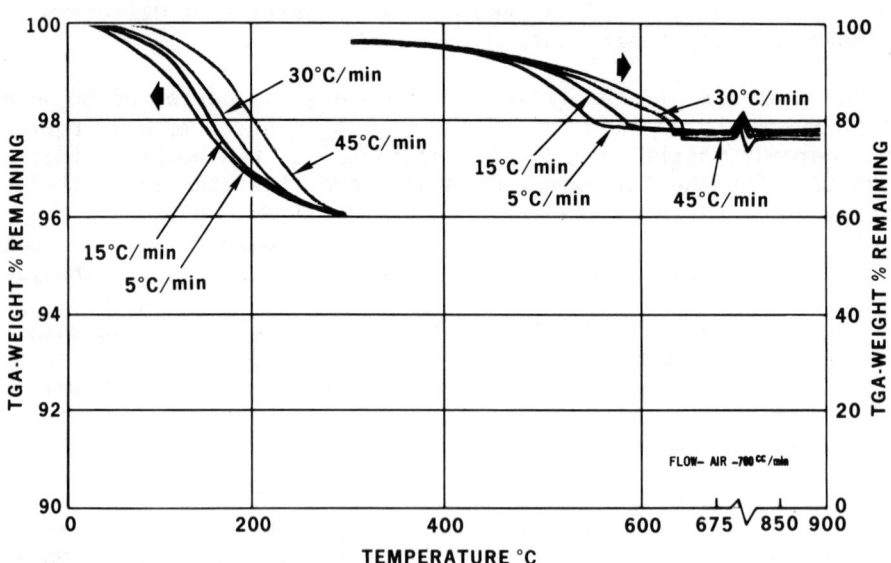

Figure 12. The results of using four different heating rates on DP 24-2/Buna N/ C 100-96.

cent, a suppression of 90 percent was used below 300°C, while above 300°C a weight scale of 0 - 100 percent was used. With this increased volatile resolution, a series of different heating rates were run (5, 15, 30, 45°C/min.) to determine the fastest rate which would give results consistent with the previously used quality control methods (Figure 12).

It was found that heating rates for the best and most consistent determinations of volatile and resin contents were 5°C/min. and 45°C/min., respectively. Subsequently, it was determined that a suppression of 60 percent could be used above 300°C, thereby increasing the resolution of the resin content determination. This procedure (5°C/min. at 90 - 100% up to 300°C, then 45°C/min. at 60-100%) was used to analyze a series of partially cured or staged prepregs (Figures 13 and 14). The integral curves (Figure 13) show that TGA can be used to determine the difference in the degree of staging of a prepreg. The volatile content decreases, as expected, with increasing staging. The derivative curves (Figure 14) show a gradual decrease in the amount of low temperature volatiles with the higher temperature volatiles being unchanged.

Analysis of a series of DP 24-2 prepregs on C 100-96 silica cloth from different sources shows the type of information the derivative TGA yields (Figure 15). These prepregs all have the same volatile content at 180°C, but different derivative curves. The

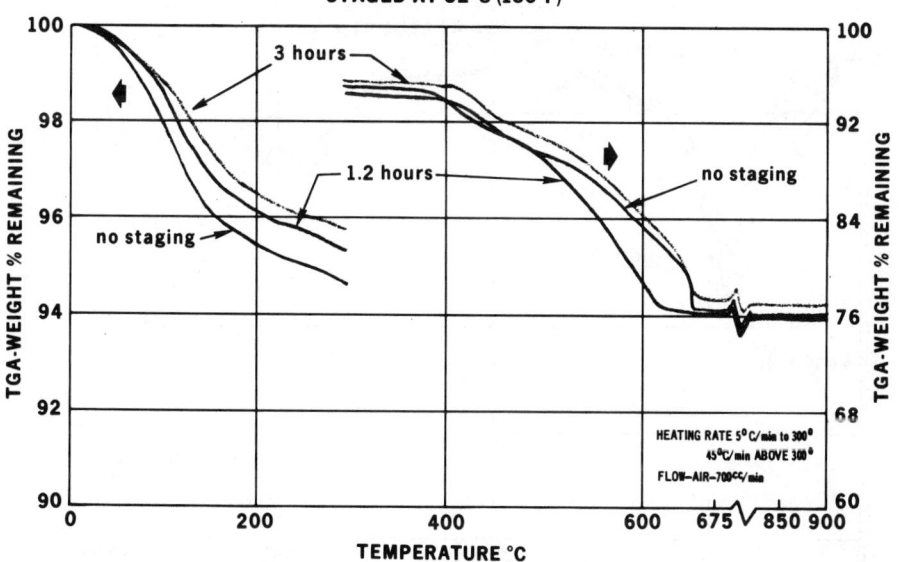

Figure 13. Integral TGA curves of DP 24-2/Buna N/C 100-96 with different amounts of staging at 82°C (180°F).

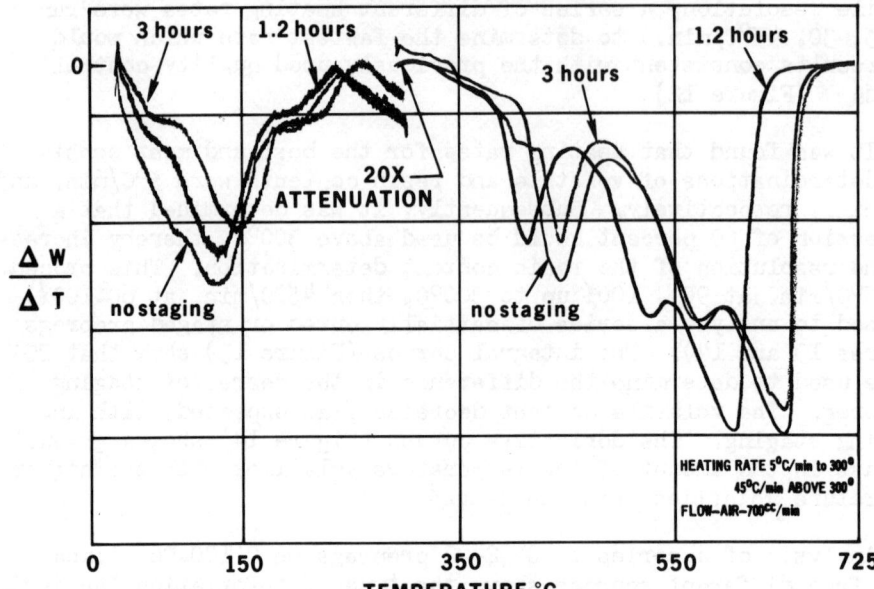

Figure 14. Derivative TGA curves of DP 24-2/Buna N/ C 100-96 with different amounts of staging at 82°C (180°F).

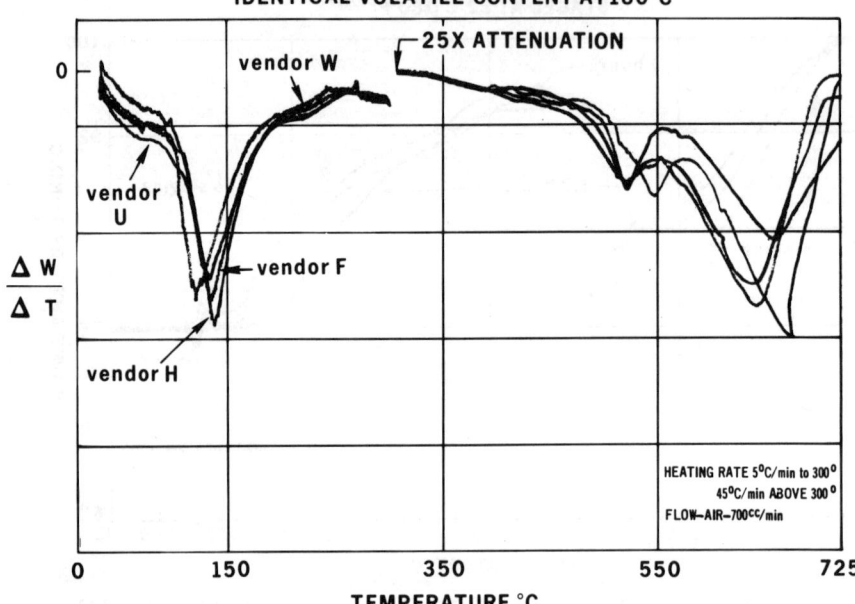

Figure 15. Derivative TGA curves from four different sources with the same volatile content.

reason for this difference is that the volatile percentage contains two types of materials: solvent and water from curing of the phenolic resin. Therefore, the shape of the derivative can show whether or not there is a difference in the ratio of solvents to uncured resins.

CONCLUSIONS

It has been shown that derivative TGA can be used to determine the ratio of curing agent to epoxy in an adhesive. With this ratio (curing agent-to-epoxy), the degree of mixing throughout the adhesive can be measured. The degree of mixing has been found to influence the sensitivity of adhesion of an epoxy. It has also been found that surface treatment can influence the sensitivity of adhesion to the degree of mixing in an epoxy since the low temperature adhesive failures occurred only with sealed anodized aluminum, whereas the unsealed anodized aluminum and bare aluminum had satisfactory bond strength with varying mix ratios. As a result, derivative TGA can be used as a quality control method for analyzing bonding problems within a mixing sensitive system and for studying adhesion as it relates to mixing and surface treatment. Although only one epoxy was discussed as an example of the use of derivative TGA, many other epoxies have been successfully analyzed, such as amine-cured and one-component catalyzed epoxies.

Phenolic-silica prepregs can be analyzed for both volatile and resin content using TGA. Thermogravimetric analysis is not only an improvement over the standard methods with increased speed and use of a smaller sample size, but yields a continuous and permanent record of volatiles and resin content versus temperature. This record allows complete freedom in choosing the temperature where volatile and resin content will be determined, and makes it possible to refer back to the material if the weight loss at another temperature is needed in the future.

The shape of the derivative TGA curve has shown that other factors are involved in the thermal analysis of phenolic-silica prepregs beyond resin and volatile content, such as solvent-to-resin ratio. Thus, the shape of the derivative curve should prove to be of value in material control and characterization in the future.

ACKNOWLEDGEMENT

The authors gratefully acknowledge the assistance, suggestions, and helpful discussion of Mr. Robert M. Washburn, Dr. Dennis W. Karle, and Mrs. Jean Anne Gordon.

BIBLIOGRAPHY

1. P. D. Garn, "Thermoanalytical Method of Investigation", Academic Press, New York, 1965.
2. W. W. Wendt, "Proceedings First Toronto Symposium on Thermal Analysis", H. G. McAdie, editor, Chemical Institute of Canada, 1965, page 101.
3. O. Vogt, V. Ivansons, H. C. Miller, and H. W. Williams. J. Macromol. Sci. - Chem., $\underline{A\ 2}$ (1) 175 (1968).
4. C. B. Murphy, Anal. Chem. $\underline{40}$, 380-R (1968).
5. I. M. Sarasohn and R. W. Tabeling, 15th Annual Pittsburgh Conference of Analytical Chemistry and Applied Spectroscopy, March 5, 1964.

A DYNAMIC DIFFERENTIAL CALORIMETRIC TECHNIQUE FOR MEASURING HEATS OF POLYMERIZATION

Jen Chiu

Plastics Department, E. I. du Pont de Nemours & Co., Inc., Experimental Station, Wilmington, Delaware 19898

INTRODUCTION

Dynamic differential calorimetry is now widely used for measuring heats of transitions and reactions in solid and in relatively nonvolatile liquid systems. Similar rapid determinations are often needed on volatile systems, such as determination of heats of polymerization of volatile monomers. Several instrument manufacturers have available so-called "hermetically" sealed sample containers for volatile samples, which, however, often rupture at relatively low pressures. More recently, other workers have devised techniques to seal the sample cells but with variable success. For instance, Horie et al. (1) modified the Perkin-Elmer sample pans for use with its Differential Scanning Calorimeter (DSC) by bonding face to face two standard aluminum pans with a heat-resistant adhesive. The sample was injected into the enclosed pan with a syringe, and the hole produced was closed with cellophane tape. Although the authors claimed good results in their polymerization studies, this sealing technique does not seem to withstand high pressures and, consequently, gas leakage may occur. Freeberg and Alleman (2) devised threaded metal cells, and Barrett and Thomas (3) sealed aluminum cylinders with polytetrafluoroethylene plugs, both for use with the Perkin-Elmer DSC. Similar metal cells have been tested in the author's laboratory. Experience in using such metal cells in either the Perkin-Elmer DSC or the Du Pont DSC unit showed poor precision in many applications, mainly because of the large thermal mass of the cell compared to the sample size, and nonreproducible heat losses from the protruding portion of the cell above the temperature sensor. Wendlandt (4) has improved this situation by sealing the sample in a glass capillary and then placing the capillary in an aluminum holder whose base makes contact with the temperature sensor and heater. In addition to reducing the weight and the protrusion of the cell, this design also allows investigation of samples corrosive to metals or of reactions catalyzed by metals. Hoyer et al. (5) reported use of sealed 4-mm glass tubes with the calorimeter attachment of the Du Pont 900 Differential Thermal Analyzer to obtain activation energies for styrene polymerization. This procedure overcomes many of the shortcomings encountered with metal containers. However, sealing of the 4-mm sample tube rapidly without damaging the sample is quite a task. Also, the sealed tube placed loosely in the sample cup of the calorimeter reduces the precision of the heat measurement because of poor heat transfer.

This paper describes a microcalorimetric technique using commercially available glass ampoules with the calorimeter module of the Du Pont 900 Thermal Analysis System. The technique is simple and suitable for use in routine determinations. High precision is generally obtained. Application of this method to determine the heat of polymerization of styrene, methyl methacrylate, acrylonitrile, and styrene-acrylonitrile mixtures is illustrated.

EXPERIMENTAL

A schematic diagram of the Du Pont Calorimeter cell (Cat. No. 900350, Instrument and Equipment Division, E. I. du Pont de Nemours & Co., Inc., Wilmington, Delaware) is shown in Figure 1. This cell is based on the Boersma design (6) with the sample and the reference material placed in silver cups located in separate air cavities in a heating block. The heat of reaction is transferred from the sample to the highly conductive metal cup. The thermal resistance of heat flow between sample cup and block is relatively large compared to that between sample and cup, thus minimizing fluctuations in heat losses. The thermocouples are welded to the bottom of the sample cups, which are rigidly positioned in the heating block to provide reproducible calibration. The differential thermogram is recorded as differential temperature (ΔT) vs. reference temperature (T) on the Du Pont 900 console. In the present work, the temperature differential signal is also fed to an external strip-chart recorder to plot simultaneously the differential temperature as a function of time to provide better precision in peak area measurement.

Because of the volatility of the monomers under investigation, a sealed ampoule technique has been established to allow rapid measurements using standard procedures. The borosilicate glass ampoules are made from the sample bulbs used in the microcell assembly for nuclear magnetic resonance measurements (Cat. No. K-897020, Kontes Glass Company, Vineland, New Jersey) which are

DU PONT CALORIMETER CELL

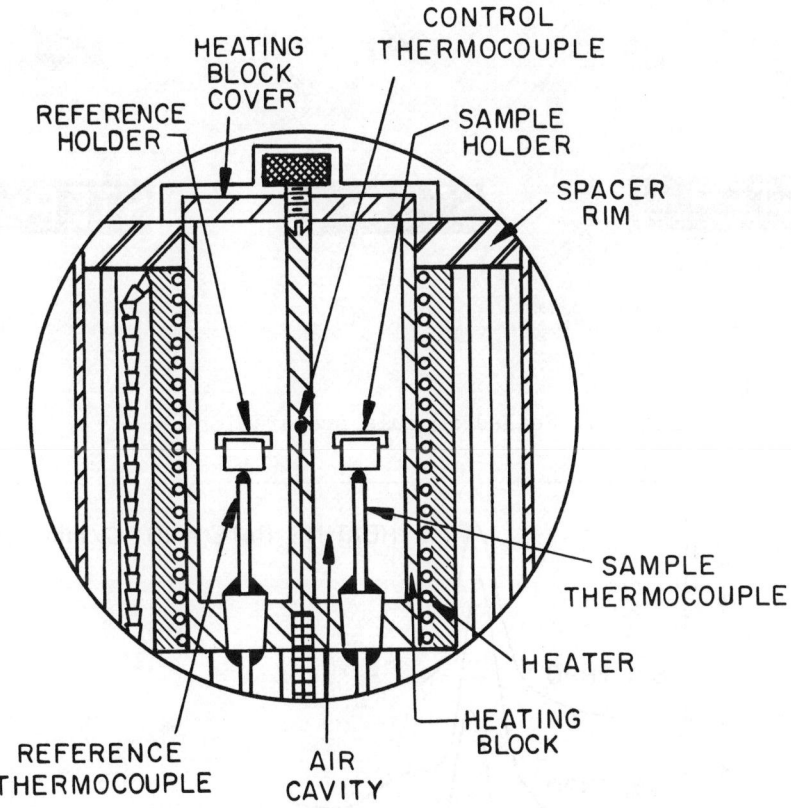

Fig. 1. Du Pont Calorimeter cell.

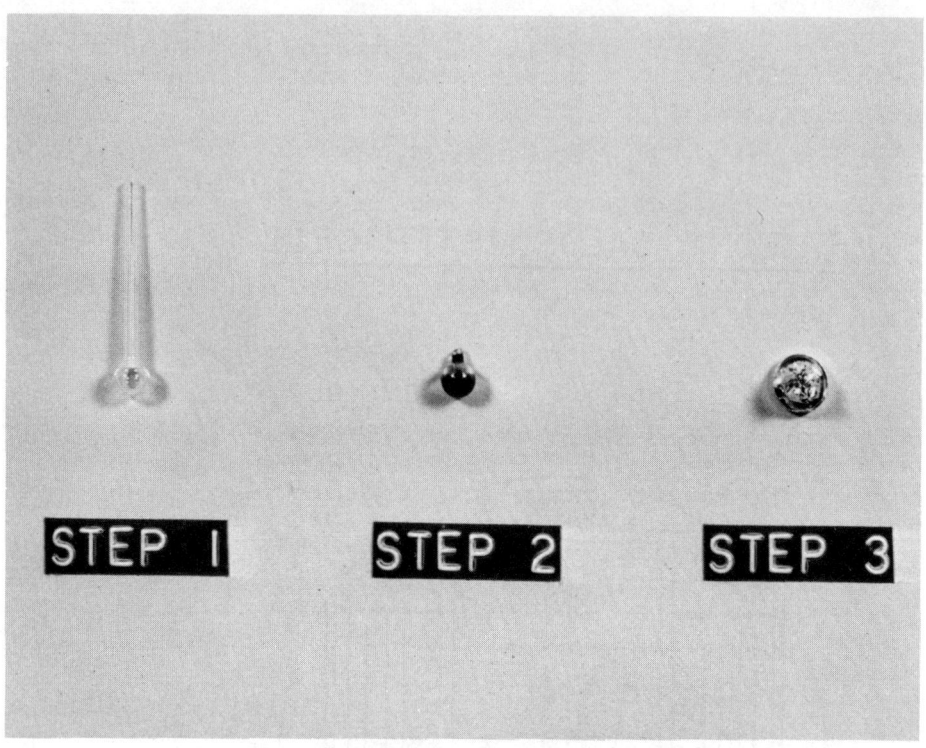

Fig. 2. Sealed ampoule preparation.

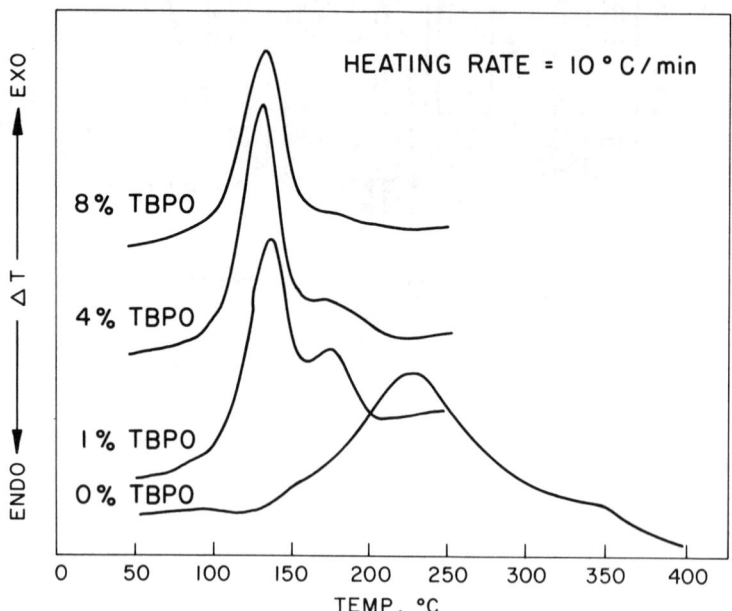

Fig. 3. Styrene polymerization with various concentrations of TBPO.

found to be reproducible in dimensions and weight. The procedure for preparation of the sample ampoule is shown in Figure 2. In step 1, the sample (ca. 10 mg) is placed in a preweighed ampoule with a microsyringe and frozen with dry ice or other suitable refrigerant. The ampoule is then sealed with a high-temperature miniature natural gas-oxygen torch, and the stem shortened, as shown in step 2. During this step, the ampoule either is positioned in a cavity of an aluminum block cooled by dry ice or is immersed in dry ice with the stem protruding through a metal cover to be sealed by the torch. The sealed ampoule is then wrapped with a measured sheet of thin aluminum foil and shaped into a plug in a metal mold to fit the inside of the sample cup of the calorimeter cell, as shown in step 3. The total weight of the plug including the sample is approximately 60 mg. A cap made from platinum or aluminum foil is placed on top of the cup to contain the heat. Another ampoule, either empty or containing Silicone DC-710, prepared in the same way, is used on the reference side. Other operational procedures are described in the instrument manual (7).

Styrene monomer was Dow Chemical Polymer Grade containing 15 ppm tert-butylcatechol as inhibitor. Acrylonitrile used was B. F. Goodrich Fiber Grade containing 32 ppm methyl ether of hydroquinone. Methyl methacrylate was inhibitor-free, redistilled, and kept refrigerated until use. The two initiators used, di-tert-butyl peroxide (DTBP) and tert-butyl peroctoate (TBPO), were from Lucidol Division of Wallace and Tiernan.

RESULTS AND DISCUSSION

Initial experiments were performed at a programmed heating rate of 10°C per minute with sample sizes varying from 4 to 15 mg. The heat of polymerization is represented by the peak area of the exotherm. Typical thermograms are shown in Figure 3 for polymerization of styrene using 0-8% of TBPO as initiator. In the absence of an initiator, the polymerization exotherm is broad and occurs at high temperatures. When an initiator is used, two polymerization steps are evident. The peak size of the high-temperature exotherm increases with decreasing initiator concentration. Presumably, this high-temperature reaction is due to thermal polymerization after the initiator has been depleted. This is supported by the fact that pure TBPO shows its decomposition exotherm at ca. 130°C, as shown in Figure 4, which is lower than the temperature for the second polymerization step.

DTBP is more thermally stable than TBPO with its decomposition exotherm at ca. 195°C (Figure 4) and, therefore, initiates polymerization at a higher temperature than TBPO. Typical thermograms are shown in Figure 4 for styrene polymerization using DTBP as initiator.

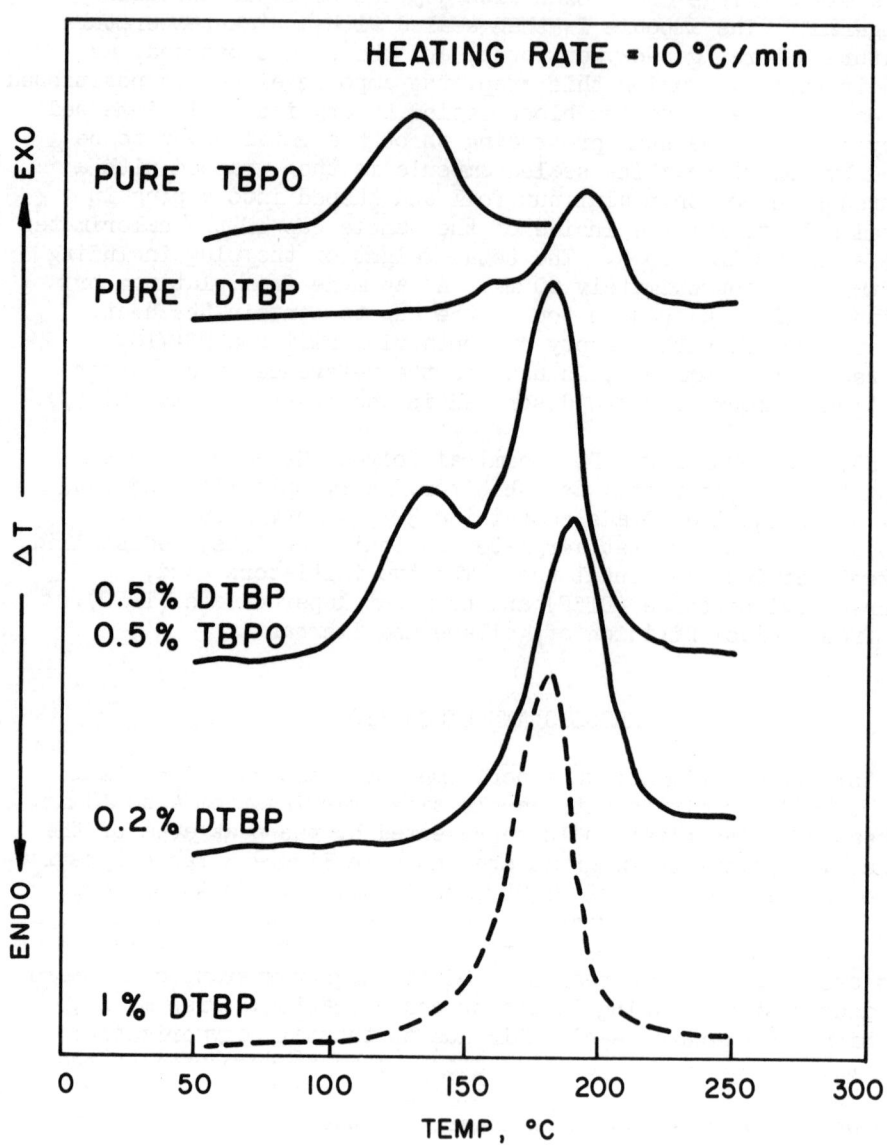

Fig. 4. Styrene polymerization with various concentrations of DTBP.

MEASURING HEATS OF POLYMERIZATION 177

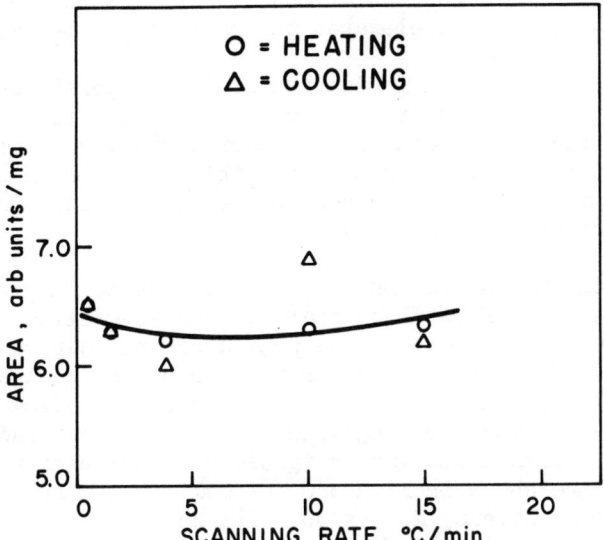

Fig. 5. Effect of scanning rate on melting and freezing of indium.

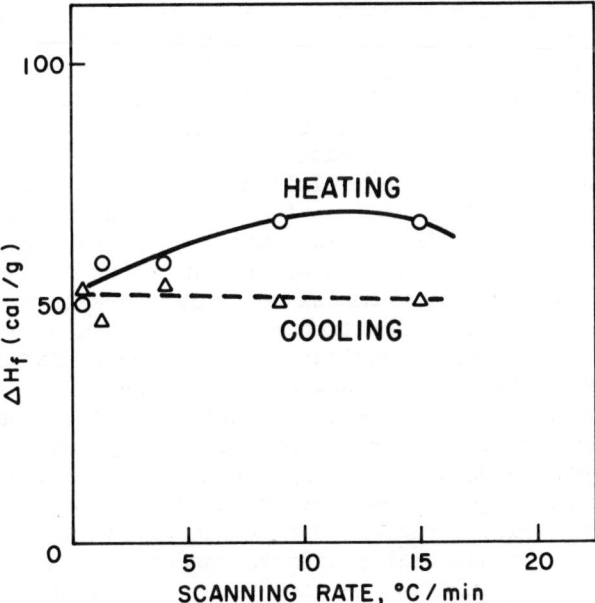

Fig. 6. Effect of scanning rate on melting and freezing of Marlex 50.

In this case, a second polymerization step does not seem to be present. Use of higher concentrations of DTBP lowers the polymerization temperature and accelerates the rate of polymerization.

To assure complete polymerization at low temperatures with a minimal amount of initiator, and to avoid a mixed mechanism of thermal polymerization, a double initiator system, such as 50:50 DTBP:TBPO, is preferred. Such a double initiator system produces a two-step polymerization corresponding to the two individual initiators as expected (Figure 4).

Calibration curves are constructed from the melting and freezing of a series of metal standards, such as gallium, indium, tin, lead, and zinc, to convert the peak areas to heat values of polymerization. The effect of scanning rate on the peak areas of the melting endotherm and the freezing exotherm of indium is shown in Figure 5. The freezing values are more scattered than the melting values. Peak areas vary slightly with scanning rate. Similar curves are shown in Figure 6 for melting and freezing of a high-density polyethylene, Marlex 50. The heat values are obtained by using indium as calibration standard. The heating curve shows a dependence of heat of fusion value on heating rate, although the cooling curve is less sensitive to cooling rate. More significantly, the two curves merge to the same point, when the scanning rate approaches zero. This variation of heat value with scanning rate is more severe with polymerization studies, because the reaction itself is both rate dependent and temperature dependent. For instance, in the case of methyl methacrylate polymerization, the amount of residual monomer in the polymer product increases with increasing heating rate. The values are 1.2, 2.8, 4.6, 14.6, and 44.3% by weight for heating rates of 0.4, 1.4, 4, 10, and 15°C per minute, respectively, with an initiator system of 0.5%:0.5% DTBP:TBPO. The residual monomer content is analyzed by gas chromatography. Dependence of the polymerization peak area on heating rate is shown in Figure 7 for both before and after residual monomer correction. Obviously, very complex mechanisms are involved at high heating rates and high reaction temperatures reached because of high heating rates. Isothermal experiments avoid such difficulties but provide lower sensitivities. Also, equilibration of the sample temperature introduces uncertainties in establishing a good baseline for peak area measurement. For all practical purposes, the heat of polymerization can be obtained fairly rapidly and precisely by the present dynamic technique, when heating rates below 2°C per minute are used. At such rates, the exothermic peak areas produced from polymerization can be calibrated with the endothermic melting peaks of metal standards scanned at the same rate. A typical calibration curve is shown in Figure 8 for a scanning rate of 0.4°C per minute. The calibration coefficient, E, is calculated according to the normal procedure (7).

MEASURING HEATS OF POLYMERIZATION 179

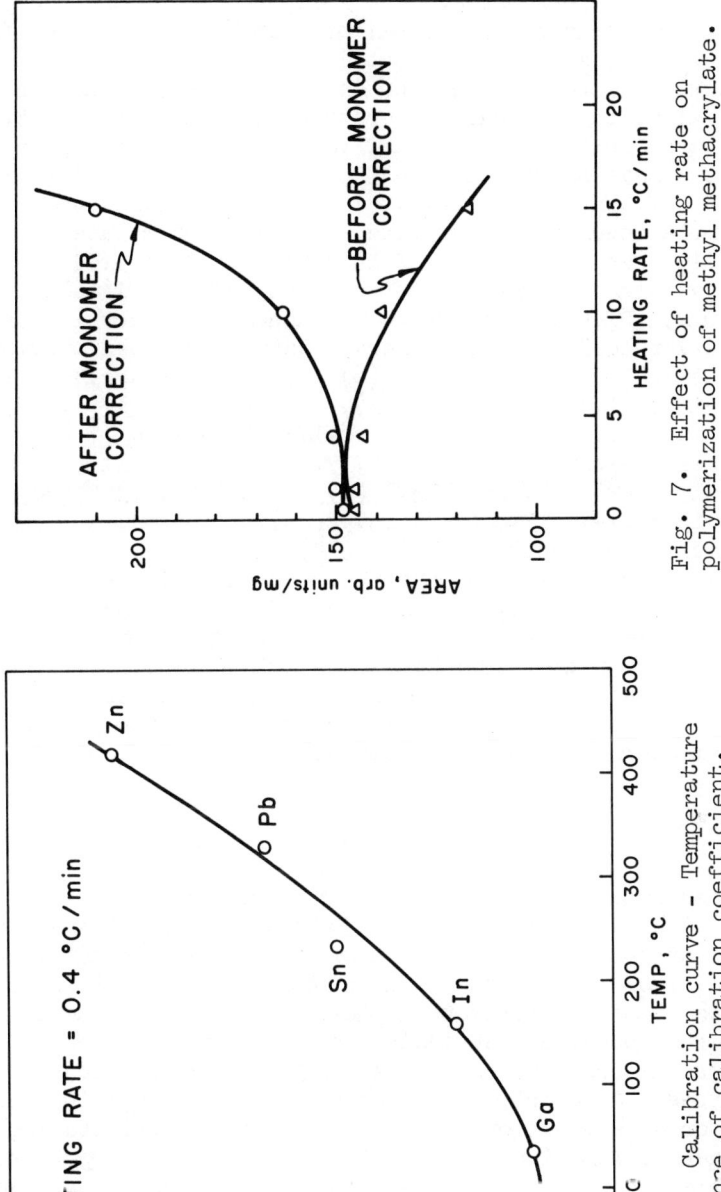

Fig. 7. Effect of heating rate on polymerization of methyl methacrylate.

Fig. 8. Calibration curve – Temperature dependance of calibration coefficient.

The heat of polymerization is obtained by incremental integration of the peak area on the strip chart to compensate for the temperature variance of the calibration coefficient. A typical scan is shown in Figure 9. The total peak is segmented into several sections, and each section is given a calibration coefficient corresponding to the average temperature of that section. By using such a calibration technique, values of heat of polymerization obtained at a heating rate of 0.4°C per minute are 151 ± 3, 149 ± 2, and 386 ± 5 cal. per gram, or 15.7 ± 0.3, 14.9 ± 0.2, and 20.5 ± 0.3 kcal. per mole monomer, for styrene, methyl methacrylate, and acrylonitrile, respectively. The initiator used is 0.5%:0.5% DTBP:TBPO by weight of monomer for styrene and methyl methacrylate, and 0.25%:0.25% DTBP:TBPO by weight of monomer for acrylonitrile and styrene-acrylonitrile copolymers. Values reported previously are as follows: styrene, 8.0-9.5 (8), 13.8-16.9 (9), 14.1 (10), 15.0 (11), 15.9 (12), 15.9-18.2 (13), 16.0 (14), 16.1 (1, 15), 16.4 (16), 16.5 (17), 16.7 (18), 16.8-17.0 (19), and 17.4 (20) kcal. per mole; methyl methacrylate, 7.8 (13), 7.9 (11), 11.6 (21), 12.9 (22), 13.0 (23, 24, 25), 13.1 (1), 13.3 (19), 13.4 (26), 13.6 (27), 13.7 (14), 13.8 (16), 13.9 (28), 16.3 (8), and 21.4 (10) kcal. per mole; acrylonitrile, 16.8 (14), 17.3 (29), 18.3 (24) and 18.4 (19) kcal. per mole. Apparently, our value for styrene agrees with most literature results, but values for methyl methacrylate and acrylonitrile are at the high end of the spectrum. A value of 205 ± 4 cal. per gram or 17.1 kcal. per mole has been obtained for a 75:25 styrene-acrylonitrile copolymer. Only one reference (14) has been discovered for such copolymer systems, which gave a value of 17.2 kcal. per mole for this composition.

CONCLUSIONS

A sealed ampoule microcalorimetric technique has been described and used for determination of the heat of polymerization for several systems. The precision of such a measurement is better than 3% relative. The procedure is simple and does not deviate greatly from standard differential thermal analysis. The sample container is disposable. An additional feature of the sealed sample container is the convenience of keeping the reaction product for other analyses. A potential certainly exists for applying this technique for determination of heat capacity, thermal stability, and heats of thermal reactions of volatile systems. By proper treatment of the sample, the kinetics of the reaction also can be studied.

ACKNOWLEDGMENT

The author wishes to thank Dr. S. Piekarski for his analysis for residual monomer and Dr. F. E. Martin for helpful discussions. He also thanks Mr. R. S. Sudol for suggesting the NMR sample bulbs and Mr. R. A. Parkinson for assistance in the experimental work.

STYRENE POLYMERIZATION

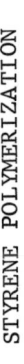

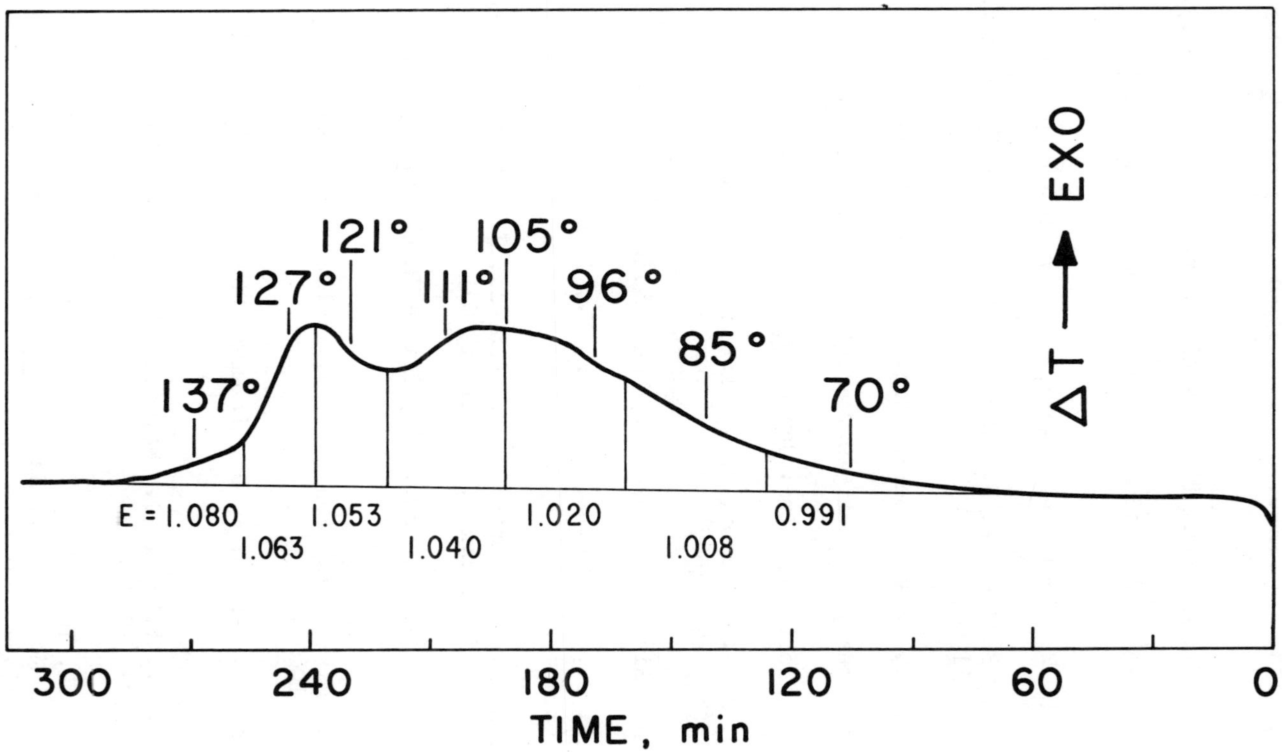

Fig. 9. A typical dynamic differential calorimetric scan for styrene polymerization. Sample weight, 6.6 mg; initiator used, 0.5% DTBP + 0.5% TBPO; heating rate, 0.4°C per minute; chart speed, 2 inches per hour; ΔT sensitivity, 0.1°C per inch. Calibration coefficients used for incremental integration are shown.

REFERENCES

1. Horie, K., Mita, I., Kambe, H., J. Polymer Sci., A-1, $\underline{6}$, 2663 (1968).

2. Freeberg, F. E., Alleman, T. G., Anal. Chem., $\underline{38}$, 1806 (1966).

3. Barrett, K. E. J., Thomas, H. R., J. Polymer Sci., A-1, $\underline{7}$, 2621 (1969).

4. Wendlandt, W. W., Anal. Chim. Acta, $\underline{49}$, 187 (1970).

5. Hoyer, H. W., Santoro, A. V., Barrett, E. J., J. Polymer Sci., A-1, $\underline{6}$, 1033 (1968).

6. Boersma, S. L., J. Amer. Ceram. Soc., $\underline{38}$, 281 (1955).

7. Instrument Manual, Du Pont 900 Thermal Analysis System.

8. Iwai, S., J. Soc. Chem. Ind. Japan, $\underline{49}$, 185 (1946).

9. Merzhanov, A. G., Abramov, V. G., Abramova, L. T., Zh. Fiz. Khim., $\underline{41}$ (1), 179 (1967).

10. Roth, W. A., Rist-Schumacher, E., Kautschuk, $\underline{18}$, 137 (1942).

11. Goldfinger, G., Josefowitz, D., Mark, H., J. Amer. Chem. Soc., $\underline{65}$, 1432 (1943).

12. Biddulph, R. H., Longworth, W. R., Penfold, J., Plesch, P. H., Rutherford, P. P., Polymer, $\underline{1}$, 521 (1960).

13. Franz, J., Mische, W., Kuzay, P., Plaste Kaut., $\underline{14}$ (7), 472 (1967).

14. Suzuki, M., Miyama, H., Fujimoto, S., Bull. Chem. Soc. Japan, $\underline{35}$, 57, 60 (1962).

15. Tong, L. K. J., Kenyon, W. O., J. Amer. Chem. Soc., $\underline{69}$, 1402 (1947).

16. Dainton, F. S., Ivin, K. J., Walmsley, D. A. G., Trans. Faraday Soc., $\underline{56}$, 1784 (1960).

17. Grikina, O. E., Tatevskii, V. M., Stepanov, N. F., Yarovoi, S. S., Vestn. Mosk. Univ., Ser. II $\underline{22}$ (4), 8 (1967); Chem. Abst., $\underline{67}$, 117437 (1967).

18. Roberts, D. E., Walton, W. W., Jessup, R. S., J. Polymer Sci., $\underline{2}$, 420 (1947).

19. Joshi, R. M., J. Polym. Sci., 56, 313 (1962).

20. Bywater, S., Worsfold, D. J., J. Polym. Sci., 58, 571 (1962).

21. Evans, A. G., Polanyi, M., Nature, 152, 738 (1943).

22. Evans, A. G., Tyrrall, E., J. Polym. Sci., 2, 387 (1947).

23. Tong, L. K. J., Kenyon, W. O., J. Amer. Chem. Soc., 67, 1278 (1945); 68, 1355 (1946).

24. Baxendale, J. H., Madaras, G. W., J. Polym. Sci., 19, 171 (1956).

25. Levin, P. I., Plast. Massy, 1959, No. 3, 29; Chem. Abst., 58, 8053 (1963).

26. Ivin, K. J., Trans. Faraday Soc., 51, 1273 (1955).

27. McCurdy, K. G., Laidler, K. J., Can. J. Chem., 42, 818 (1964).

28. Ekegren, S., Öhrn, O., Granath, K., Kinell, P., Acta Chem. Scand., 4, 126 (1950).

29. Tong, L. K. J., Kenyon, W. O., J. Amer. Chem. Soc., 69, 2245 (1947).

THERMAL METHODS FOR DETERMINATION OF DEGREE OF CURE OF THERMOSETS

James P. Creedon

E. I. du Pont de Nemours & Company (Inc.), Photo Products Dept., I&E Division, Wilmington, Delaware 19898

INTRODUCTION

Most thermosetting resins are infusible, insoluble, covalently crosslinked, thermally stable, network polymer structures. The formation of this network structure, during curing, is responsible for the desirable physical properties which are typical of thermosetting resins over a broad temperature range.

The use of thermosetting resins dates back to ancient times. Until Baekeland[1] demonstrated in the early part of this century that through the use of the proper catalyst, pressure and temperature, one could form thermosets in reasonable lengths of time, the synthetic resin industry was nonexistent. Over the past sixty years, a wide variety of thermosetting resins have been introduced and used commercially. The good physical properties of these materials are primarily due to their chemical structure and the amount of crosslinking. This latter property is governed in large part by their degree of cure. For this reason considerable effort has been spent to measure and to establish the amount of cure needed for the necessary property level.

Chemical, physical, electrical and thermal properties have been examined to characterize the degree of cure of thermosets. Complete cure, resulting in the maximum number of chemical crosslinks possible, is usually never obtained even under laboratory conditions. The methods for studying cure have taken the empirical character of maximizing a particular property versus the conditions of cure.

Two general types of tests have been employed to characterize the degree of cure of thermosets. The first type determines conversion, which is the disappearance of reactive groups or the appearance of a particular structure. Typical tests for determining conversion involve infrared or chemical analyses. The second type determines crosslinking, which is the coupling of the molecule into three dimensional networks through the reactive sites to form the desired resin. Those methods which follow only the degree of conversion do not completely characterize the degree of cure since they give no indication of the extent of crosslinking. Hence, the determination of degree of crosslinking is a positive mode of characterization, rather than using the negative aspects of conversion. For this reason, the present paper will be concerned with methods for determining degree of crosslinking.

One of the most common techniques employed to characterize crosslinking has been the determination of physical properties of the resin. Such properties as tensile, compressive and flexural strength, impact resistance, Vicat softening point and deflection temperature have all been used to indicate a relative degree of cure in a resin system[2]. In general, as the degree of cure increases in a given system, the values determined for these properties increase to a maximum value, since the material becomes harder, more rigid and softens at a higher temperature. The determination of tensile properties as a function of curing conditions is most commonly employed in the characterization of epoxies. The limitations of this technique regarding sensitivity and ability to discriminate between small differences in degree of cure are obvious, since the technique really can see only reasonably large differences in material properties. Chemical properties such as moisture resistance, resistance to acids, bases and solvents, and weathering and corrosion resistance are all somewhat sensitive to the degree of crosslinking. The general behavior reflects greater resistance to change by these agents as the cure increases. Electrical properties have also been studied, and it has been found that changes in volume resistivity, dielectric constant and dissipation factor as a function of cure can be determined for various systems[3, 4].

The purpose of this paper is to survey existing thermoanalytical techniques which have been successfully applied to the determination of differences in the degree of cure in thermosetting resins, and to point out, where applicable, the potential use of these techniques for both research and quality control in the thermosetting resin industry.

THERMAL METHODS

Thermogravimetric Analysis (TGA)

The thermal degradative stability of an epoxy resin has been related to its degree of cure by Lee and Levi[5]. These workers employed thermogravimetry in both the dynamic and isothermal modes to obtain the overall activation energy for the thermal decomposition of an epoxy resin. The samples consisted of cured and uncured blends of DER-332 (diglycidyl ether-bisphenol A) and m-phenylenediamine. They determined the overall activation energy for thermal decomposition as a function of curing temperature. Their results are shown in Figure 1, which illustrates a proportional relationship between activation energy and curing temperature. They have also shown (Figure 2) that under identical dynamic conditions there were marked differences in the thermal stability of this resin as a function of curing conditions. They concluded that this data was consistent with the view that higher curing temperatures resulted in a higher degree of crosslinking. This work strongly suggested the potential value of TGA as a tool for investigating proposed curing cycles and curing agents in other resin systems.

Dudley, Smith and Youren[6] have also successfully applied TGA to the characterization of the degree of cure of thermosetting resins. They calculated overall activation energies for thermal decomposition for ethylene-propylene rubbers. Their data (Figure 3) demonstrated that the overall activation energy for thermal decomposition increases with cure time and therefore, with degree of cure. These results, they concluded, were expected, since an increase in the number of bonds per unit mass of polymer will cause an increase in the energy required to degrade a given mass into its volatile products.

The preceding examples illustrate that kinetic data can be obtained on a thermogravimetric analyzer. This data should be correlated with that from other methods of testing for degree of cure and ultimately replace these other methods. For example, activation energies could be correlated with the much less sensitive tensile properties of a given epoxy, and possibly replace tensile testing.

Differential Scanning Calorimetry (DSC)

Differential Scanning Calorimetry (DSC) has also been employed to measure differences in degree of cure in thermosetting resins. Fava[7] used this technique to study the cure of DER-332 with hexahydrophthalic

OVERALL ACTIVATION ENERGY FOR THERMAL DEGRADATION OF CURED AND UNCURED DER 332[a]

Curing Temperature, °C	Activation Energy (Kcal/mole)	
	Isothermal	Dynamic
200	32	32
175	28	30
150	25	26
125	25	26
100	22	20
UNCURED	13	14

(a) Lee and Levi, J. Appl. Polym. Sci., 13, 1703 (1969)

FIGURE 1.

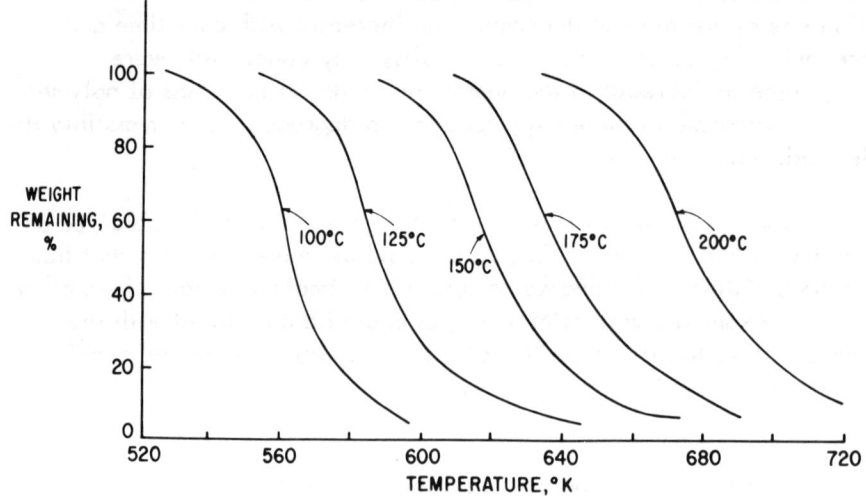

THERMOGRAMS AT A RATE OF HEATING 10°C/MIN FOR EPOXIDES CURED AT THE INDICATED TEMPERATURE[a]

(a) Lee and Livi, J. Appl. Polym. Sci., 13, 1703 (1969)

FIGURE 2.

anhydride as the curing agent. His results (Figure 4) demonstrated that as curing temperature was varied the character of the DSC scan changed. The scans in Figure 4 are inverted so that an exotherm appears as a downward deflection of the pen. These scans were obtained on samples which had first been cured in the DSC cell at the indicated temperature until no further reaction was evident. This was accomplished by having heated the sample quickly to the curing temperature, having held that temperature constant and then having recorded the exothermic cure until the pen returned to the baseline. The bottom scan was obtained on a sample which had been cured at 330°K, cooled and then scanned at 8°C/minute. The other samples were cured under similar conditions at the indicated temperatures and scanned at the same rate. Figure 4 illustrates the effect of curing temperature on the DSC curve. As the curing temperature was increased, the glass transition temperature was seen to increase and in fact to coincide with the curing temperature. Also the residual curing exotherm diminished in size as the initial curing temperature was increased. After 400°K, there was no apparent residual exotherm, which indicated that at this temperature the reaction was indeed complete at least as far as the limits of detectability of the DSC was concerned. In addition, the upper limit of the glass transition appeared to be 400°K. Fava reasoned that, if a material were cured at some temperature below its limiting glass transition temperature, the reaction would be extremely slow once the glass transition temperature coincided with the temperature of cure. This behavior resulted from the fact that diffusion inhibits further reaction in such a case. This data demonstrated the importance of curing a resin above its limiting glass transitions temperature if optimum physical properties were to be attained. If curing was conducted as in this example, below 400°K, the reaction would cease when the material became glassy. This behavior was demonstrated in Figure 5, where he showed DSC curves of samples cured at 360° and 330°K for up to 2000 hours, and there was no appreciable decrease in the exotherm size after the first 20 hours in either case.

This type of data can be obtained for any resin whose cure involves the evolution or absorption of heat. While DSC can effectively be applied to the analysis of degree of cure and to the selection of proper cure time, temperature or most effective curing agent, it can also be used in comparing curing agents (Figure 6). These scans were obtained on samples of DER-332 with DMP-30 [Tris (dimethyl-aminomethyl) phenol] activator. The procedure employed was to mix a batch of resin and activator according to specifications (100 parts DER-332 plus 5 parts DMP-30) in a glass container and then transfer the liquid epoxy to the DSC sample pans. The scans were run on a Du Pont 900 Thermal Analyzer in the DSC module. The curve labeled 'uncured' was obtained on a filled sample pan which had been kept in a 70°C oven for 3 hours, while the curve marked 'cured' was obtained on a

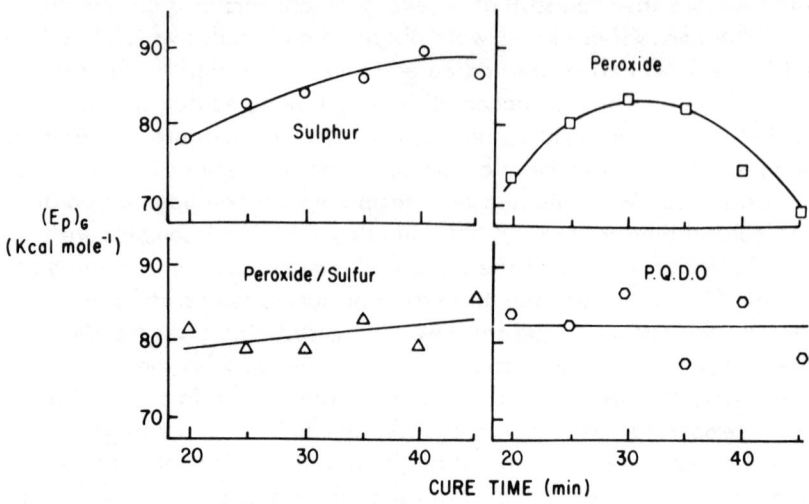

OVERALL ACTIVATION ENERGIES FOR THERMAL DEGRADATION OF ETHYLENE-PROPYLENE RUBBERS VS CURE TIME [a]

(a) Dudley, Smith and Youren, Thermal Analysis, Vol. 1, Schwenker and Garn, Eds., Academic Press 1969 pp. 643-665

FIGURE 3.

DSC SCANS OF DER-332 WITH HEXAHYDROPHTHALIC ANHYDRIDE CURING AGENT PREVIOUSLY CURED TO APPARENT COMPLETION AT INDICATED TEMPERATURE [a]

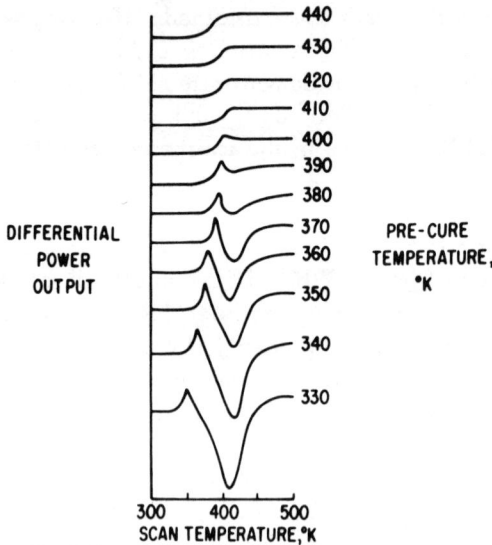

(a) Fava, R.A., Polymer, 9, 137 (1969)

FIGURE 4.

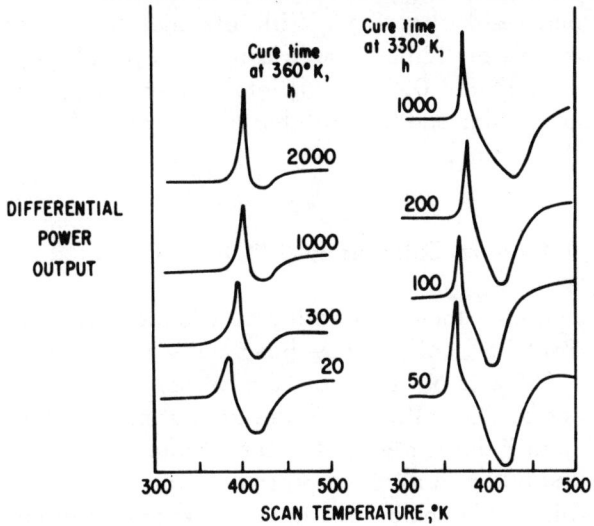

FIGURE 5.

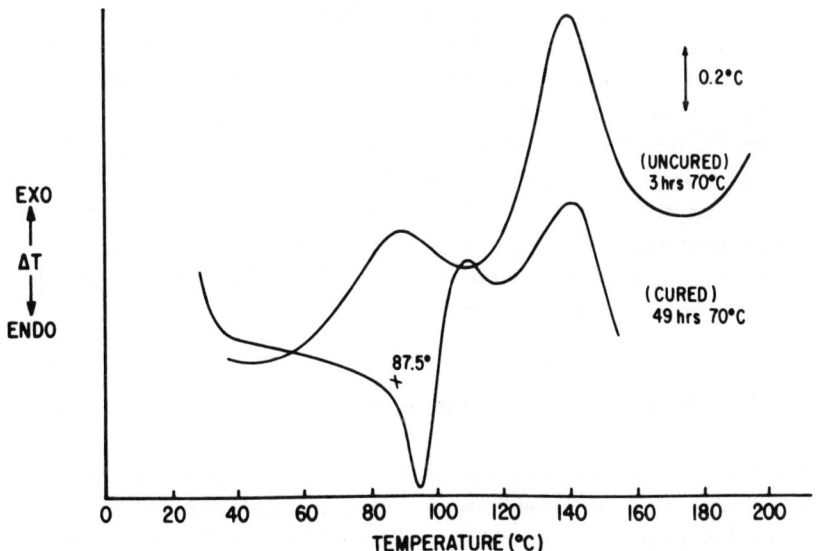

FIGURE 6.

filled sample pan which had been kept at the curing temperature for 49 hours. The uncured sample shows two exotherms indicating a characteristic two-stage curing reaction while the cured curve indicates a transition occuring at 87.5°C followed by some residual curing. With reference to Fava's work [7], these data indicate that this particular sample is far from completely cured. The main point, however, is that DSC is completely amenable to the study of curing reactions in epoxy resins and the path has been cleared for application of this technique to other resins.

High Pressure Calorimetry (HPC)

Another method available for investigating cure is high pressure calorimetry. A modified Du Pont DSC cell has now been made available to operate at pressures as high as 1000 psig or as low as 10 microns. Some reactions are not amenable to analysis by DSC at ambient pressure but can be studied at elevated pressure. Figure 7 shows a sample of a phenolic resin analyzed by DSC at both ambient and 800 psig nitrogen pressure. The scan at ambient pressure was totally without the characteristic curing exotherm of an uncured resin. The reason for this strange curve was that an endothermic water loss peak masked the exotherm of curing. When this same material was run under 800 psig nitrogen, the water endotherm was shifted out of field and two distinct exotherms for the cure of the phenolic were observed which indicated a two-stage curing reaction[8].

Even in those cases where the curing exotherms can be seen at ambient pressure, it is sometimes helpful to study the reaction under pressure. A curve was obtained for a sample of DER-332 with DMP-30 activator at 800 psig nitrogen. This curve is shown in Figure 8 and has a somewhat different character than the curve of the same uncured material in Figure 6. The conditions of sample preparation were identical in both cases for these curves and the indications are that the samples behave differently when at ambient and elevated pressures. This pressure technique can be employed as a means of elucidating reactions which cannot be studied at ambient and as an extension of data obtained from other techniques.

Thermomechanical Analysis (TMA)

Cuthrell[9] investigated the factors affecting the cure of epoxy polymers and employed expansion measurements as a means of determining the extent of cure. Figure 9 shows the expansion curve of DEA-828 [Epichlorhydrin-bisphenol A (Epon-828) cured with diethanolamine (DEA)]. Two phase transition temperatures are evident by this technique. Cuthrell demonstrated

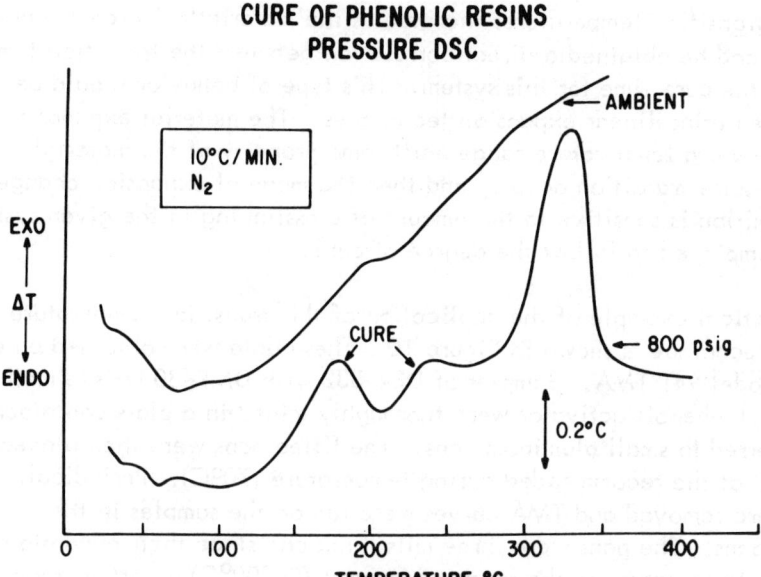

FIGURE 7.

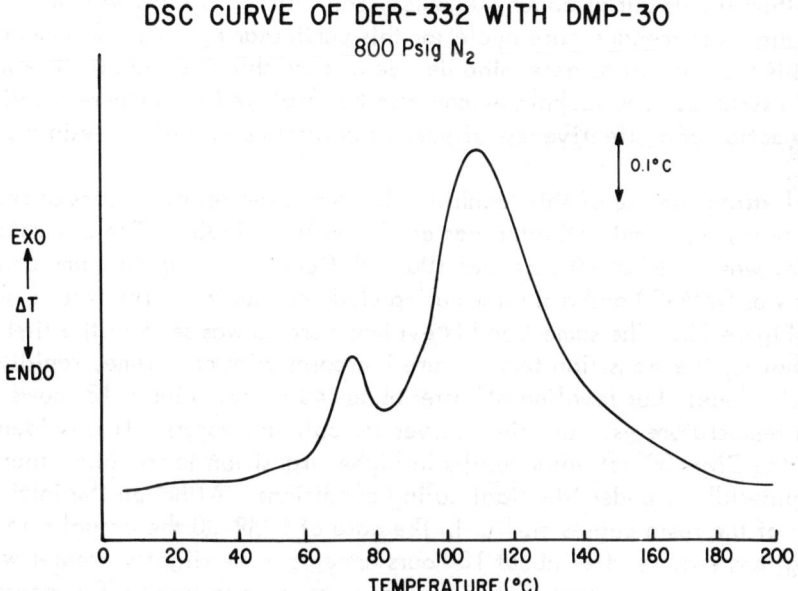

FIGURE 8.

that these transition temperatures increase as the material's degree of cure increases, and he obtained a direct correlation between the transition temperatures and the cure time for this system. This type of behavior should be typical when using linear expansion techniques. The material expands at a given rate over a temperature range until some property of the material changes or some transition occurs, and then the mode of expansion changes. If this transition is sensitive to the amount of crosslinking in the given system, it can be employed to follow the degree of cure.

A practical example of the application of this transition temperature vs. cure time technique is shown in Figure 10. These data were obtained on a Du Pont Model 941 TMA. Samples of DER-332 with DMP-30 [Tris(dimethylaminomethyl)phenol] activator were thoroughly mixed in a glass container and transferred to small aluminum pans. The filled pans were then placed in an oven set at the recommended curing temperature (70°C). Periodically samples were removed and TMA curves were run on the samples in the aluminum pans. The pans were three mils thick and since their expansion is linear with temperature in the range of interest (0-200°C) no attempt was made to remove the pans prior to analysis of the epoxy. Transition temperatures were determined as shown in Figure 10 and then these temperatures were plotted vs. cure time as shown in Figure 11. It is interesting to note that the transition temperature increases rapidly with time during the early stages of cure and then levels out at about 40 hours. This data correlates well with the fact that the recommended cure cycle for this particular epoxy is 40 hours at 70°C. This type of test to determine degree of cure should be applicable to other resin systems. The technique may also be employed to compare relative rates of reaction and effectiveness of various activators in a given resin matrix.

To illustrate the use of this technique in comparing relative rates of reaction in a resin, a second activator was employed in DER-332. The activator, piperidine, was mixed at 10 parts per 100 with DER-332 under the same conditions as was DMP-30 and a similar curing study conducted. The results are shown in Figure 12. The same trend is evident here as was seen in the first system, that is, the transition temperature increases with cure time, rapidly in the early stages, but leveling off after about 40 hours. Figure 13 shows transition temperature vs. cure time curves for both activators. It is evident here that the DMP-30 activator results in higher transition temperature than does the piperidine, under identical curing conditions. Although the initial hardening of the resin begins earlier in the case of DMP-30 the actual rate of increasing temperature after about 10 hours appears to be slightly greater with piperidine. It is also evident that both activators seem to level off at approximately the same time. This type of information could be of great importance in the processing of this resin and in the selection of the best curing agent and conditions for a given application. It is also of interest that this data can be

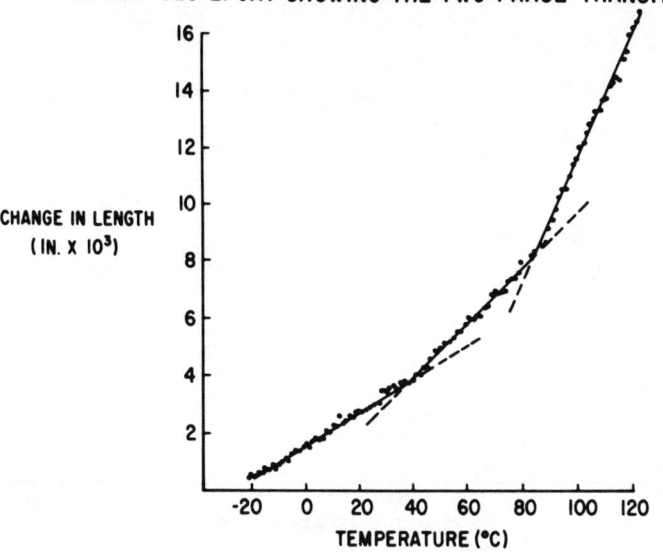

FIGURE 9.

(a) Cuthrell, J. Appl. Polym. Sci., 12, 955 (1968)

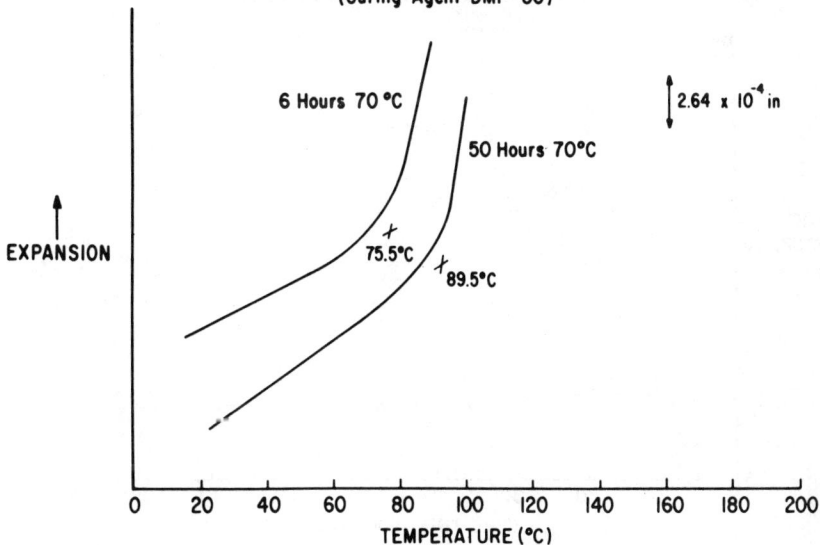

FIGURE 10.

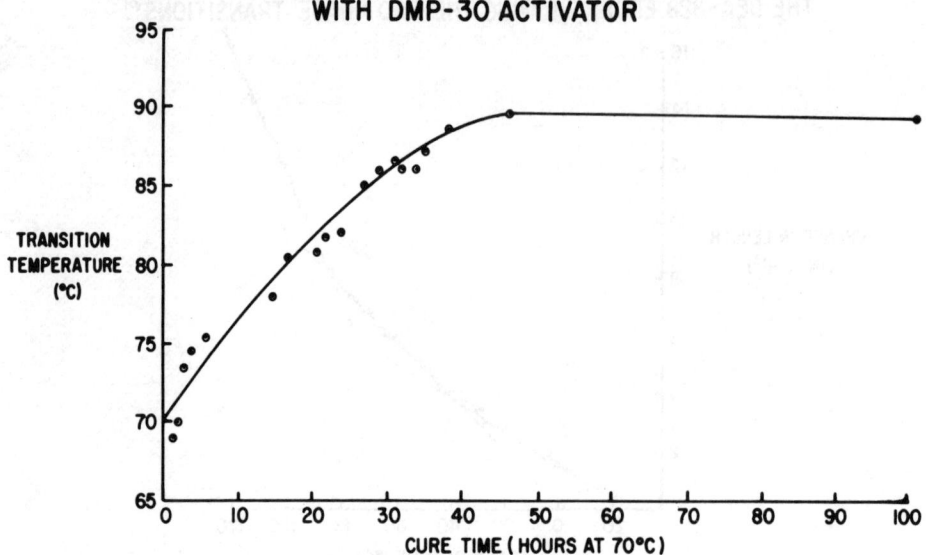

FIGURE 11.

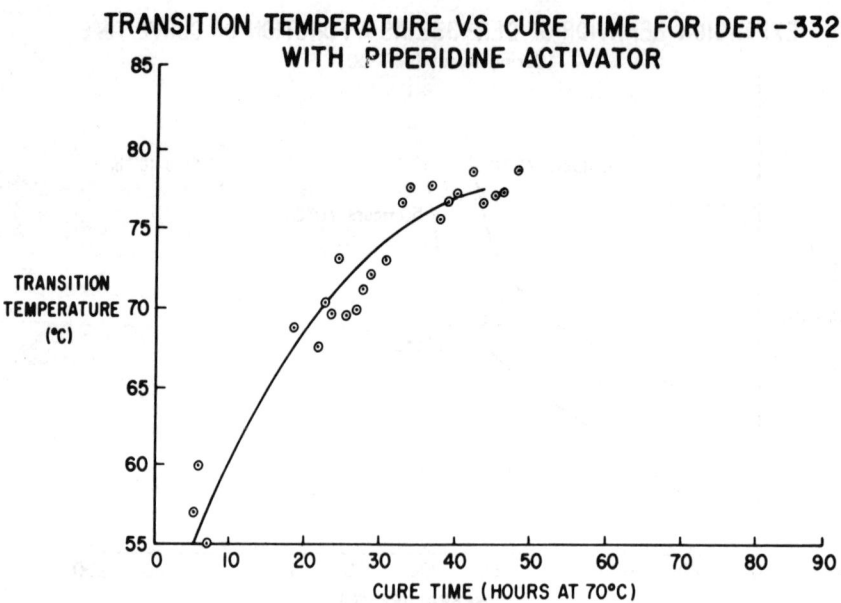

FIGURE 12.

obtained on very small samples either in the research lab or out in the manufacturing plant.

A third epoxy resin examined by thermoanalysis for degree of cure was a low coefficient of expansion adhesive resin. This material was submitted for evaluation without any identification and the data is merely shown to demonstrate that the same technique does not apply to every epoxy. Figure 14 shows the behavior of this system by TMA as a function of cure time. In this example the uncured resin mixture had a high viscosity and was not easily transferred to the aluminum pans employed above. The resin and activator were mixed well in a glass container and pressed into a sandwich 50 mils thick, between two layers of 2 mil aluminum foil. The sandwich was then placed in an oven set at the recommended curing temperature (93°C). Periodically, pieces were cut from the sandwich and analyzed directly without removing the aluminum. Again, the expansion behavior of aluminum is linear with temperature and would not interfere with transition behavior. As is seen in Figure 14 the expansion profile of the material changes with cure time but not the same way as was evident with DER-332. The final stages of cure of this material can be studied readily since one of the specifications for the final epoxy is that after cure its expansion coefficient should be $25.2 \times 10^{-6} (°C)^{-1}$. This allows curing conditions to be set which will give the final material this specified expansion coefficient.

CONCLUSIONS

The application of thermal analysis to the determination of the degree of cure of thermosetting resins has been discussed. A survey of some existing thermoanalytical techniques which have been successfully applied in this area has been given and some new data on the application of TMA and DSC to investigating cure have been presented. It is concluded that these four techniques (TGA, DSC, HPC and TMA) have been successfully applied to the solution of only a small percentage of the problems to which they are amenable.

ACKNOWLEDGMENT

The author wishes to thank P. F. Levy and G. W. Miller for their constructive criticism and guidance in the preparation of this paper.

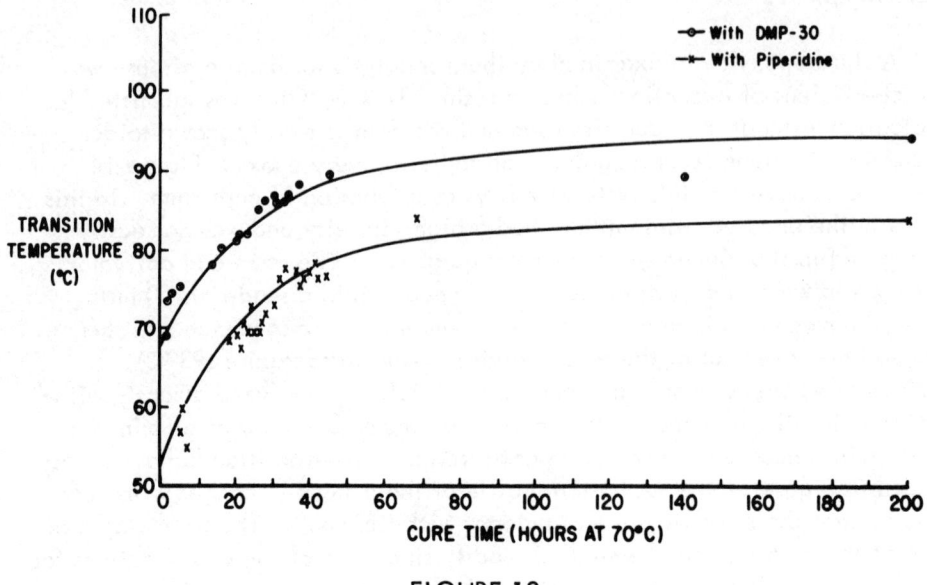

FIGURE 13.

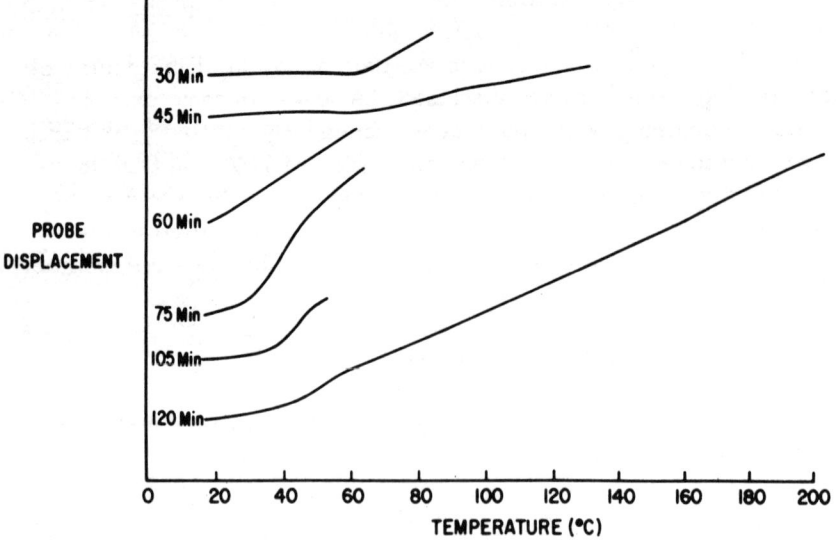

FIGURE 14.

REFERENCES

1. Baekeland, L. H., Chem. Ztg., 33, 317, 326, 347, 353, 1268 (1909), 36, 1245 (1912).

2. Lee, H. and Neville, K., Handbook of Epoxy Resins, McGraw-Hill, 1967, pp. 6-3, 6-61.

3. Warfield and Petree, The Temperature Dependence and Activation Energy of Electrical Conduction in High Polymers, NAVORD Report, 6246, (1958).

4. Delmonte, SPE Symposium, Minneapolis, Minnesota, October 1958.

5. Lee, H. T. and Levi, D. W., J. Appl. Polym. Sci., 13, 1703 (1969).

6. Dudley, Smith and Youren, Thermal Analysis, Vol. I, Schwenker and Garn Eds., Academic Press 1969, pp. 643-665.

7. Fava, R. A., Polymer, 9, 137 (1968).

8. Levy, P. F., Nieweboer, G. and Semanski, L. C., Pressure Differential Scanning Calorimetry, paper presented at ACS Meeting, Houston, Texas Feb. 22-27, 1970.

9. Cuthrell, R. E., J. Appl. Polym. Sci., 12, 955 (1968).

DYNAMIC CURE ANALYSIS OF THERMOSETTING POLYMERS

R. Bruce Prime

International Business Machines Corporation

Systems Development Division, Endicott, N. Y.

INTRODUCTION

A number of articles have been published which deal with the dynamic differential thermal analysis of chemical reaction kinetics.[1,2,3] In addition, there are publications which treat the polymerization kinetics of epoxy resins by dynamic differential scanning calorimetry (DSC).[4,5] One of these[5] contains a detailed treatment for obtaining the kinetic parameters from dynamic DSC data. The determination of epoxy cure kinetics from adiabatic calorimetry measurements has also been reported.[6] Inherent in all of the previous treatments is the tacit assumption that the isothermal rate is equivalent to the temperature variant or dynamic rate; i.e.

$$\left(\frac{\partial \alpha}{\partial t}\right)_T \equiv \frac{d\alpha}{dt} \qquad (1)$$

where α is the fraction reacted, t the time, and T the absolute temperature.

This paper describes for the first time the relationship between isothermal cure kinetics and dynamic cure kinetics. Dynamic kinetics are those which govern programmed heating rate (PHR) experiments. A new expression was developed for the dynamic rate of cure.[7] Results of dynamic DSC cure experiments on a model epoxy system prove that it is now possible to obtain isothermal kinetic parameters from programmed heating rate experiments.

THEORY

Recently, Draper pointed out that the assumptions embodied in Equation 1 are incorrect.[8] Since α is a function of time and temperature, the total derivative, $d\alpha$, must be used as follows:

$$d\alpha = \left(\frac{\partial \alpha}{\partial t}\right)_T dt + \left(\frac{\partial \alpha}{\partial T}\right)_t dT \quad (2)$$

From this equation the dynamic rate is obtained,

$$\frac{d\alpha}{dt} = \left(\frac{\partial \alpha}{\partial t}\right)_T + \left(\frac{\partial \alpha}{\partial T}\right)_t \frac{dT}{dt} \quad (3)$$

Note that the dynamic rate is related to the isothermal rate, $(\partial \alpha/\partial t)_T$; the general expression for the isothermal rate is

$$\left(\frac{\partial \alpha}{\partial t}\right)_T = f(\alpha) \, A \, \exp[-E/RT] \quad (4)$$

where A and E are the Arrhenius pre-exponential and activation energy, respectively; R is the gas constant; and $f(\alpha)$ is a function of the extent of reaction, and it is often assumed that

$$f(\alpha) = (1-\alpha)^n \quad (5)$$

where n is the order of the reaction.

Integration of Equation 4 followed by differentiation with respect to time and temperature yields the new dynamic rate expression[7]

$$\frac{d\alpha}{dt} = Z \left(\frac{\partial \alpha}{\partial t}\right)_T \quad (6)$$

where Z is a variable factor which relates the isothermal rate to the dynamic rate and is expressed as

$$Z = \left(1 + \frac{tE}{RT^2} \frac{dT}{dt}\right) \equiv \left(1 + \frac{E \cdot \Delta T}{RT^2}\right) \quad (7)$$

The factor Z depends on three constants — heating rate (dT/dt), activation energy, and gas constant — and on one independent variable — time or temperature. ΔT is the difference between the instantaneous temperature and the initial reaction temperature.

CALCULATIONS

For an exothermic reaction, the DSC measures the rate of heat production, $d\Delta h/dt$, as a function of temperature and time (see Figure 1).

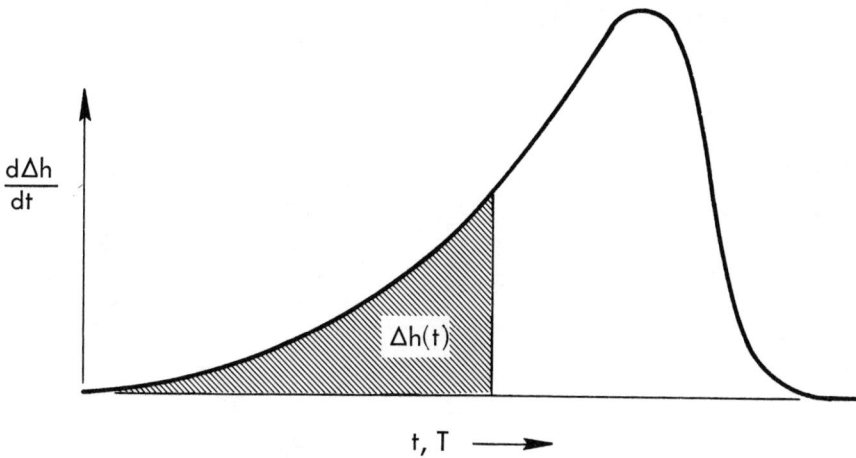

Fig. 1. DSC trace for an exothermic reaction.

The rate of cure is related to the DSC experiment by Equation 8

$$\frac{d\alpha}{dt} = \frac{1}{\Delta H} \cdot \frac{d\Delta h}{dt} \tag{8}$$

where ΔH is the total heat of reaction and $\Delta h(t)$ is the heat produced up to and including time t. Equations 5 and 6 were used to analyze dynamic DSC cure data. A relationship between n and E is needed to reduce Equation 6 to two unknowns. This relationship is obtained from the conditions at the peak exotherm (where $d^2\alpha/dt^2 = 0$) as follows:

$$n \approx \frac{E}{RT_p^2} \cdot \frac{(1-\alpha)_p}{(d\alpha/dt)_p} \cdot \frac{dT}{dt} = cE \tag{9}$$

Subscript p refers to values at the peak in the exotherm. The combination of Equations 5, 6, and 9 results in

$$\ln\left(\frac{d\alpha/dt}{Z}\right) = \ln A + bE \tag{10}$$

where

$$b = -1/RT + c \ln(1-\alpha).$$

A plot of $\ln\left(\frac{d\alpha/dt}{Z}\right)$ versus b yields ln A and E from the intercept and slope, respectively (see Figure 2). Once E is obtained, n is calculated from Equation 9. For comparison, calculations were also made without the Z factor.

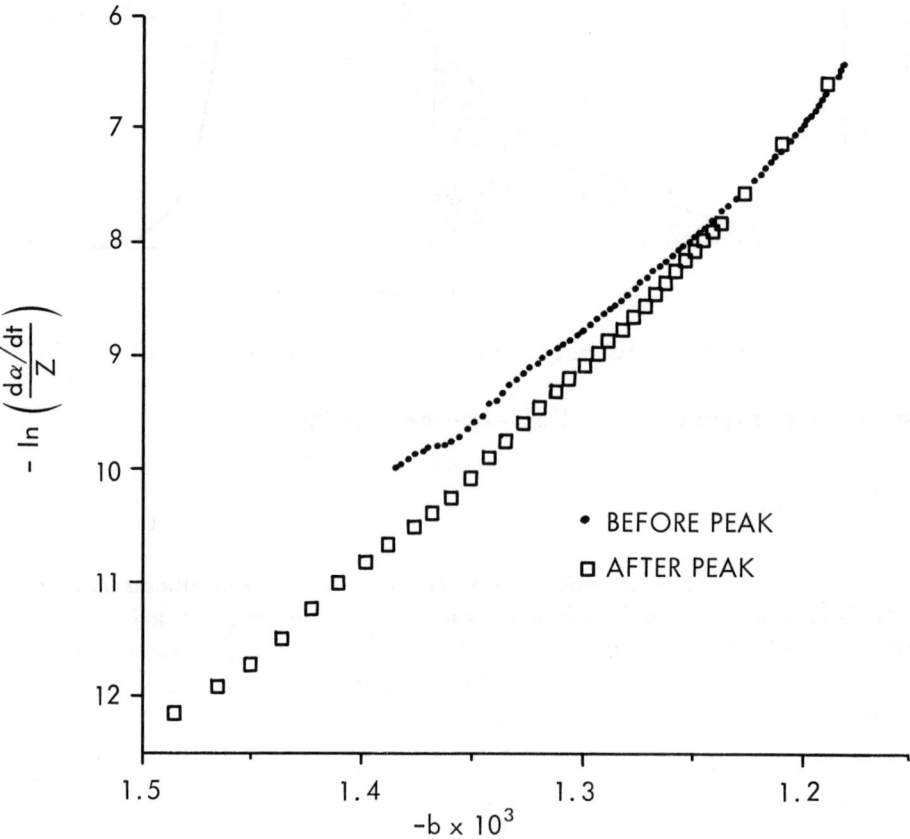

Fig. 2. Plot of Equation 10 variables. The slope is equal to E and the intercept is equal to ln A. This plot was made from the output shown in Figure 3.

EXPERIMENTAL

Materials

The comonomers used in this study were the chemically pure diglycidyl ether of bisphenol A (DER 332; Dow Chemical Co.) and m-phenylene diamine (m-PDA; Aldrich Chemical Co.; 99% pure). Methylene chloride solutions of these compounds were mixed and the solvent removed by freeze drying. The comonomer mixture contained 20 parts by weight m-PDA to 100 parts DER 332, or 1.3 equivalents amine to 1.0 equivalent epoxy. The mixture was stored at $-20°C$ and discarded after two weeks.

DSC — Computer Interface

The instrument used in this research was a differential scanning calorimeter (Perkin-Elmer DSC-1) coupled to an IBM 1800 Data Acquisition and Control System computer. The calorimeter was modified to give finer ZERO control by replacing the 500 ohm, single-turn, ZERO potentiometer with a ten-turn, Bourns potentiometer of the same resistance. Modifications were also made to the readout amplifier card of the calorimeter to increase the range of the ZERO control to -10 to +16 millivolts.

The principle of the DSC is to keep two sample holders at approximately the same temperature during a reaction by adjusting the supply of electrical power to the respective heaters. One of the holders is a reference while the other contains the epoxy sample. The differential power (ΔP) signal is recorded on a stripchart along with time and temperature signals. In addition, ΔP is recorded on a logical tape of the computer. This is accomplished by first amplifying* the differential power signal 1000 times and then digitizing the signal with an eight-range ohmmeter (Non-Linear Systems, Inc. Series X-1); associated equipment interfaces with the computer. (More sophisticated equipment which also allows the temperature signal to be recorded has since been installed.)

Data Collection

The data collection program was initiated by an IBM 2260 Video Display Unit which allows the operator to set the time interval between data points. Actual data collection was initiated and concluded by means of a pushbutton console. The DSC, digital voltmeter and associated hardware, and the 2260 Video Display Unit were all located in the same, temperature- and humidity-controlled room. At the conclusion of an experiment, the

---

* Dana Model 2000 DC Amplifier which also serves as an isolation amplifier.

data (which could be viewed on the video display unit) were smoothed by a nine point smoothing routine and stored. At the conclusion of a series of experiments, the data were punched on cards and read into the memory of an IBM System/360 computer for analysis with a terminal based language.

Before the data could be analyzed according to Equation 10, the beginning and end of the reaction had to be established and a baseline mathematically interpolated. The baseline is simply a cubic polynomial with the following four constants: the first deviation from the baseline, the leading slope, the return to the baseline, and the trailing slope. A computer controlled plot of the baseline and data are shown in Figure 3. The leading and trailing baselines were eliminated from this plot (compare with Figure 1). After the baseline was subtracted from the data, the appropriate calculations were made.

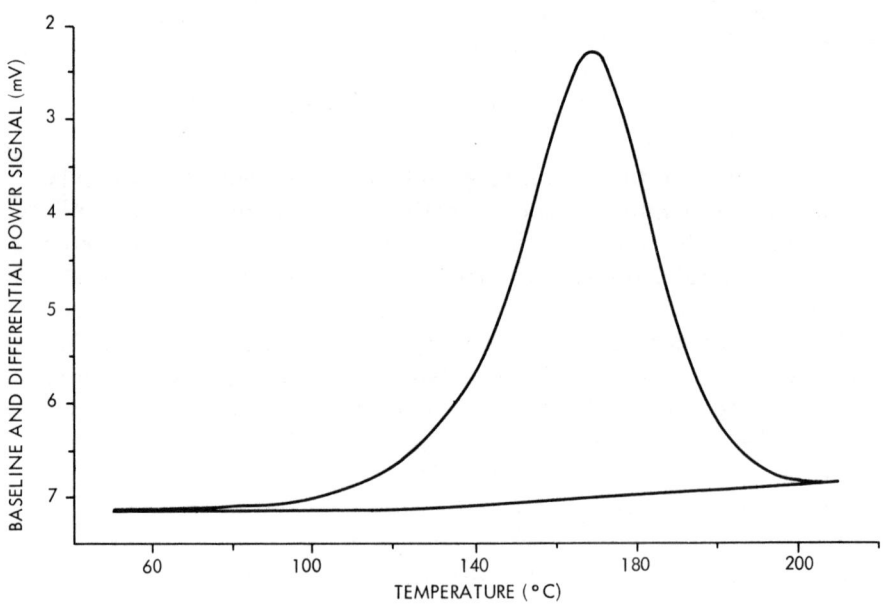

Fig. 3. Computer-controlled plot of differential power signal and baseline versus true temperature; heating rate: 20°C/min; scale: 8 mcal/sec.

Procedure

The comonomer mixture was removed from the freezer, and 5-20 mg was quickly weighed in a sample pan containing a disk of clean glass fabric. Two more such disks and an aluminum disk were added and the sample pan was crimped, placed in the sample holder, and covered with a dome cover. The reference was an empty pan with a dome cover. The glass fabric was used to evenly distribute the sample and prevent sample motion during cure. Experiments were run at heating rates of 5, 10, and 20 degrees per minute; each experiment started at 5°C. Data were collected at the rate of one per degree. At the conclusion of a series of experiments at each heating rate, a calibration run was made by using Indium (99.999% minimum purity). This information was coupled with previous calibration data of pure metals and zone-refined organic standards to correct the recorded temperature. Corrections were also made for the nonisothermal state of the DSC during a transition.

RESULTS

The general results of the dynamic cure experiments are listed in Table I. The peak temperatures (T_p) and the temperatures at the end of the reaction (T_f) increase with the increasing heating rates because of the kinetic nature of the reaction. The initial reaction temperatures (T_i) are the temperatures at which the reaction is fast enough to be detected by DSC. T_i is the least reproducible measurement and therefore is responsible for much of the difficulty in the kinetic analysis.

Table II presents the results of dynamic cure calculations (with and without the Z factor) along with the results of the isothermal experiments.[9] Dynamic calculations were made for values up to and including the DSC peak values (see Figure 2). Calculations from the other side of the peak led to higher values for each of the kinetic parameters. Three independent measuring techniques (DSC, infrared, and volume resistivity) were used in the isothermal study, and the results were calculated from the isothermal rate equation (Equation 4).

DISCUSSION

A new method for obtaining the cure parameters of thermosetting polymers has been presented. Cure studies on an epoxy system have demonstrated that use of a new dynamic rate equation allows the measurement of isothermal kinetic parameters from experiments conducted at programmed heating rates. In addition the successful coupling of the DSC with the computer has been reported.

TABLE I
General Results from Dynamic Cure Experiments

Heating Rate (°C/min)	ΔH (cal/g)	T_i (°C)	T_p (°C)	T_f (°C)
5	117	64	130	181
10	122	65	148	208
20	113	51	170	229

TABLE II
Kinetic Parameters from Dynamic and Isothermal Experiments

	Heating Rate (°C/min)	E (kcal/mole)	ln A	n
Dynamic without Z	5	23.2	23.5	0.9
	10	19.4	18.1	0.7
	20	15.0	12.3	0.6
Dynamic with Z	5	15.3	12.2	1.0
	10	12.4	6.6	0.8
	20	12.4	8.2	0.8
Isothermal	—	12.5	10.1	0.9

To the author's knowledge this is the first report of the proper relationship between isothermal and dynamic kinetics. Previous publications[1-6] have assumed the validity of Equation 1, and only one of these[5] presents a method for measuring all of the kinetic parameters. By using the method presented here, it is possible to obtain all the isothermal kinetic parameters (E, ln A, and n) from one PHR experiment. This approach can be applied to other kinetic analyses, for example, dehydration and decomposition.

Results of dynamic cure experiments on an epoxy system have demonstrated the validity of the new dynamic rate expression. At heating rates of 10 and 20°C per minute the results are within one standard deviation of the isothermal results. At 5°C per minute the activation energy is within three to four standard deviations of the isothermal value. The value of ln A is the least accurate of the kinetic parameters because of the long extrapolation to obtain it (see Figure 2).

Dynamic calculations were also made without the Z factor (i.e. assuming the validity of Equation 1). For these results there is a striking dependence on heating rate, and only at 20°C per minute do they approach the isothermal results; also both E and ln A are too large and both increase as the heating rate decreases. The observation was also made that the Z factor brings post-peak values into closer agreement with pre-peak values.

The nature of the cure reaction is also important to this discussion. Isothermal measurements of cure were made over a wide temperature range (23 to 160°C) by three experimental techniques.[9] From these measurements it was concluded that after the gel point is reached the chain extension through primary amine reaction cannot be distinguished from crosslinking through secondary amine reaction. Cure can be considered as simply one reaction: network formation by both primary and secondary amines where the reaction is limited in its extent of completion by the mobility of the reacting groups.

Coupling of the DSC with the computer offers new dimensions in differential scanning calorimetry. Among these dimensions are more sophisticated baseline interpretation, increased precision and accuracy (which may necessitate more sensitive instruments), and the possibility of computer simulated DSC experiments to further a better understanding of how the DSC functions. These ideas are presently being explored at IBM.

ACKNOWLEDGMENTS

The author gratefully acknowledges the assistance of Messrs. D. G. Sedor, R. C. Lasky, and F. S. Strock. The author also appreciates the contribution of Mr. J. E. Riedy in preparing the manuscript.

LITERATURE CITED

1. G. O. Piloyen, I. D. Ryabchikov, and O. S. Novikova, Nature, 212, 1229 (1969).

2. H. J. Borchardt and F. Daniels, J. Am. Chem. Soc., 79, 41 (1957).

3. H. E. Kissinger, Anal. Chem., 29, 1702 (1957).

4. R. A. Fava, Polymer, 9, 137 (1968).

5. O. R. Abolafia, SPE AnTech, 15, 610 (1969).

6. I. N. Sokol'nikova, I. M. Gurma, Y. M. Sivergin, and M. S. Akutin, Plast. Massy, (English translation: RAPRA 43E-55 CAm, 34), 9, 32 (1967).

7. R. B. Prime, to be published.

8. A. L. Draper, Proceedings of the Third Toronto Symposium on Thermal Analysis (H. G. McAdie, editor), 63 (1969).

9. M. A. Acitelli, R. B. Prime and E. Sacher, to be published.

DIFFERENTIAL THERMAL ANALYSIS OF THERMALLY REVERSIBLE GELS

Howard C. Haas, Monis J. Manning, Stanley A. Hollander

Polaroid Corporation Research Laboratories

Cambridge, Massachusetts 02139

INTRODUCTION

A thermally reversible gel is one which on the application of heat reverts to a fluid. The latter, on cooling, again undergoes gel formation. Thermally reversible gels are not only scientifically interesting but are also of commercial importance. Gelatin gels form the basis of photographic silver halide emulsion preparation. Aqueous gels of agar-agar are used as media for bacterial cultures, and pectin and gelatin for the manufacture of certain gelling food products.

A large amount of theoretical and experimental research has been carried out on polymeric gels in which the three dimensional gel network involves only covalent bonding. These studies have dealt with statistics of gel formation, the thermodynamics of swelling of crosslinked polymers, and the rubber-like mechanical behavior of these swollen networks. Generally speaking, the behavior of these systems is reasonably well understood. Much less emphasis has been placed on understanding thermally reversible polymeric gels. These gels pose problems not normally encountered in systems such as swollen vulcanized rubber or a poly-functional vinyl monomer polymerized beyond the gel point. One such problem is that it is difficult to determine when a thermally reversible gel is in true thermodynamic equilibrium. It is easy to quench a thermally reversible gel into a state which does not represent a minimum free energy situation particularly when crystallization and gelation are concurrent phenomena. The nature and structure of thermally reversible crosslinks are much less well defined than covalently crosslinked junction points. The reversible crosslinks may consist of hydrogen bonds, hydrophobic bonds, electrostatic interactions, dipole interactions, or combinations of all of these. A crosslink may be simple and involve only one group from each of two polymer chains or be complex and crystalline and involve many groups from two or more chains. The strength of a thermally reversible crosslink will be solvent dependent because of solvation effects on the groups involved. In many cases, the properties of the gel will be determined by its past thermal history. The study of the dilute solution behavior of polymer-solvent systems which form gels can involve experimental difficulties. The solvent must necessarily be a poor one for the system to gel. At the gel point, polymer-polymer contacts are preferred over polymer-solvent interactions and the system gels (or severely aggregates) before the θ temperature is reached. The θ temperature, therefore, must exist somewhere below the gel point and its onset is probably noted by the buildup of haze within the gel network.

There are numerous examples of aqueous thermally reversible gels which are based on synthetic polymers. Copolymers of methacrylic acid and methacrylamide[1], copolymers of acrylic acid and acrylonitrile[2], solutions of poly (vinyl alcohol) (PVA) containing various gelling agents[3,4], and partial acetoacetates of PVA with dihydrazides as gelling agents[5] show thermal reversibility. At certain pH's, phthaloylated ethyl cellulose and copolymers of acrylic acid and ethyl or methylacrylate also form thermally reversible gels[6], like some of the above. At high concentrations of polymer, poly (vinyl alcohol) forms a thermally reversible gel in water. If a small amount of vinyl fluoride is introduced into the PVA backbone, thermally reversible aqueous gels result at a very low copolymer concentration[7]. Some years ago in our laboratories, we synthesized the first vinyl homopolymers which yield thermally reversible gels in water which are somewhat akin to gelatin gels[8]. These materials are the poly-(acrylyl-) and -(methacrylylglycinamides). We have carried out a large amount of research on the properties of the dilute solutions and gels of these and other polymers and the results are currently being reported in the Journal of Polymer Science, Part A-2 as a series of papers titled "Synthetic Thermally Reversible Gel Systems". During this continuing effort, we employed differential thermal analysis to study the calorimetric behavior of thermally reversible gels. Surprisingly, we could find no other literature references reporting the use of DTA for this purpose. It is with this subject that this paper is concerned.

THEORETICAL

In 1954, Eldridge and Ferry[9] considered the thermodynamics of thermally reversible gelation. It can be shown that at the gel point, the point at which a three dimensional network infinite in extent first appears, where C is the concentration of polymer

$$fm_{c1} = c/2\bar{M}_w \qquad (1)$$

in grams per liter, $\bar{M}_w$ is the weight average molecular weight before crosslinking, m_{c1} is the total concentration of crosslinks in moles per liter at the gel point and f is the fraction of crosslinks that form non-cyclical structures. If m_1 is defined as the molar concentration of free crosslinking loci on a polymer molecule and we assume that single crosslinks are formed by binary association between chains, we can write the following equilibrium constant:

$$m_{c1}/m_1^2 = K(T, \bar{M}_w) \qquad (2)$$

Equations (1) and (2), on taking logarithms, differentiating with respect to temperature at constant $\bar{M}_w$ and combining with van't Hoff's law yield

$$\left[\frac{d(\ln C)}{dT}\right]_{\bar{M}_w} \left[1 - \frac{2\,\partial \ln m_1}{\partial \ln C}\right] = \frac{\Delta H^\circ}{RT^2} + \left[\frac{d(\ln f)}{dt}\right]_{\bar{M}_w} + \frac{2\,\partial(\ln m_1)}{\partial T} \qquad (3)$$

If it is assumed that $(d\ln f/dT)_{\bar{M}_w} = 0$, $\partial \ln m_1/\partial \ln C = 1$ and $\partial \ln m_1/\partial T = 0$, equation (3) reduces to

$$-\left[\frac{d(\ln C)}{dT}\right]_{\bar{M}_w} = \frac{\Delta H^\circ}{RT^2} \qquad (4)$$

which on integration gives

$$\log_{10} C = \Delta H^\circ / 2.303 RT + \text{Constant} \qquad (5)$$

Here, C is the concentration of polymer in the gel of g/liter, T is the absolute melting point of the gel and ΔH° is the heat of the reaction for the process: m_1 (one mole) + m_1 (one mole) ⟶ m_{c1} (one mole). Eldridge and Ferry applied equation (5) to melting point data on aqueous gelatin gels, and obtained the predicted linearity of log C vs. 1/T plots and values for ΔH° ranging from -50 to -220 kcal per mole of crosslinks. The high values of ΔH° were interpreted as meaning that each crosslink might involve between 10 and 45 hydrogen bonds assuming a value of -5 kcal/mole for the H bond. Annealed gels had higher melting points and yielded higher values for $\Delta H°$ than those formed by quick chilling. The crystalline nature of gelatin gels has also been demonstrated by X-ray[10], light scattering[11], and specific volume measurements. The high value of $\Delta H°$ implies that the gel junction joints involve numerous peptide groups and are also probably small crystallites. Similar measurements on poly(acrylylglycinamide)[8,12], copolymers and terpolymers of acrylylglycinamide[12], poly(methacrylylglycinamide)[13] and gels of ethylacrylate-acrylic acid and phthaloylated ethyl cellulose[6] have yielded values between -5 and -12 kcal/mole of crosslinks for the heat of crosslinking. This implies that a crosslink probably consists of only one group from each of two chains. The complex nature of the crosslinks in gelatin and the simple nature of the crosslinks in poly(acrylylglycinamide) gels has been verified by equilibrium swelling measurements and moduli measurements on the equilibrium-swelled gels[12] and application of the Flory-Rehner and rubber-like elasticity theories. Annealing gels of poly(acrylylglycinamide) has little effect on either the melting points of the gels or the value for $\Delta H°$ which is consistent with their non-crystalline character[14].

EXPERIMENTAL

Measurements

A Model 900 duPont Thermal Analyzer modified to increase the signal gain was employed. The gels were contained in small sealed aluminum cups. Heating rates ranged from 1 to 10°C per minute. Some measurements were also made using the penetrometer probe and expansion dilatometer attachments.

Materials and Thermally Reversible Gel Preparation

Gelatin. A photographically inert Rousselot limed ossein gelatin was used to prepare an aqueous gel containing 18.4% gelatin by weight (19.3g of gelatin per 100 ml of gel). The gel was conditioned for many days at 3°C prior to measurement.

Poly(acrylylglycinamide). The polymer was obtained from a persulfate-initiated polymerization of the monomer in water containing a small amount of isopropanol at 75°C. The intrinsic viscosity $[\eta]$ in 2M NaCNS at 25°C was 0.64 corresponding to an $\overline{M}_n$ of 200×10^3. A gel was prepared by allowing a dry film to swell to equilibrium in water at 25°C. The gel contains 19.65 percent polymer by weight.

6,6 Nylon. A commercial sample of poly(hexamethyleneadipamide) was used to prepare a 8.8% by weight thermally reversible gel in 5/1 meta-cresol/dimethylformamide by volume. The gel was conditioned at room temperature overnight.

Poly-γ-Benzyl-L-Glutamate. The polymer having a molecular weight between 200 and 400×10^3, was obtained from Pierce Chemical Co. A thermally reversible gel in xylene (Mallinckrodt, Analytical Reagent) was prepared containing 5% PBLG by weight. The gel was conditioned overnight at room temperature.

Vinyl Alcohol-Vinyl Fluoride Copolymer. The polymer was prepared by an AIBN-initiated copolymerization of 20g of vinyl acetate and 2g of vinyl fluoride at 60°C in 30g of ethanol followed by an acid catalyzed conversion to the polyol. The copolymer contained 10 mole percent vinyl fluoride by fluorine analysis. An aqueous gel containing 13.4g of polymer per 100 ml was prepared and conditioned at room temperature for several days prior to measurement.

Poly (Vinyl Chloride). A commercial sample of poly (vinyl chloride) having a $\overline{M}_n$ of 60×10^3 was used. A dibutylphthalate gel containing 0.6g of polymer per 10 ml was prepared. The gel was conditioned for several days at room temperature.

EXPERIMENTAL RESULTS AND DISCUSSION

Gelatin

The first gel to be studied was gelatin. A DTA curve is presented in Figure 1 for a heating rate of 2°C/min at a gain level of 0.01°C/in. An exotherm (15-25°C) precedes two overlapping endotherms, one starting at 32°C, and the second shortly thereafter. The premelt exotherm corresponds to approximately -0.11 cals/g of gel. The measured heat for the combined endotherms is 1.35 cals/g of gel. The heat observed during the first endotherm corresponds to about 20% of the total of the two endotherms. We believe the premelt exotherm to be a heat of gel reorganization resulting in larger and more perfect crystallites in the gel structure. In essence, this is similar to an annealing operation just below the melting point. Since the macroscopically observed melting point of the gel is very close to 32°C, the first endotherm probably corresponds to breaking the gel network and this is followed by (second endotherm) melting of the larger crystallites within the now soluble aggregates. A penetration probe measurement (Figure 2a) also confirms that network breakdown occurs at about 32°C and corresponds to the lower endotherm. Figure 2b represents the type of curve we obtain using the duPont expansion dilatometer with a silaceous material as a packing between the gel and the flat probe. A helix-coil transition may also be contributing to the magnitude of the second endotherm[15]. This interpretation is consistent with X-ray diffraction results which show that a relatively sharp diffraction pattern exists for

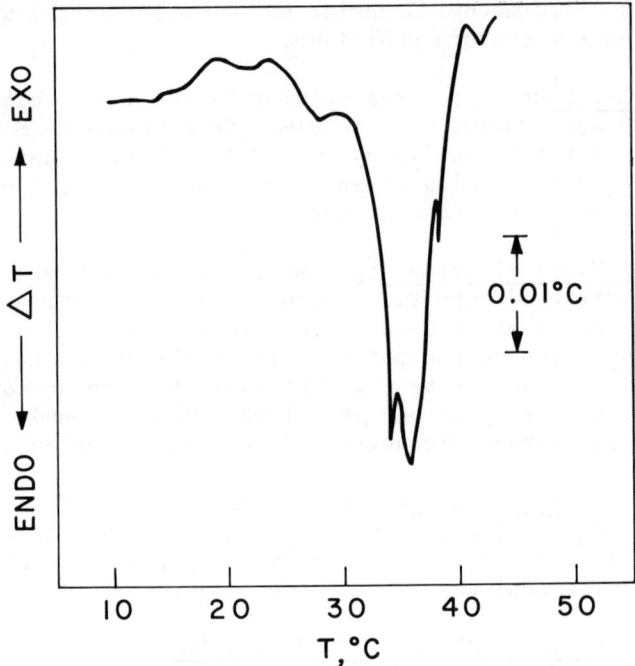

Figure 1. Gelatin Gel; 18.4% in H_2O; Sample 18.8 mg; Heating Rate 2°C/min.

some time after a gelatin gel has melted. In an article now in press[16] we made an approximate calculation from thermodynamic measurements which indicates that breaking the network required about 20% of the total endothermic heat observed by DTA.

Poly(Acrylylglycinamide)

Measurements suggest that these gels are non-crystalline and the measured heat of crosslinking is only -5 to -12 kcal/mole of crosslinks[12]. This would place an upper limit on the magnitude of the DTA endotherm of only 0.03 cals/g of gel. The melting point of this particular gel is between 80 and 90°C as determined on a Fisher-Johns apparatus. A typical DTA thermogram (Figure 3) shows no signal in the region of the gel melting point. This behavior is consistent with our prior measurements and conclusions which would place the magnitude of the endotherm outside the range of DTA sensitivity at least at the gain level of 0.05°C/in used for this particular gel.

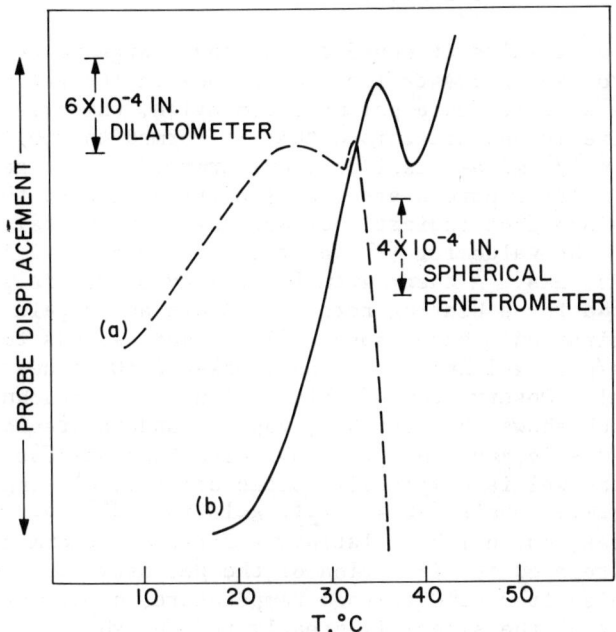

Figure 2. Gelatin Gel; 18.4% in H_2O; Curve (a) ---Spherical Penetrometer Probe, 20 mg load; Curve (b) —— Dilatometer Probe, 2g load, silaceous filler used; Heating Rate 1°C/min.

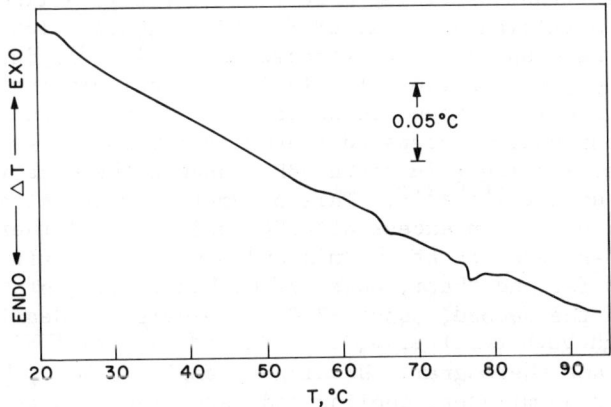

Figure 3. Poly(acrylylglycinamide) Gel; 19.65% w/w in H_2O; Sample 6.1 mg; Heating Rate 5°C/min.

6,6 Nylon

From the above results it would appear that large values for $\Delta H°$ are related to having crystalline crosslinks in the gel networks. Therefore we selected a polymer, 6,6 nylon, which is readily crystallizable and determined that it formed thermally reversible gels in 5/1 meta-cresol/dimethylformamide. The gels are hazy and the melting points are very sensitive to past thermal history which are good indications that they are crystalline. In spite of this, the value for $\Delta H°$ was found to be only -5.5 kcal/mole of crosslinks. A thermogram for the nylon gel is presented in Figure 4a for a heating rate of 5°C/min and a gain of 0.01°C/in. The first endotherm, onset (63°C), corresponds to a heat of 0.13 cals/g of gel and the second, onset 68°C, a heat of 0.19 cals/g of gel. Observation of this gel on a Fisher-Johns melting point block shows that at 66°C, vapors condense rapidly on the glass cover slip, and the gel flows with "stringiness" at 68°C. At 74°C, the gel is completely molten and free flowing. The first endotherm accounts for 40% (cf. gelatin = 20%) of the total absorbed heat, but unlike gelatin the first endotherm does not appear to be related to disruption of the gel network. If the sample is cooled from 92°C to room temperature, held there for 80 minutes and rerun, the signal is greatly reduced showing that reconstruction of the gel structure is slow (Figure 4b). Although we know $\Delta H°$ for this nylon gel, we do not know the crosslink density in the gel and therefore cannot predict the size of the endotherm to be expected for network fusion.

Poly-γ-Benzyl-L-Glutamate (PBLG)

It is apparent that something more than just crystallinity is necessary for obtaining high values for $\Delta H°$. While nylon crystallites are made up of fully extended chains[17,18], during the gelation and crystallization of gelatin, helical winding of the gelatin random coil is known to occur[19]. Assuming that this might be important, we prepared thermally reversible gels of PBLG in xylene, a solvent in which PBLG assumes the α-helical form of Pauling and Corey[20,21]. This polymer-solvent gel system led to values of $\Delta H°$ in excess of -300 kcal/mole. A thermogram for the PBLG-xylene gel at 5°C/min and 0.01°C/in is given in Figure 5a. The first endotherm, onset 72°C, has a value of 0.13 cals/g of gel and the second, onset 82°C, 0.10 cals/g. Heating the sample through both endotherms, cooling and rerunning yields essentially the same thermogram. Heating a sample to 74°C, holding it at 74°C for 30 minutes, cooling and rerunning results in disappearance of the first and intensification of the second endotherm (Figure 5b). A test tube melting point of 82°C and a dilatometer measurement, Figure 5c, indicate that network fusion

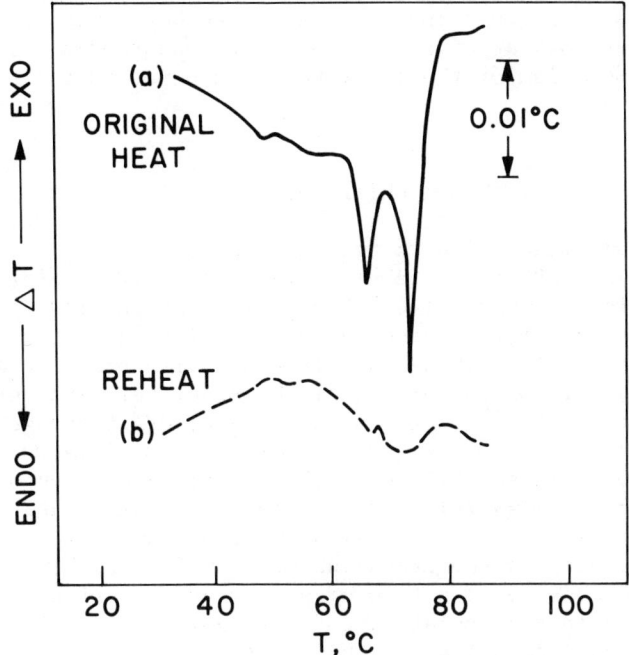

Figure 4. 6,6 Nylon Gel; 8.8% w/w in 5/1 m-cresol/dimethylformamide; Sample 13.2 mg; Heating Rate 5° C/min.

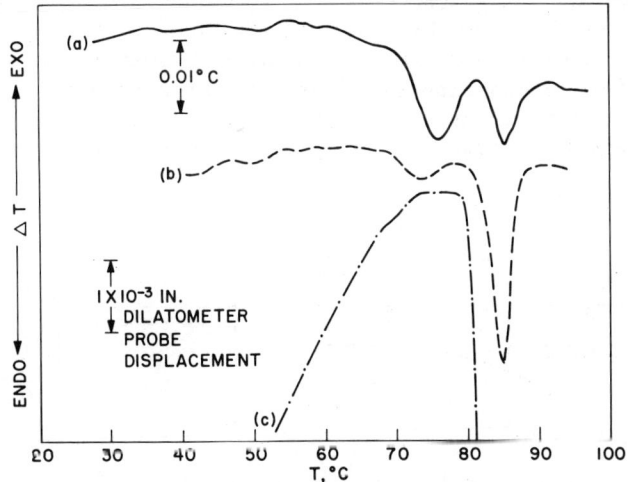

Figure 5. PBLG Gel; 5% w/w in xylene; Samples (a) and (b) both 11.2 mg; Sample (b) previously held at 74°C for 30 minutes, then cooled and reheated; Heating Rate 5° C/min.

is related to the second endotherm. This result is similar to that observed with 6,6 nylon. As yet, with both nylon and PBLG, we are not sure of the origin of the lower endotherms. They may involve polymorphism or the existence of a paracrystalline phase (22,23).

Poly(Vinyl Chloride); PVA-F

A limited amount of work has been carried out on poly(vinyl chloride) and PVA-F gels. We have measured $\Delta H°$ for poly(vinyl chloride) gels in dioxane (-8.2 kcal/mole) and in dibutylphthalate (-5.8 to -6.3 kcal/mole depending on conditioning). The melting points of the gels in dibutylphthalate are also somewhat dependent on the thermal history. Poly(vinyl chloride) gels are believed to be crystalline[24,25]. In spite of this, DTA thermograms do not show much signal in the region of the gel melting point. One run at 10°C/min and 0.02°C/in (Figure 6) gave an endotherm corresponding to 0.029 cals per g of gel. The small heat involved indicates that the crystal bonding energy is low if crystallites are present and the crosslink density is probably also low. The crosslink density of a film of poly(vinyl chloride) equilibrium-swelled in dioxane at 25°C (17.1% polymer) involves approximately only one crosslink per chain[26].

The measured $\Delta H°$ for poly(vinyl alcohol-fluoride) copolymer gel in water varies from -28 to as high as -80 kcal/mole. This may imply that this copolymer exists in water in a helical conformation which is stabilized by -O-H-F- bonds. The thermogram (Figure 7, 5°C/min, 0.02°C/in) has many endotherms. They may result from the combination of heterogeneities of not only molecular weight but also fluorine contents and the resulting broad variety of sequence distributions present in the copolymer backbones.

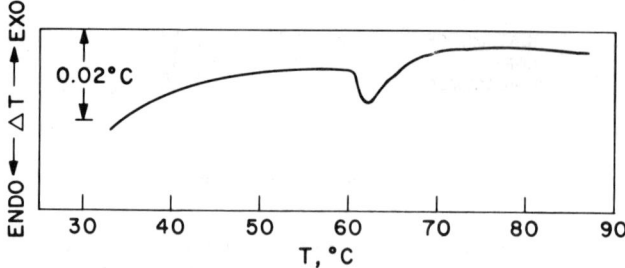

Figure 6. PVC Gel; 8% w/w in dibutylphthalate; Sample 12.8 mg; Heating Rate 10°C/min.

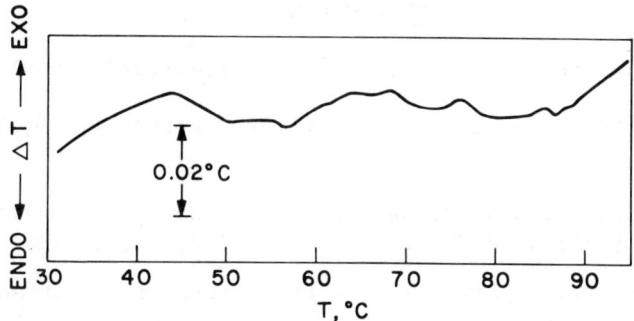

Figure 7. PVA-F gel; 13% w/w in H_2O; Sample 12.6 mg; Heating Rate 5°C/min.

CONCLUSIONS

1. Differential thermal analysis is a useful tool for studying the thermal behavior of certain thermally reversible gels.

2. Non-crystalline gels like poly(acrylylglycinamide) having low heats of crosslinking are at the limit of DTA sensitivity as we now know it.

3. Endotherms are readily detectable by DTA at temperatures near the melting point for crystalline gels like gelatin -H_2O, 6,6 nylon-m-cresol/DMF and PBLG-xylene. The DTA curves of these gels are dependent on past thermal history, as observed for other crystalline polymers.

4. For gelatin-H_2O gels, the heat required for network fusion probably lies close to 20% of the total DTA endotherms.

5. Nylon and PBLG also exhibit at least two endotherms but unlike gelatin, the endotherm at the lower temperature does not appear to be related to network breakup but probably with paracrystallinity, polymorphism, or some other phenomenon.

6. High heats of crosslinking in thermally reversible gels are probably the result of crystallinity involving helical conformations.

7. DTA has been used for studying the denaturation kinetics of biopolymers in solutions[27] and similar types of transitions in solid polypeptides[28]. Thermally reversible gels of biopolymers might be useful new media for studying their conformational changes and related energies.

REFERENCES

1. C. E. Barnes and W. O. Ney, Jr., U.S. 2,461,023 (1949)

2. J. Pouradier and A. C. A. Clavier, Ger. Patent 1,040,370 (1958)

3. W. H. McDowell and W. O. Kenyon, U.S. 2,234,186 (1941); 2,249,536 (1941); 2,249,537 (1941) and 2,249,538 (1941).

4. W. G. Lowe, U.S. 2,286,215 (1942); 2,311,058 (1943), and 2,311,059 (1943)

5. G. D. Jones, J. Appl. Polymer Sci., $\underline{6}$, 15 (1962)

6. J. S. Yudelson and R. E. Mack, J. Polymer Sci., A $\underline{2}$, 4683 (1964)

7. B. W. Howk and L. Plambeck, Jr., U.S. 2,499,097 (1950)

8. H. C. Haas and N. W. Schuler, Polymer Letters $\underline{2}$, 1095 (1964); H. C. Haas, R. D. Moreau and N. W. Schuler, J. Polymer Sci., A-2, $\underline{5}$, 915 (1967)

9. J. E. Eldridge and J. D. Ferry, J. Phys. Chem., $\underline{58}$, 992 (1954)

10. K. Herrmann, O. Gerngross and W. Abitz, Z. Phys. Chem., $\underline{\text{B } 10}$, 371 (1930)

11. H. Boedtker and P. Doty, J. Phys. Chem., $\underline{58}$, 968 (1954)

12. H. C. Haas, C. K. Chiklis and R. D. Moreau, J. Polymer Sci., paper III, in press

13. H. C. Haas, R. L. MacDonald and A. N. Schuler, paper VII, submitted to J. Polymer Sci.

14. H.C. Haas, R. L. MacDonald and A. N. Schuler, J. Polymer Sci., paper VI, in press

15. S. V. Pudenko and S. M. Tevi, Vysokomol. Soendin., Ser. A, $\underline{10(3)}$ 647 (1968)

16. H. C. Haas, M. J. Manning and M. H. Mach, J. Polymer Sci., paper V, in press.

17. C. W. Bunn and E. V. Garner, Proc. Roy. Soc. (London) A $\underline{189}$, 39 (1947)

18. D. R. Holmes, C. W. Bunn and D. J. Smith, $\underline{17}$, 159 (1955)

19. A. Veis, The Macromolecular Chemistry of Gelatin, Academic Press, New York, 1964, pp 268-9

20. L. Pauling and R. B. Corey, Proc. Natl. Acad. Sci., U.S.A. $\underline{37}$, 241,729 (1951)

21. P. Doty, J. H. Bradbury and A. M. Holtzer, J. Am. Chem. Soc. $\underline{78}$, 947 (1956)

22. E. T. Samulski and A. V. Tobolsky, Macromolecules $\underline{1(6)}$, 555 (1968)

23. C. Robinson, Mol. Cryst., $\underline{1}$, 467 (1966)

24. T. Alfrey, N. Wiederhorn, R. Stein and A. Tobolsky, Ind. Eng. Chem., $\underline{41}$, 701 (1949)

25. H. Morowetz, Proc. Paint Res. Inst., $\underline{38}$, 59 (1966).

26. Unpublished results.

27. H. W. Hoyer, J. Am. Chem. Soc., $\underline{90\text{-}10}$, 2480 (1968)

28. D. B. Green, F. Happey and B. M. Watson, European Polymer Journal, $\underline{6}$, 7 (1970)

THERMAL BEHAVIOR OF AQUEOUS GELATIN SOLUTIONS

S.E.B. Petrie and R. Becker

Research Laboratories, Eastman Kodak Company, Rochester
New York 14650

ABSTRACT

The thermal behavior of aqueous gelatin solutions prepared from deionized bone gelatin has been studied by differential scanning calorimetry. The peak temperatures of the endotherms associated with the loss of gel structure were found to be a function of the gelation temperature and time. For 10% solutions, the transitions for material gelled for 48 hours at 4, 10, and 22.5°C were observed at 32, 33, and 37°C, respectively. In order to achieve the sensitivity required for evaluating the transition enthalpy, a Du Pont Differential Thermal Analyzer was coupled with a Keithley Microvolt Amplifier. The enthalpy changes calculated from the dsc endotherms for 10% solutions varied from 4 to 7 cal/g of gelatin depending on the gelation time and temperature. Preliminary data obtained for 5% solutions indicate that the thermal behavior is similar to that of the 10% solutions.

A study has been made of the influence of heating rate and various annealing procedures on the thermal behavior of developed gel structure. The increase in the transition temperature with increasing annealing temperature and the development of multiple endotherms are similar to those observed for first-order phase transitions in polymers. For a 10% gelatin solution, a limiting transition temperature of 45°C was estimated from an extrapolation of the data.

INTRODUCTION

Because the thermal energy associated with the gel-sol transition in aqueous gelatin solutions is small,[1,2,3,4] calorimetric studies of the transition are difficult to carry out. Extensive information regarding the factors that influence the amount and thermal stability of gel structure developed in gelatin solutions, however, has been derived from measurements of other physical properties that are sensitive to gelatin conformational changes. These investigations,[2,3] based mainly on measurements of rheological and optical rotatory properties, indicate that the nature of the gel structure is influenced by several variables involving the conditions of gelation and the thermal history, as well as the type and concentration of gelatin. Information concerning the transition enthalpy has been based primarily on estimates made from indirect measurements.[1,4,5,6]

With the sensitivity and ease of measurement available with differential thermal techniques,[7,8] a quantitative study of the effect of influential variables on the thermal behavior of gelatin solutions would seem to be possible. Nevertheless, until recently, attempts in this laboratory to measure directly the transition energy associated with the loss of gel structure by the differential thermal method have not been successful. The sensitivities of the commercial, differential scanning calorimeters (dsc) employed, the Perkin-Elmer Model dsc-1B and the Du Pont Model 900 with dsc cell, were found to be inadequate. Transition temperatures, however, could be detected readily in solutions having gelatin concentrations of 1% or more. Purcell et al.[9] have reported similar qualitative differential thermal data for aqueous collagen solutions. Except for the quantitative studies with differential adiabatic calorimeters by Andronikashvili et al.[10] and Privalov and Tiktopulo[11] on collagen solutions, other differential thermal studies of gelatin and collagen have been made on material in either the solid[6,12,13,14,15] or swollen state.[16]

Several authors[17,18] have succeeded in measuring with adiabatic calorimetry the enthalpy changes related to the helix-to-random coil transition in synthetic polypeptides in dilute solutions. In order to retain the ease of measurement provided by dsc, however, an attempt was made to improve the sensitivity of the Du Pont 900, which was found to have good temperature stability in the temperature interval 0 to 50°C. With additional amplification of the differential signal, and with careful matching of the heat capacities of the sample and reference materials, the response required for making a quantitative investigation was achieved.

In this paper, preliminary studies are reported of the influence of gelation conditions and thermal history on the thermal behavior of gel structure, developed in 5 and 10% aqueous solutions of deionized bone gelatin.

EXPERIMENTAL

A Du Pont 900 Differential Thermal Analyzer equipped with a dsc cell and with an x-y recorder, suitably modified, was used in all the experiments. The block diagram of the modified analyzer system in Fig. 1 traces the path of the differential temperature signal (ΔT) measured by the differential thermocouple system. In the Du Pont 900 Analyzer, the signal goes from the thermocouples to a high gain, low noise, differential temperature preamplifier, and then through the ΔT range control of the y-axis recorder input. The sensitivity amplification was made by inserting a Keithley, model 150 AR, microvolt-ammeter into the circuitry immediately preceding the y-axis recorder input and after the ΔT preamplifier output. This modification increased the sensitivity by 10- to 100-fold. Normally, a 10X amplification was sufficient.

The gelatin solutions and samples were handled in the following manner. Aqueous, 5 and 10% solutions, 0.1 $\underline{M}$ in KCl and at pH 4.8, were prepared from deionized bone gelatin and distilled water and then stored in a refrigerator. In the preparation of the samples, the stock solution was melted in a water bath maintained at 47 $\pm$ 0.5°C. A microliter gas syringe, also heated to 47°C, was used to transfer about 15 mg of the solution to the sample holders, which were pressure-sealable, aluminum pans suitable for volatile liquids.

In order to erase any thermal history that might have developed during sample preparation and weighing, the samples were heated in the 47°C bath for 15 min before being placed at the chosen gelation or annealing temperature. Three gelation temperatures, 4, 10 and 22.5°C, were used. The 4°C temperature was maintained in a refrigerator. Both the 10° and 22.5°C temperatures were maintained in water baths.

Because of the limited temperature interval between the freezing temperatures of these aqueous solutions and the transition temperatures, it was necessary to minimize the initial baseline displacement. Careful matching of the heat capacities of the reference and sample was required. Distilled water of the same weight as the aqueous gelatin samples proved to be adequate as a reference material. Much of the initial baseline displacement associated with the experiment was eliminated provided a good match was achieved.

Since the temperature range of the transition was between 10° and 40°C, gallium was chosen for the calibration of the temperature and differential temperature signals. The melting temperature and heat of fusion of gallium are 29.6°C and 19.1 cal/g, respectively.[19]

Error incurred in the determination of the transition energy, ΔH, as a result of error in the calibration coefficient was about 1%,

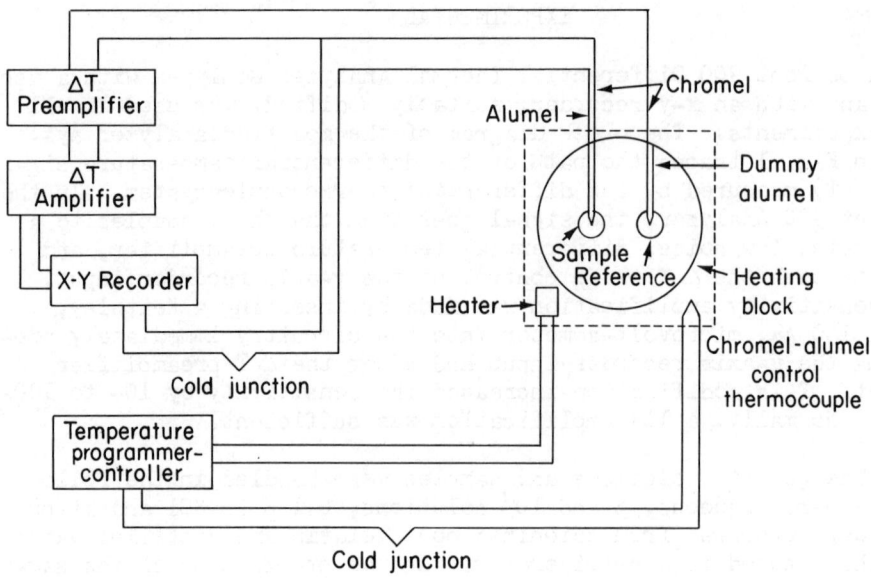

Fig. 1. Diagram of modified Du Pont 900 Differential Thermal Analyzer system with dsc cell.

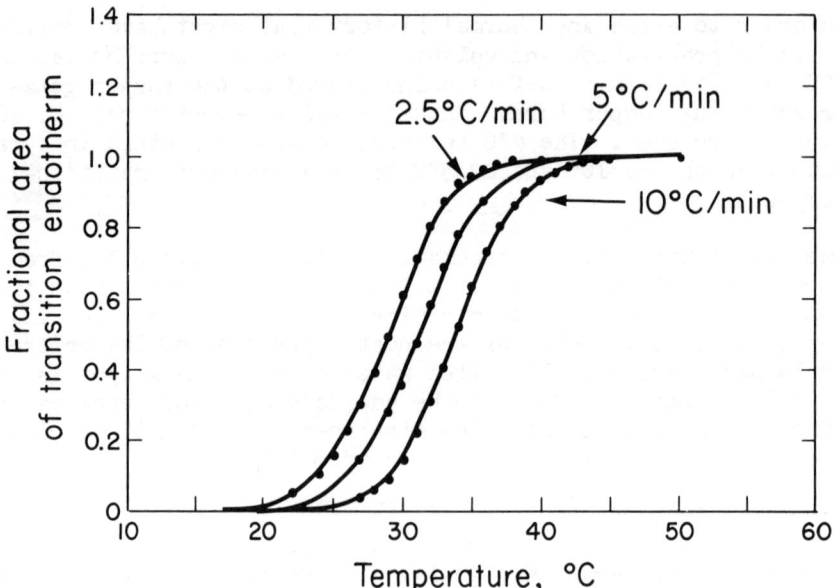

Fig. 2. Effect of heating rate on the thermal behavior of the transition in a 10% aqueous gelatin solution gelled at 10°C.

while that resulting from weighing errors was found to be 0.3%. The largest error in calculating ΔH was associated with the determination of the endotherm area. It was important to obtain thermograms in which the initial baseline displacement did not distort the transition endotherm. Even when this precaution was taken, the error introduced into the ΔH calculation by discrepancies in area determinations was 5%. Because of the difficulties encountered in matching sample and reference materials, multiple runs were made for each determination, and thermograms with excessive baseline displacement were rejected.

The range of heating rates over which quantitative data could be obtained was limited to about 2.5 to 10°C/min. At slower rates, the signal associated with the transition was too small for quantitative analysis, and amplification of the signal was limited by the signal-to-noise ratio. Uncertainties in the baseline were large at rates greater than 10°C/min, because sufficient separation was not obtained between the initial thermal imbalance of the system and the transition signal.

A comparison of the normalized, integrated plots of thermal energy absorbed during the transition indicated that there was relatively little change in the amount of annealing taking place during the heating cycle over this range of heating rates (Fig. 2). The temperature range of the transition was elevated by about 2.5°C with each increase in the heating rate. This increment was considerably greater than that observed for the melting of indium. An increase of about 1°C, which was observed between the melting points of indium measured at 2.5°C/min and at 10°C/min, agrees with data reported by Strella and Erhardt.[20] Before a conclusion can be reached regarding superheating of the gel structure, information about the effect of heating rate on the melting associated with a sharp transition involving a small enthalpy change is required.

Since the matching of sample and reference heat capacities was critical at 10°C/min, runs were made at 5°C/min.

RESULTS AND DISCUSSION

The effects of gelling period and temperature on the thermal behavior of 10% aqueous gelatin solutions are illustrated in Figs. 3-7. For comparisons of the transition data, the peak temperature was selected as the transition temperature. As anticipated from other studies,[2,3] it increased with increasing annealing temperature. Only at the highest gelling temperature, 22.5°C, was there observed an appreciable elevation of the transition temperature with increasing periods of incubation extending to about 250 hr (Fig. 3).

Although the transition energy measured for material gelled at 4°C was higher than those observed for higher temperature gels, there

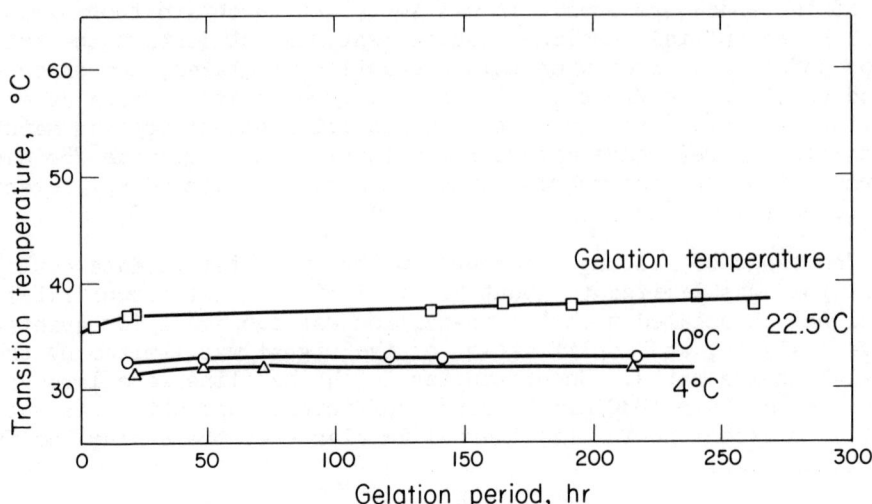

Fig. 3. Transition temperature as a function of gelation period and gelation temperature for 10% aqueous gelatin solutions.

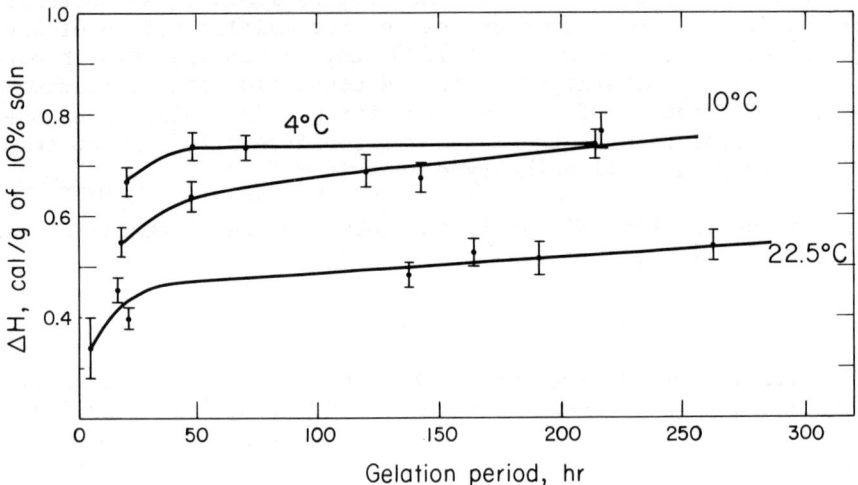

Fig. 4. Transition energy as a function of gelation period and gelation temperature for 10% aqueous gelatin solutions.

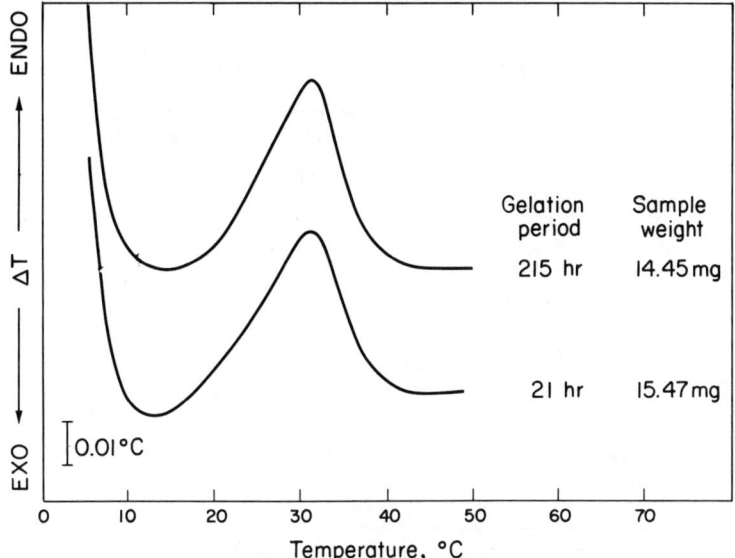

Fig. 5. Influence of gelation period at 4°C on the thermal behavior of 10% aqueous gelatin solutions.

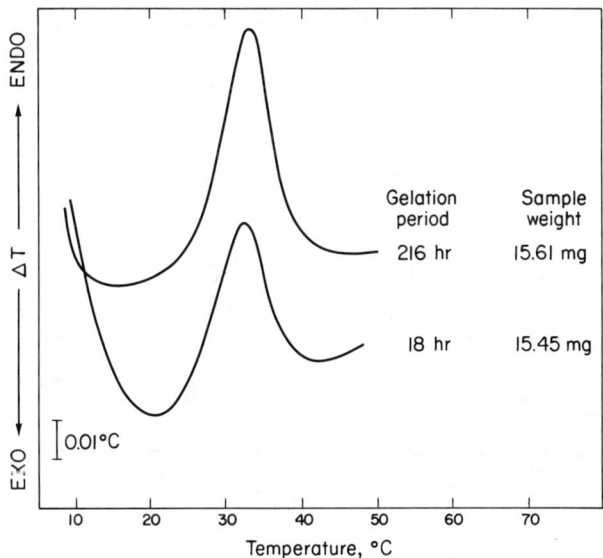

Fig. 6. Influence of gelation period at 10°C on the thermal behavior of 10% aqueous gelatin solutions.

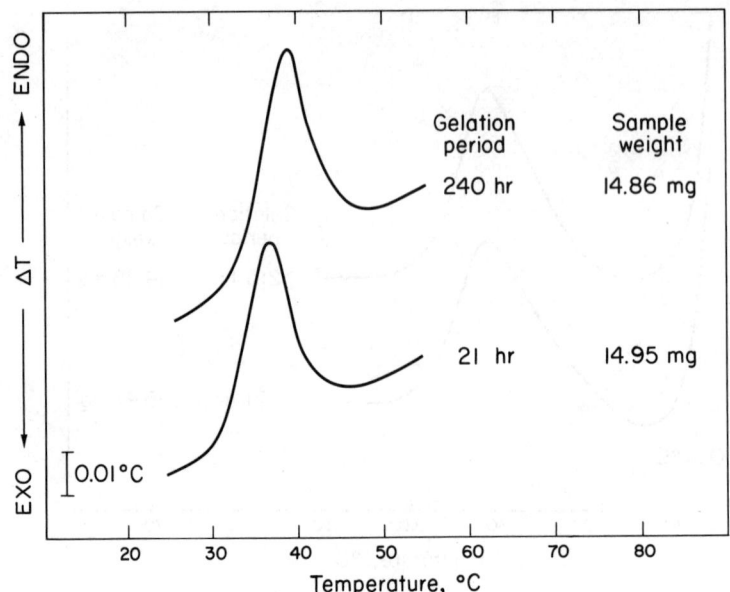

Fig. 7. Influence of gelation period at 22.5°C on the thermal behavior of 10% aqueous gelatin solutions.

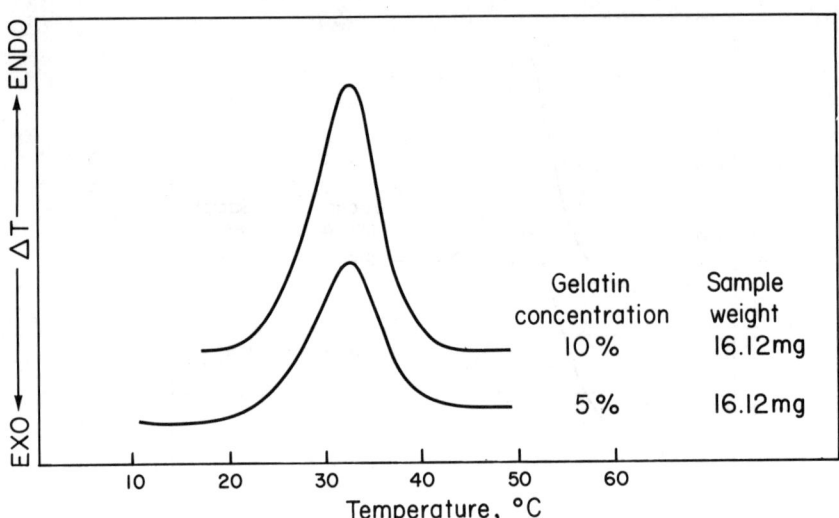

Fig. 8. Influence of gelatin concentration on the thermal behavior of gelatin structure developed in 48 hr at 10°C in aqueous solutions.

was no observable increase after 48 hr in the transition energy with increased gelling time up to about 200 hr. As the plots in Fig. 4 illustrate, annealing effects were observed in both of the other gels. For gelling periods of about 200 hr, the transition enthalpy of the gel prepared at 10°C had increased from 6.5 cal/g after 48 hr to a level comparable to that of the lower temperature gel, 7.5 cal/g of gelatin. The transition enthalpy of the gel prepared at 22.5°C, however, did not approach that of the lower temperature gels for these annealing periods. In all of the gels, the major portion of the structure had formed within the first day.

The effects of extended gelling times and of gelling temperature on the transition temperature interval are illustrated by typical dsc thermograms, reproduced in Figs. 5, 6, and 7, representing short and long annealing periods. Over the temperature range studied, the temperature interval of the transition increased with decreasing temperature. Some narrowing of the temperature interval occurred with increased gelling time at the lower temperatures. At 22.5°C, the shape of the endotherm, initially narrower than the others, was not altered by increased gelling time. The temperature of the transition, however, was elevated during the extended annealing.

Although the thermal data obtained for 5% aqueous gelatin solutions is incomplete, the data indicate that the thermal behavior is similar to that of the 10% solutions. As in the case of the 10% solutions, transition temperatures of 32, 33, and 37°C were observed for 5% gelatin solutions gelled for 48 hr at 4, 10, and 22.5°C, respectively. Also, there was no significant difference in the temperature interval of the transition with the change in concentration (Fig. 8). The transition enthalpy measured for these gels varied from 7.5 cal/g of gelatin for a solution gelled at 4°C to 5 cal/g for that gelled at 22.5°C.

In an attempt to establish the effect of isothermal annealing on the thermal behavior of gel structure developed at a lower temperature, some preliminary annealing studies were undertaken. Gel structure formed at 10°C was used, and it was annealed 30 min at various temperatures after having been programmed at 5°C/min to the designated temperature. The dsc traces of the 10% solutions, cooled to 4°C immediately following the annealing period, are reproduced in Fig. 9. Gel structure of greater thermal stability was formed during the annealing. While the transition temperature of this material increased with increasing annealing temperature, the amount formed decreased. During the cooling cycle, structure having a lower transition temperature reformed. Similar observations were made with the 5% solutions.

These observations parallel those of annealing studies of polymer crystalline structure.[21,22] Evidently, the annealing

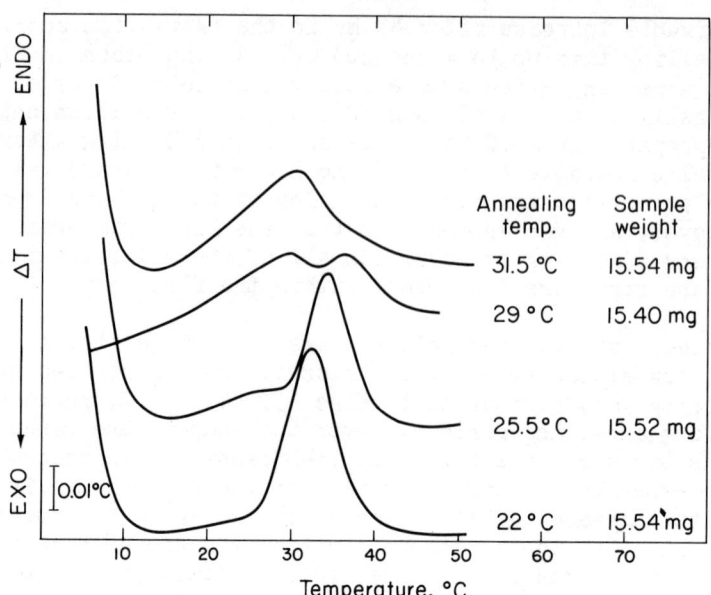

Fig. 9. Effect of additional isothermal annealing on the thermal behavior of gelatin structure formed at 10°C in 10% aqueous gelatin solutions.

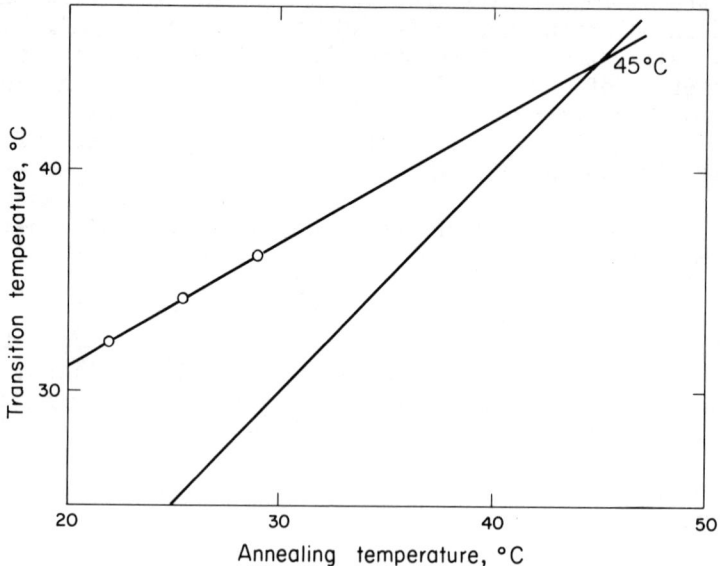

Fig. 10. Extrapolated equilibrium transition temperature for 10% aqueous gelatin solutions gelled at 10°C and annealed 30 min at various temperatures.

processes permit the formation of better ordered and/or more extensive structures. An estimate of an equilibrium transition temperature was made by extrapolating the data for the 10% solutions in the manner employed in estimating equilibrium melting points of polymer crystalline structures. An equilibrium transition temperature of 45°C was obtained from the extrapolated data (Fig. 10).

Secondary annealing periods of 30 min were considered adequate for the extrapolation, since the thermal behavior of a gel annealed for 15 min was the same as that observed for a gel annealed under similar conditions for the longer time (Fig. 11).

For secondary annealing temperatures higher than 35°C, a higher temperature gel was used. In Fig. 12, the thermal behavior of a gel formed at 22°C and annealed 30 min at 37°C is illustrated. A small amount of a more thermally stable structure was formed.

SUMMARY

The thermal behavior of 10% aqueous gelatin solutions prepared from deionized bone gelatin has been studied by dsc. For a heating rate of 5°C/min, the temperature and enthalpy of the gel-sol transition observed during the heating cycle were found to be a function of the gelation time and temperature. The major portion of the gel structure formed in less than 24 hr.

For material gelled 48 hr, the transition temperatures observed increased from 32°C for a gelling temperature of 4°C to 37°C for a gelling temperature of 22.5°C. The transition enthalpy measured for these gels decreased from 7.5 cal/g of gelatin for the gel developed at 4°C to 4.5 cal/g for the high temperature gel. Transition enthalpies measured after annealing periods of about 250 hr did not have such a large temperature dependence, since the structure developed at higher temperatures was influenced more by the gelling period.

In the concentration range from 5 to 10%, it appears that the effect of gelatin concentration on the thermal behavior of the solutions is small.

Preliminary studies of the effect of various annealing procedures on the thermal behavior of developed gel structure indicate that the process parallels that associated with the crystalline structure in polymers. It appears that the processes taking place during the annealing permit the formation of more highly ordered and/or larger ordered regions.

Because of a possible superheating of the gel structure at the heating rates employed in dsc, a direct comparison of the transition behavior based on dsc thermal data with that obtained from other

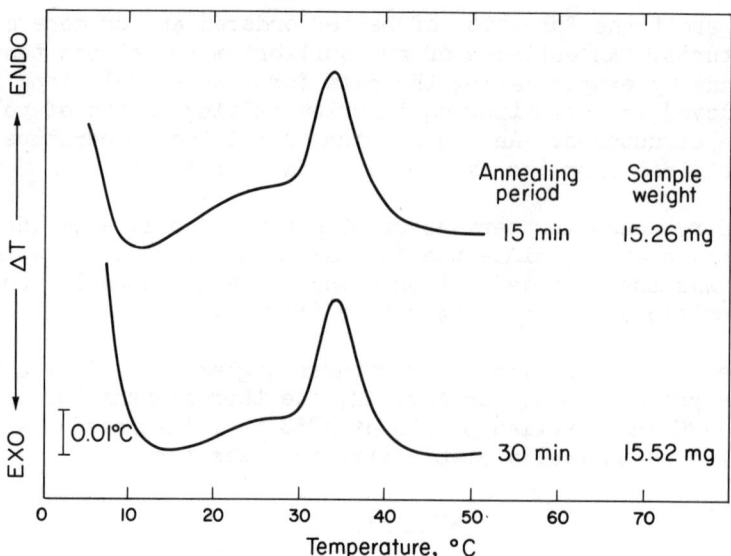

Fig. 11. Effect of duration of additional isothermal annealing at 25.5°C on the thermal behavior of gelatin structure formed at 10°C in 10% aqueous gelatin solution.

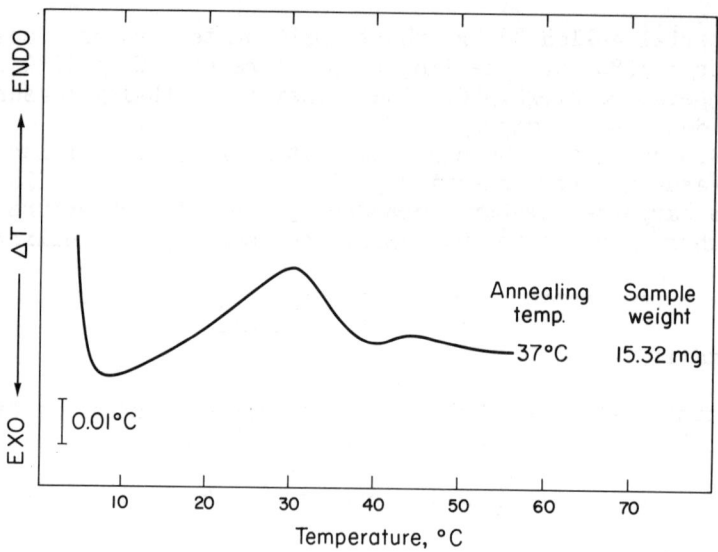

Fig. 12. Effect of additional isothermal annealing on the thermal behavior of gelatin structure formed at 22°C in 10% aqueous gelatin solution.

techniques, such as optical rotation, ought to be made with caution. The preliminary studies indicate that such a comparison ought to be made with data obtained under conditions involving comparable temperature programs.

REFERENCES

1. P. J. Flory and R. R. Garrett, J. Amer. Chem. Soc., 80, 4836 (1958).

2. A. Veis, "The Macromolecular Chemistry of Gelatin," Academic Press, New York, N.Y., 1964.

3. P. H. von Hippel in "Treatise on Collagen," Vol. I, G. N. Ramachandran, Ed., Academic Press, New York, N.Y., 1967, Chapter 6.

4. W. F. Harrington and N. V. Rao in "Conformation of Biopolymers," Vol. II, G. N. Ramachandran, Ed., Academic Press, New York, N.Y., 1967.

5. L.W.J. Holleman, H. G. Bungenberg de Jong, and R. S. Tjaden Modderman, Kolloid Beihefte, 40, 211 (1934).

6. J. E. Jolley, Tutorial paper given at the Annual Conference of the Soc. Photogr. Sci. Eng., Los Angeles, 1969.

7. W. J. Smothers and Y. Chiang, "Handbook of Differential Thermal Analysis," Chemical Publishing Company, New York, N.Y., 1966.

8. B. Ke in "Newer Methods of Polymer Characterization," B. Ke, Ed., Interscience, New York, N.Y., 1964, Chapter 9.

9. A. W. Purcell, A. S. Jahn, and L. P. Witnauer, J. Amer. Leather Chem. Ass., 61, 273 (1966).

10. E. L. Andronikashvili, N. G. Bakradze, G. V. Madzhagaladze, D. R. Monaselidze, G. M. Mrevlishvili, Z. I. Chanchalashvili, Zh. Eksp. Teor. Fix., Pisma Red., 8, 311 (1968).

11. P. L. Privalov and E. I. Tiktopulo, Biopolymers, 9, 127 (1970).

12. Y. Okamoto and K. Seki, Kolloid-Z. Z. Polym., 194, 124 (1964).

13. J. B. Yannas and A. V. Tobolsky, J. Macromol. Chem., 1, 723 (1966).

14. D. Puett, Biopolymers, 5, 327 (1967).

15. I. V. Yannas, Europ. Polym. J., 4, 257 (1968).

16. J. Naghski, A. Wisnewski, E. W. Harris, Jr., and L. P. Witnauer, J. Amer. Leather Chem. Ass., 61, 64 (1966).

17. F. E. Karasz, J. M. O'Reilly, and H. E. Bair, Biopolymers, 3, 241 (1965).

18. T. Ackermann and E. Newmann, Biopolymers, 5, 649 (1967).

19. D. E. Gray, Ed., "American Institute of Physics Handbook," 2nd ed, McGraw-Hill, New York, N.Y., 1963.

20. S. Strella, and P. F. Erhardt, J. Appl. Polym. Sci., 13, 1373 (1969).

21. J. Coste, Ind. Plastiques Mod., 9, 37 (1957).

22. J. D. Hoffman and J. J. Weeks, J. Res. Nat. Bur. Stand., 66A, 13 (1962).

SOME CALORIMETRIC MEASUREMENTS ON ADDUCTS OF ORGANOTIN HALIDES

WITH NITROGEN BASES

Edward W. Kifer and Charles H. Van Dyke

Koppers Co., Inc. and Carnegie-Mellon University

Considerable interest has developed over the past few years concerning the nature and constitution of complexes of organotin halides. The use of infrared, nuclear magnetic resonance, Mössbauer, and X-ray crystallographic techniques has provided much information about the unique structures and bonding of these complexes. Pyridine was used as the donor ligand in most of the early work with these compounds, however, gradually other adducts which involve more complex ligands have been studied. Surprisingly, little work has been done with the methyl, amino, and other related substituted pyridine series of donor ligands. These systems offer some excellent examples of how steric, electronic, and other fundamental factors influence the formation, stability and structures of the complexes. We have initiated a study of some of these systems and the various structural and stability factors which go along with them. We now report the results of our calorimetric measurements on various organic nitrogen base adducts of $(CH_3)_3SnCl$ and $(CH_3)_2SnCl_2$. Bases chosen for study include: pyridine, 2-picoline, 3-picoline, 4-picoline, 2-aminopyridine, 3-aminopyridine, 4-aminopyridine, 4-cyanopyridine, aniline, and 2,2'dipyridyl

The enthalpies of formation of a number of 1:1 complexes of $(CH_3)_3SnCl$ have been measured in non-polar solvents.[1,2] As far as we can ascertain, no detailed calorimetric studies of adducts of $(CH_3)_2SnCl_2$ have been reported. The method developed for use in this investigation was essentially a very simple, direct calorimetry method. The method is rather unique since only very small amounts of reactants were needed as opposed to the much larger amounts used for most investigations on reaction heats. In addition, the determinations were extremely fast and convenient to perform. The total time for one experimental run, including sample preparation and product characterization, was usually less than 90 minutes. There are some disadvantages to studying the reactions in the manner done in this work. Since, for example, the reaction studied was

$$\text{Acceptor}_{(S)} + n\text{Donor}_{(L)} \rightarrow \text{Acceptor} \cdot n\text{Donor}_{(S)}$$

one must be rather careful concerning the conclusions drawn. It must be kept in mind that the heat of reaction is not equivalent to the bond energy of the coordinate linkage, nor does it represent the heat of formation of the simple (gaseous) adduct molecule. Rather, it is a direct measure of the tendency of the reactants to form an isolable solid adduct. This consideration was, in fact, the prime motivation for carrying out this study. In separate investigations the failure to isolate solid adducts of trimethyltin chloride with certain bases was not well understood. For example, it was known that pyridine and 4-picoline form isolable adducts with $(CH_3)_3SnCl$, yet all attempts to prepare 2-picoline and 3-picoline adducts of the halostannane were unsuccessful.[3] It had not been established whether or not the failure to isolate the complex was due to some experimental problem (solvent, temperature, etc.) or whether it was due to an inherent inertness of the particular systems. It was thought that calorimetric investigations might answer some of the unresolved questions. Also, in the ideal case, the relative values of the heat of reaction become a good approximation of the relative values of heat of formation. Therefore, besides being of value in predicting isolable complexes, the measured heats of reaction allow some conclusions to be made concerning the <u>relative</u> strengths of the coordinate bond.

EXPERIMENTAL DETAILS

All reagents used in this work were purified by standard techniques. Purities were established by melting points, boiling points, and in certain cases infrared methods. Reaction heats for complex formation were measured by using a DuPont thermal analysis system with a calibrated calorimeter cell accessory.

The areas of the transition and reaction peaks were determined by drawing a straight line from the point where the thermogram departed from the baseline to the point where it returned. The enclosed area was measured either by planimetry or by weight. The peak area is related to the ΔH value by the equation

$$\Delta H(cal/g) = E \frac{A \Delta T_s T_s}{Ma}$$

where E represents the calibration coefficient in cal/°C-min, A represents the peak area in square inches, T_s represents the Y-axis sensitivity setting in °C/in, T_S represents the X-axis sensitivity setting in °C/in, M represents the sample mass in mg, and a represents the heating rate in °C/min. From an inspection of the thermograms, it was obvious that the results were represented by a resultant of two thermal contributions. The first contribution is the fusion endotherm of the base, while the second contribution is the exothermic reaction leading to the adduct. The only complicating factor was that the two contributions were not completely resolved due to the fact that the reaction ensued as soon as melting of the base commenced. Thus, much of the melting endotherm was counterbalanced by the reaction exotherm and the resulting thermogram also had a smaller exotherm than it would otherwise have had. Fortunately, since the magnitude of the endotherm that should have resulted was known, the fact that the two contributions were not resolved did not inject any difficulty into the calculations.

As an example of the calculation procedure, a reaction between $(CH_3)_2SnCl_2$ and 2,2'-dipyridyl will be considered here. Preliminary experiments had established the value of the heat of fusion of 2,2'-dipyridyl to be 33.0 cal/g. When 13.2 mg of 2,2'-dipyridyl and 16.5 mg of $(CH_3)_2SnCl_2$ were reacted, the total endothermic area was only 0.09 square inches which corresponds to 94.5 millicalories. The expected endotherm would be 13.2 x 33.0 or 435.6 millicalories. Thus, the remaining 341.1 millicalories was offset by the exothermic contribution which was, in turn, diminished by that same amount. The exothermic area was 0.68 square inches which equates to 719.4 millicalories. The total exothermic contribution then was 341.1 + 719.4 or 1060.5 millicalories. For the 0.0845 millimoles of product the heat of reaction calculates out to be 12.6 kcal/mole.

In order to study the heat effects during reactions of the type

$$(CH_3)_3SnCl_{(s)} + Base_{(1)} \rightarrow (CH_3)_3SnCl \cdot Base_{(s)}$$

$$(CH_3)_2SnCl_{2(s)} + 2\ Base_{(1)} \rightarrow (CH_3)_2SnCl_2 \cdot 2\ Base_{(s)}$$

it was necessary to combine the two reactants in such a way that no reaction took place before the system was positioned in the thermal analysis equipment for measurement. This made it necessary to develop a general operating procedure which would insure that all the heat accompanying the reaction in question was measured; and also insure that the heat effects measured were indeed those originating in the nominal reaction.

In the case of the solid organic bases which were studied, it was found through preliminary experiments that no heat was produced when the two solid reactants were mixed at room temperature under dry nitrogen. Thus, the only sample preparation necessary for these samples was the thorough admixing of the stoichiometric amounts of the two reactants.

For those organic bases which were liquid at room temperature, sample preparation was more demanding. The procedure which proved most effective was to freeze the organic base to a solid at approximately -135° and then quickly add the precooled organotin chloride in a finely divided state. This operation was done under a flow of dry nitrogen.

After the reactants were combined and in the calorimeter cell, the temperature was programmed to increase at the desired heating rate and the complete thermal profile of the reaction was recorded.

Solid reactants were always weighed accurately into aluminum pans. Liquid reactants were either weighed accurately or measured into the pan with a microliter syringe. In some cases both techniques were used as a check on procedure. At no time did the two techniques lead to values that differed by more than 0.1 mg. Water was rigorously excluded at all steps in the experimental procedure.

In order to make certain that the actual product was the expected adduct, the infrared spectrum and melting point of the product were compared to the infrared spectrum and melting point of an alternate sample of the same adduct which had been synthesized by solution techniques and the purity of which had been established by analysis. The spectra were identical and the melting points were within 1° of each other in all cases.

<u>Trimethyltin Chloride and 4-Picoline</u>. The necessary preliminary runs were made to determine the heat of fusion of 4-picoline. The heat of reaction was measured for the reaction

$$(CH_3)_3SnCl_{(s)} + CH_3C_5H_4N_{(l)} \rightarrow (CH_3)_3SnCl \cdot CH_3C_5H_4N_{(s)}$$

Experimental details and results of all the experiments carried out are summarized in Table I.

<u>Dimethyltin Dichloride and 4-Picoline</u>. In these experiments the heat of reaction was determined for the reaction

$$(CH_3)_2SnCl_{2(s)} + 2\ CH_3C_5H_4N_{(l)} \rightarrow (CH_3)_2SnCl_2 \cdot 2\ CH_3C_5H_4N_{(s)}$$

Results are summarized in Table II.

TABLE I

Experimental Details and Results of $(CH_3)_3SnCl$ Reactions with 4-Picoline

Run Number	$(CH_3)_3SnCl$ (mg)	4-Picoline (mg)	Molar Ratio	$-\Delta H$ Observed
1	-	8.9	-	-28.4* cal/g
2	-	13.3	-	-28.6* cal/g
3	12.9	6.0	1:1	8.04 kcal/mole
4	19.0	17.8	1:2	8.34 kcal/mole
5	12.4	5.8	1:1	8.16 kcal/mole
6	7.8	3.6	1:0.99	8.13 kcal/mole
7	7.8	3.6	1:0.99	7.46 kcal/mole

*Literature value, 29.7 ± 1.7 cal/g[4]

TABLE II

Experimental Details and Results of $(CH_3)_2SnCl_2$ Reactions with 4-Picoline

Run Number	$(CH_3)_2SnCl_2$ (mg)	4-Picoline (mg)	Molar Ratio	$-\Delta H$ Observed (kcal/mole)
1	8.9	7.6	1:2	16.0
2	8.1	6.9	1:2	16.2
3	12.9	10.95	1:2	16.3
4	6.6	5.6	1:2	16.4
5	12.5	5.3	1:1	8.0

Trimethyltin Chloride and 3-Picoline. Preliminary experiments were conducted to establish a value for the heat of fusion of 3-picoline. Then, two experiments were made to determine the value of the heat of reaction for the hypothetical reaction

$$(CH_3)_3SnCl_{(s)} + CH_3C_5H_4N_{(l)} \rightarrow (CH_3)_3SnCl \cdot CH_3C_5H_4N_{(s)}$$

The product in this case was an ill-defined mull which could not be solidified upon cooling to 0°. No solid complex could be isolated using solution techniques. Thus, the observed "heat of reaction" should probably be more accurately described as the "heat of interaction" since the product was not solid $(CH_3)_3SnCl \cdot CH_3C_5H_4N$ in this instance. The purity of the product could not be determined. The pertinent data are presented in Table III.

TABLE III

Experimental Details and Results of $(CH_3)_3SnCl$ Reactions with 3-Picoline

Run Number	$(CH_3)_3SnCl$ (mg)	3-Picoline (mg)	Molar Ratio	$-\Delta H$ Observed
1	-	8.4	-	-29.0 cal/g*
2	-	8.6	-	-30.8 cal/g**
3	11.1	5.2	1:1	2.48 kcal/mole
4	13.7	6.4	1:1	2.51 kcal/mole

*Literature value, 26.5 ± 0.8 cal/g[4]
**Sample was from a different source than run number 1.

Dimethyltin Dichloride and 3-Picoline. In these experiments the heat of reaction was measured for the reaction

$$(CH_3)_2SnCl_{2(s)} + 2\ CH_3C_5H_4N_{(l)} \rightarrow (CH_3)_2SnCl_2 \cdot 2\ CH_3C_5H_4N_{(s)}$$

Experimental data and results are presented in Table IV.

TABLE IV

Experimental Details and Results of $(CH_3)_2SnCl_2$ Reactions with 3-Picoline

Run Number	$(CH_3)_2SnCl_2$ (mg)	3-Picoline (mg)	Molar Ratio	$-\Delta H$ Observed (kcal/mole)
1	9.4	8.0	1:1	16.5
2	25.0	21.15	1:2	16.8
3	12.0	10.2	1:2	16.8
4	7.1	6.0	1:2	16.3

Trimethyltin Chloride and 2-Picoline. After preliminary experiments to determine the heat of fusion of 2-picoline, the reaction of 2-picoline with trimethyltin chloride was studied. However, the system was found to undergo no reaction. This was demonstrated by the fact that the only peak in the entire thermal profile was the melting endotherm for 2-picoline. The heat of fusion of 2-picoline was measured as 22.2 cal/g and 22.7 cal/g in two separate experiments. When warmed together with trimethyltin chloride the fusion endotherm was measured to be 24.9 cal/g. The infrared spectrum of the mixture showed no peak shifts indicative of interaction.

Dimethyltin Dichloride and 2-Picoline. In two experiments the heat of reaction was determined for the reaction

$$(CH_3)_2SnCl_{2(s)} + 2\ CH_3C_5H_4N_{(l)} \rightarrow (CH_3)_2SnCl_2 \cdot 2\ CH_3C_5H_4N_{(s)}$$

The pertinent data are summarized in Table V.

TABLE V

Experimental Details and Results of $(CH_3)_2SnCl_2$ Reactions with 2-Picoline

Run Number	$(CH_3)_2SnCl_2$ (mg)	2-Picoline (mg)	Molar Ratio	$-\Delta H$ Observed (kcal/mole)
1	14.3	12.1	1:2	8.5
2	21.3	18.1	1:2	9.0

Trimethyltin Chloride and Pyridine. Preliminary experiments were carried out to evaluate the heat of fusion of pyridine. Then reaction heats were measured for duplicate runs of the reaction

$$(CH_3)_3SnCl_{(s)} + C_5H_5N_{(l)} \rightarrow (CH_3)_3SnCl \cdot C_5H_5N_{(s)}$$

Results are summarized in Table VI.

TABLE VI

Experimental Details and Results of Trimethyltin Chloride Reactions with Pyridine

Run Number	$(CH_3)_3SnCl$ (mg)	C_5H_5N (mg)	Molar Ratio	$-\Delta H$ Observed
1	-	9.6	-	-22.4 cal/g*
2	-	20.3	-	-22.3 cal/g*
3	24.8	9.8	1:1	6.9 kcal/mole
4	14.0	5.8	1:1	6.7 kcal/mole

*Literature value, 22.4 ± 0.6 cal/g[4]

Dimethyltin Dichloride and Pyridine. In these duplicate runs, the reaction heats were measured for the reaction

$$(CH_3)_2SnCl_{2(s)} + 2\ C_5H_5N_{(l)} \rightarrow (CH_3)_2SnCl_2 \cdot 2\ C_5H_5N_{(s)}$$

Results are presented in Table VII.

TABLE VII

Experimental Details and Results of $(CH_3)_2SnCl_2$ Reactions with Pyridine

Run Number	$(CH_3)_2SnCl_2$ (mg)	C_5H_5N (mg)	Molar Ratio	$-\Delta H$ Observed (kcal/mole)
1	22.4	16.1	1:2	13.2
2	8.3	6.0	1:2	12.7

Dimethyltin Dichloride and 2-Aminopyridine. A preliminary determination was made of the heat of fusion of 2-aminopyridine. Then the calorimetry of the reaction of interest was studied:

$$(CH_3)_2SnCl_{2(s)} + 2\ NH_2C_5H_4N_{(1)} \rightarrow (CH_3)_2SnCl_2 \cdot 2\ NH_2C_5H_4N_{(s)}$$

A summary of the data is presented in Table VIII.

TABLE VIII

Experimental Details and Results of $(CH_3)_2SnCl_2$ Reactions with 2-Aminopyridine

Run Number	$(CH_3)_2SnCl_2$ (mg)	$NH_2C_5H_4N$ (mg)	Molar Ratio	$-\Delta H$ Observed
1	-	20.9	-	-41.1 cal/g
2	10.5	9.0	1:2	15.3 kcal/mole
3	18.7	16.0	1:2	15.1 kcal/mole
4	18.4	15.8	1:2	15.2 kcal/mole

Dimethyltin Dichloride and 3-Aminopyridine. A preliminary experiment was carried out in order to determine the heat of fusion of 3-aminopyridine. Then duplicate runs were made to measure the heat of reaction of the reaction

$$(CH_3)_2SnCl_{2(s)} + 2\ NH_2C_5H_4N_{(1)} \rightarrow (CH_3)_2SnCl_2 \cdot 2\ NH_2C_5H_4N_{(s)}$$

The pertinent data are listed in Table IX.

TABLE IX

Experimental Details and Results of $(CH_3)_2SnCl_2$ Reactions with 3-Aminopyridine

Run Number	$(CH_3)_2SnCl_2$ (mg)	3-Aminopyridine (mg)	Molar Ratio	$-\Delta H$ Observed
1	13.0	11.1	1:2	14.8 kcal/mole
2	3.1	2.7	1:2	14.3 kcal/mole
3	-	2.0	-	-39.8 cal/g

Dimethyltin Dichloride and 2,2'-Dipyridyl. A preliminary experiment was carried out to determine the heat of fusion of 2,2'-dipyridyl and then duplicate runs were made to measure the heat of reaction of the reaction

$$(CH_3)_2SnCl_{2(s)} + (C_5H_4N)_{2(l)} \rightarrow (CH_3)_2SnCl_2 \cdot (C_5H_4N)_{2(s)}$$

It was also determined that 2,2'-dipyridyl forms no adduct with $(CH_3)_3SnCl$. The data are summarized in Table X.

TABLE X

Experimental Details and Results of $(CH_3)_2SnCl_2$ Reactions with 2,2'-Dipyridyl

Run Number	$(CH_3)_2SnCl_2$ (mg)	$(C_5H_4N)_2$ (mg)	Molar Ratio	$-\Delta H$ Observed
1	-	7.8	-	-33.0 cal/g
2	13.2	9.4	1:1	12.7 kcal/mole
3	16.5	13.2	1:1	12.6 kcal/mole

Dimethyltin Dichloride and 4-Cyanopyridine. A preliminary experiment established a value for the heat of fusion of 4-cyanopyridine. Two experiments were then carried out to determine the heat of reaction of the reaction

$$(CH_3)_2SnCl_{2(s)} + 2\ NCC_5H_4N_{(l)} \rightarrow (CH_3)_2SnCl_2 \cdot 2\ NCC_5H_4N_{(s)}$$

Experimental details and results are given in Table XI.

TABLE XI

Experimental Details and Results of $(CH_3)_2SnCl_2$ Reactions with 4-Cyanopyridine

Run Number	$(CH_3)_2SnCl_2$ (mg)	4-Cyanopyridine (mg)	Molar Ratio	$-\Delta H$ Observed
1	-	11.3	-	-45.7 cal/g
2	4.3	4.1	1:2	12.9 kcal/mole
3	7.3	6.9	1:2	12.5 kcal/mole

Dimethyltin Dichloride and Aniline. Preliminary experiments were carried out to establish the value of the heat of fusion of aniline. Duplicate runs were then made to measure the heat of reaction of the reaction

$$(CH_3)_2SnCl_2(s) + 2\ C_6H_5NH_2(l) \rightarrow (CH_3)_2SnCl_2 \cdot 2\ C_6H_5NH_2(s)$$

The pertinent data are listed in Table XII.

TABLE XII

Experimental Details and Results of $(CH_3)_2SnCl_2$ Reactions with Aniline

Run Number	$(CH_3)_2SnCl_2$ (mg)	$C_6H_5NH_2$ (mg)	Molar Ratio	$-\Delta H$ Observed
1	-	7.5	-	-27.8 cal/g
2	-	22.9	-	-27.8 cal/g
3	10.1	8.6	1:2	7.6 kcal/mole
4	10.3	8.7	1:2	7.9 kcal/mole

DISCUSSION

The stereochemistry of the adducts prepared in this investigation will not be discussed in this report. It has been well established in other studies that the 1:1 adducts of $(CH_3)_3SnCl$ are molecular and possess trigonal bipyramidal configurations with the methyl groups in the equatorial positions.[5,6] The geometry of the 2:1 adducts of $(CH_3)_2SnCl_2$ has been shown to be basically octahedral with the two methyl groups occupying positions trans to each other.[7,8] The relative positions of the two chlorines and the two remaining ligands have not been determined with certainty and may vary from one complex to another.[8]

The numerical results of the reaction heat measurements for the halostannanes and nitrogen bases investigated are summarized in Table XIII.

TABLE XIII

Summary of Results of Calorimetric Measurements

Acceptor (1)	Donor (1)	$-\Delta H$ Observed (kcal/mole)
$(CH_3)_3SnCl$	4-Picoline	8.2 ± 0.2
$(CH_3)_3SnCl$	3-Picoline	2.5 ± 0.2
$(CH_3)_3SnCl$	2-Picoline	0
$(CH_3)_3SnCl$	Pyridine	6.8 ± 0.2

Acceptor (1)	Donor (2)	$-\Delta H$ Observed (kcal/mole)
$(CH_3)_2SnCl_2$	4-Picoline	16.2 ± 0.2
$(CH_3)_2SnCl_2$	3-Picoline	16.6 ± 0.4
$(CH_3)_2SnCl_2$	2-Picoline	8.7 ± 0.3
$(CH_3)_2SnCl_2$	Pyridine	12.9 ± 0.3
$(CH_3)_2SnCl_2$	2-Aminopyridine	15.2 ± 0.2
$(CH_3)_2SnCl_2$	3-Aminopyridine	14.5 ± 0.3
$(CH_3)_2SnCl_2$	2,2'-Dipyridyl	12.6 ± 0.2
$(CH_3)_2SnCl_2$	4-Cyanopyridine	12.7 ± 0.2
$(CH_3)_2SnCl_2$	Aniline	7.7 ± 0.3

In the series of $(CH_3)_3SnCl$ adducts the values determined for the heat of reaction decrease in the order: 4-picoline > pyridine > 3-picoline > 2-picoline. In order to explain this behavior it is necessary to consider the nature of the coordinate link and those factors which may have an effect on its strength. The molecular orbital description of the situation has been discussed by Mulliken.[9] In very simplified terms the coordinate donor (D) - acceptor (A) linkage can be regarded as a resonance hybrid of two structures; the "no-bond" structure (D,A), and the electron transfer structure, (D^+-A^-). The wave function then for the ground state can be represented by:

$$\psi_N = a\psi_0(D,A) + b\psi_1(D^+-A^-).$$

In this equation, ψ_N is the overall wave function, ψ_0 is the "no-bond" wave function, ψ_1 is the donor wave function representing complete electron transfer of an electron from D to A, and $\underline{a}$ and $\underline{b}$ are the mixing coefficients. Obviously, the larger $\underline{b}$ is, the more stable the adduct molecule is. Since this is the case it is reasonable to conclude that the most important factors influencing the strength of the D→A bond are very fundamental properties of the donor and acceptor molecules, such as the ionization potential and base strength of the donor and the electron affinity or effective electronegativity of the acceptor molecule. Also of prime importance are the various inductive and steric effects of substitution on the donor and acceptor atoms.

The order of base strengths (toward hydrogen ion!) for the series of picolines is reasonably well established. Table XIV lists the pK_a values of a series of substituted pyridines, most of which were determined in aqueous solution.

TABLE XIV

pK_a Values of Some Substituted Pyridines[10,11]

Base	pKa	Base	pKa
4-aminopyridine	9.12	pyridine	5.22
4-picoline	5.98	aniline	4.58
2-picoline	5.96	4-cyanopyridine	1.86
3-picoline	5.63		

It is quite apparent that the order of reaction heats with $(CH_3)_3SnCl$ does not parallel the order of basicities for the series 4-picoline, 3-picoline, 2-picoline, pyridine as given by their pK_a values. The measured heats of reaction do parallel exactly the chemical behavior observed in the solution syntheses of the complexes, however. In the solution experiments, 4-picoline and pyridine readily formed isolable complexes, while 3-picoline formed only an oily material which could not be isolated. There was no evidence of any interaction in the case of 2-picoline.

The answer to the apparent anomalous behavior is most likely found in two significant pieces of information. First, the steric effect of the methyl group in the 2 position is undoubtedly the overwhelming factor which causes the lack of interaction between 2-picoline and $(CH_3)_3SnCl$. This steric effect has been noted before,[5] and is of great importance when the steric requirement of the acceptor is large also, as is the case with $(CH_3)_3SnCl$.

In the case of 3-picoline the reaction heat would be expected to be greater than that for pyridine. The measured values (6.8 kcal/mole for pyridine, 2.5 kcal/mole for 3-picoline) are at odds with the predicted results. Once again, however, the results are not unprecedented. In measurements carried out on the thermochemistry of transition metal complexes of pyridine and the three picolines, the heat of dissociation of the 3-picoline complex (CoL_2X_2) was nearly 7 kcal/mole less than the value obtained for the pyridine complex.[5] Similar behavior has been noted in other investigations.[12,13]

Another possibility which must be considered as a complicating factor is the relative lattice energies of the series of complexes. The relative lattice energies cannot be predicted and have not been measured. However, due to the similarity of the series of donors, it is not likely that any gross differences would result in the lattice energies of the adducts. This conclusion is substantiated by the fact that the value measured in this work for the reaction heat between $(CH_3)_3SnCl$ and pyridine (6.8 ± 0.2 kcal/mole) compares very well with the value obtained by Bolles and Drago[1] for the same reaction using solution techniques.

Many interesting comparisons result from examining the thermal data for the series of $(CH_3)_2SnCl_2$ adducts. The observed heat of reaction for 4-picoline (16.2 kcal/mole) is almost exactly twice the value measured for the $(CH_3)_3SnCl$ adduct with the same base (8.2 kcal/mole). One possible interpretation would be to suggest that the coordinate bond (N→Sn) energy is nearly constant and that the only difference is that two bonds are formed in the 2:1 adduct with $(CH_3)_2SnCl_2$. The values observed for the pyridine adducts follow an identical trend (12.9 ± 0.3 kcal/mole for the $(CH_3)_2SnCl_2$ adduct vs. 6.8 ± 0.2 kcal/mole for the $(CH_3)_3SnCl$ adduct). It is also interesting to note the similarity of the values obtained for the $(CH_3)_2SnCl_2$ adducts of pyridine (12.9 ± 0.3 kcal/mole) and 2,2'-dipyridyl (12.6 ± 0.2 kcal/mole).

Comparing the series of picolines is again a somewhat arduous task. The values for the reaction heats are in the order: 3-picoline ≅ 4-picoline > pyridine > 2-picoline. The position of 2-picoline can again be attributed to steric factors. The only disparity, then, is the near equivalence observed for the reaction heats of 4- and 3-picoline. This was unexpected based on the results obtained with $(CH_3)_3SnCl$.

The remaining bases which were used in the series of adducts with $(CH_3)_2SnCl_2$ were investigated primarily to demonstrate the flexibility of the experimental method. With the exception of aniline all the additional free bases were solids at room temperature. This made the experimental procedure very simple and the speed of determination very rapid.

Very little is known about the manner in which the various aminopyridines coordinate to an acceptor molecule.[14] It was assumed here that they were acting as monodentate ligands, coordinating most likely through the ring nitrogen. This being the case, the results observed for the 2-aminopyridine and 3-aminopyridine adducts appear to be those expected on the basis of the respective pK_a values. The value for 4-aminopyridine could not be measured because of its high melting point which resulted in the $(CH_3)_2SnCl_2$ being volatilized before the 4-aminopyridine melted.

The value obtained for the interaction of aniline and $(CH_3)_2SnCl_2$ is quite reasonable and readily explainable on the basis of the low pK_a value (4.58) for aniline. The measured value for the heat of reaction of $(CH_3)_2SnCl_2$ with 4-cyanopyridine is quite high in view of the pK_a value of 4-cyanopyridine (1.86) and the lack of important steric considerations. The complexes of 4-cyanopyridine are again thought to coordinate through the ring nitrogen.[14]

The results of the calorimetric investigations illustrate the variety of factors which must be considered in attaching significance to the experimentally determined trends in the heats of reactions. It is obvious that steric and basicity factors cannot be more than partly responsible for the results. Other contributions to the stability of acid-base adducts of organotin halides such as crystal packing effects deserve much more careful examination.

BIBLIOGRAPHY

1. Bolles, T. F. and Drago, R. S., J. Am. Chem. Soc., 88, 3921 (1966).

2. Bolles, T. F. and Drago, R. S., J. Am. Chem. Soc., 87, 5015 (1965).

3. Van Dyke, C. H., Nasta, M. A. and Neilson, R. H., unpublished results.

4. Biddiscombe, D. P., Coulson, E. A., Handley, R. and Herington, E. F. G., J. Chem. Soc., 1957 (1954).

5. Beech, G., Mortimer, C. T. and Tyler, E. G., J. Chem. Soc. (A), 925 (1967).

6. Hulme, R., J. Chem. Soc., 1524 (1963).

7. Beattie, I. R. and McQuillen, G. P., J. Chem. Soc., 1519 (1963).

8. Clark, J. P. and Wilkins, C. J., J. Chem. Soc. (A), 871 (1966).

9. Mulliken, R. S., J. Am. Chem. Soc., 74, 811 (1952).

10. Andon, R. J. L., Cox, J. D. and Herington, E. F. G., Trans. Farraday Soc., 50, 918 (1954).

11. Fischer, A., Galloway, W. J. and Vaughn, J., J. Chem. Soc., 3591 (1964).

12. Cabral, J. deO., King, H. C. A., Nelson, S. M., Shepherd, T. M. and Koros, E., J. Chem. Soc. (A), 1348 (1966).

13. Nelson, S. M. and Shepherd, T. M., Inorg. Chem., 4, 813 (1965).

14. Frank, C. W. and Rogers, L. B., Inorg. Chem., 5, 615 (1966).

THERMAL ANALYSIS STUDIES OF THE DECOMPOSITION OF HYDRATED AND DEUTERATED ROCHELLE SALTS

Alfred C. Glatz[**], Irving Litant [*], and Bernard Rubin[*]

[**] Voland Corporation, New Rochelle, N.Y. 10802

[*] NASA-Electronics Research Center, Cambridge, Mass.

I. INTRODUCTION

Rochelle Salt is the common name given to the double tartrate of sodium and potassium, which crystallizes as the tetrahydrate. The important dielectric and ferroelectric properties of hydrated and deuterated Rochelle Salts have been well documented (1,2,3). These materials are important commercially for their piezoelectric properties as acoustic transducers. Although there have been many reported investigations of the physical properties of Rochelle Salts, the thermal decomposition has not been investigated above 60 °C.. It is the purpose of this paper to report studies of the thermal decomposition of single crystals of these salts by the thermal analytical techniques: Differential Scanning Calorimetry (DSC) andThermogravimetry (TGA).

The structural formulas for hydrated and deuterated Rochelle Salts are shown below,

Hydrated Rochelle Salt   Deuterated Rochelle Salt

The molecules are, as indicated, enantiomorphic. The most important physical properties of these salts are that they exhibit two Curie Points, with the ferroelectric range being confined to the region between these Curie Points. Hydrated Rochelle Salt has a lower Curie Point of - 18 °C. and an upper Curie Point of 24 °C., whereas the deuterated salt has a lower Curie Point of - 21 °C. and an upper Curie Point of 35 °C. (4). These Curie

Points are characterized by large anomalies in the dielectric constants and very small enthalpic effects. The latter having specific heats of less than one calorie/mole/degree Centigrade. It is believed that the upper and lower Curie Points in Rochelle Salts are second order transitions. The ferroelectric transition is also accompanied by a lowering of the crystal symmetry from orthorhombic to monoclinic. However, the monoclinic deformation from the orthorhombic structure is very small. For example, in hydrated Rochelle Salt at 11 °C. the deformation is only 3 minutes in the angle beta (5).

II. EXPERIMENTAL METHODS

A. Crystal Growth - The preparation of single crystals of Rochelle Salts was carried out by deposition from super-saturated solutions of these salts using the apparatus shown in Figure 1. For hydrated Rochelle Salt, a saturated solution of the salt (Fisher Certified Reagent Grade) was initially prepared, and suitable seed crystals made by repeated watchglass evaporations. Using this technique seed crystals (3mm x 6mm x 6mm) of hydrated Rochelle Salt were obtained. A hole was made in the seed crystal by careful water etching and drilling. The seed crystal was then suspended on a platinum wire and placed in a jar into a saturated solution of Rochelle Salt as shown in Figure 1. The glass jar was immersed in a bath thermostated at 40 $\pm$0.1 °C. and allowed to come to equilibrium. The crystals were grown by reducing the bath temperature 0.5 °C./day until a temperature of 29 °C. was reached (21 days). In this manner single crystals of Rochelle Salt (19mm x 19mm x 50mm) were obtained.

Single crystals of deuterated Rochelle Salt were grown in the same manner, but required the initial preparation of a saturated solution of the deuterated salt. The deuterated salt was prepared in the following manner: Hydrated Rochelle Salt powder was heated in an all glass system to 110 °C. and simultaneously pumped down. The powder was removed and weighed at room temperature. The process was repeated until the powder reached constant weight. The weight loss corresponded to 99.8 % of the theoretical weight loss for four molecules of water. The resulting powder was added to the exact amount of D_2O (99.9%) to form a saturated solution. After being equilibrated at 40 °C. the supernatent solution was used to prepare seed crystals of the deuterated salt. These seed crystals were used to grow crystals of this salt in the same manner as was described for the hydrated salt above.

The crystals of hydrated and deuterated Rochelle Salts grown in this manner were transparent, had well-formed faces, and were stored in atmospheres to maintain them in their appropriate hydrated states.

B. Differential Scanning Calorimetry (DSC) - The DSC used for these studies was the Perkin-Elmer Model DSC-1B, which was operated at a heating rate of 2.5 °C./minute and at recorder speeds of

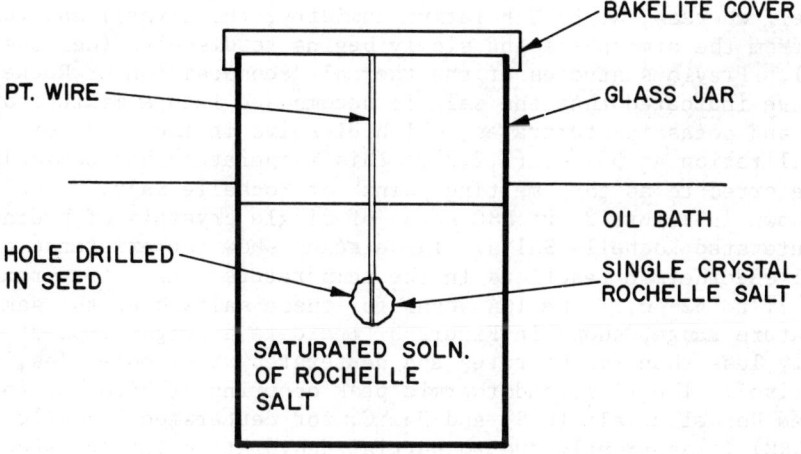

Figure 1 - Apparatus for Single Crystal Growth of Rochelle Salts

0.5 - 1.0 inches/minute. The samples were placed in aluminum sample holders with loose fitting covers to allow water-evaporation. Nitrogen gas was flowed through the system at a rate of 30 cc./min..

C. <u>Thermogravimetric Analysis (TGA)</u> - The TGA equipment used for these studies was the Perkin-Elmer Thermobalance, Model TGS-1, which was operated at a heating rate of 2.5 °C./minute and at a recorder speed of 0.25 inches/minute. The samples were placed in open platinum sample holders and nitrogen gas was flowed through the system at a rate of 30 cc./min..

III. EXPERIMENTAL RESULTS

It is well known that at room temperature the tetrahydrate of Rochelle Salt may be unstable, depending upon the atmospheric humidity (2). For example, at 25 °C. and a relative humidity of less than 40 %, the crystal loses water of crystallization (efflorescence); whereas, at 85 % relative humidity, the crystal absorbs water from the atmosphere and slowly begins to dissolve (deliquescence). Previous studies of the thermal decomposition of Rochelle Salt have indicated that the salt is decomposed into a mixture of sodium and potassium tartrates, which dissolve in the water of crystallization at 56 °C. (1,2,3). This temperature has commonly been referred to as the "melting point" of Rochelle Salt.

Shown in Figure 2 are DSC scans of single crystals of hydrated and deuterated Rochelle Salts. These scans show the existence of several endothermic reactions in the temperature range of approximately 31 to 52 °C.. The TGA scans for these salts over the same temperature range, shown in Figure 3, indicate a weight loss of slightly less than two hydrated and deuterated water molecules, respectively. The first endothermic peak occuring at 34.8 °C. for hydrated Rochelle Salt (HRS) and 34 °C. for deuterated Rochelle Salt (DRS) is apparently due to partial dehydration (efflorescence) of the Rochelle Salt as given by equation 1.

$$RS \cdot 4H_2O \xrightarrow[\Delta]{34°C} RS \cdot (4-x)H_2O \qquad (1)$$

It would be expected that under the conditions of this experiment, i.e. the flowing of nitrogen gas through the system, that dehydration of the Rochelle Salts should occur as the temperature is raised above room temperature. The vapor pressure of Rochelle Salt has been reported as 35 mm. of Hg. at 34 °C. (6), which is probably significantly greater than that of the system being flushed with nitrogen gas. Therefore, an endothermic dehydration reaction probably explains the endothermic peak observed at approximately 34 °C. for these salts.

It may be observed from the DSC scans (Figure 2) that after the peak at 34 °C., there is a slowly increasing endotherm terminating in a peak at 50 °C. for both the hydrated and deuterated Rochelle Salts. This endothermic behavior is apparently due to the decomposition of the Rochelle Salt into a mixture of sodium and potassium tartrates and the solubility of these tartrates in the water of crystallization as given by equation 2.

$$2RS \cdot yH_2O \text{ (s)} \xrightarrow{36°C} K_2Tar \cdot 1/2H_2O + Na_2Tar \cdot 2H_2O \text{ (solids)}$$

$$\begin{array}{c} K_2Tar \cdot 1/2H_2O \text{ (}\ell\text{)} \\ + \\ Na_2Tar \cdot 2H_2O \text{ (}\ell\text{)} \end{array} \xleftarrow{50°C} \qquad (2)$$

It may be observed from the DSC scans of HRS and DRS (Figure 2) that after the endotherm at 34 °C. is complete, a slow endothermic reaction beginning at about 36 °C. is observed. This endotherm is apparently the decomposition of the Rochelle (double) Salt into the corresponding single salts, $Na_2Tar \cdot 2H_2O$ and $K_2Tar \cdot 1/2H_2O$. According to LeChatelier's Principle this decomposition is endothermic, because the product single salts (sodium and potassium tartrates) are the stable species at the higher temperature. It has been reported that Rochelle Salt is unstable at temperatures greater than 40 °C. (6,7). At about this temperature the solution of a solid phase, probably the more highly soluble $K_2Tar \cdot 1/2H_2O$, was experimentally observed in an alternate experiment. As the temperature is further raised the solution becomes more dilute, because more Rochelle Salt is decomposing, and the solubility of the tartrates increases. At 52 °C. only a liquid phase is present. The solution of the sodium and potassium tartrates have negative heats of solution (8), and are exothermic reactions. It may be observed for the DSC scans of Rochelle Salt, shown in Figure 4, that there is an exothermic reaction between the two endothermic peaks, which is probably due to the solution of the sodium and potassium tartrates in the water of crystallization. It would be expected that the exothermic reactions could be observed if the partial pressure of water has been restrained from escaping from the environment of the sample, thereby preventing further decomposition of the Rochelle Salt as shown by equation 2. Under certain conditions the endothermic decomposition reaction apparently masks the exothermic reaction for the solution of the tartrates and the latter are not observed, as shown by Figure 2.

On the DSC scan for HRS (Fig. 2-a) there is a small endothermic effect occuring between 52 and 59 °C.. This enthalpic effect is not observed for DRS. This endotherm is probably a non-equilibrium condition, because other DSC scans for HRS did not show this effect.

It may be observed from Figure 4 that when the temperature is reversed from 60 to 30 °C., no enthalpic effects are observed. This verifies that the decomposition of Rochelle Salt is irreversible.

As the sample is heated above 60 °C. there is a slow loss of water as shown on the TGA curves in Figure 3. The endothermic

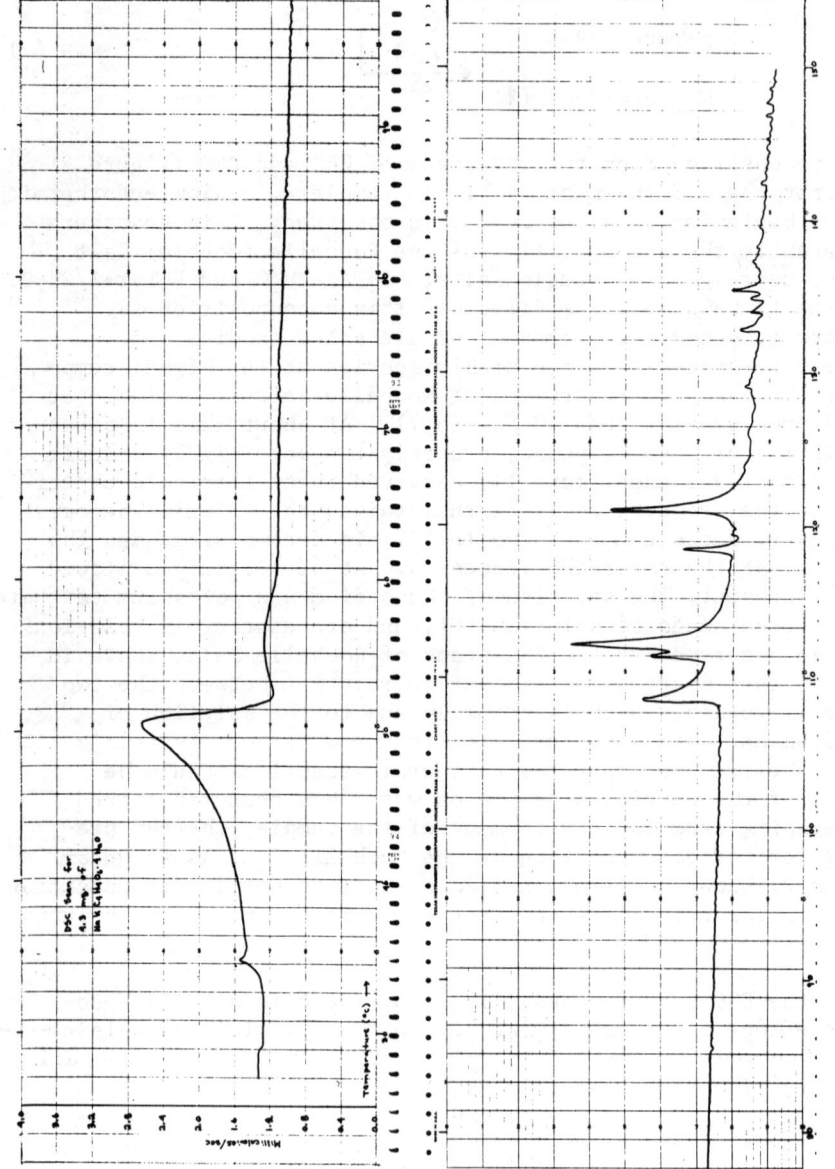

Figure 2-a – DSC Scan for Single Crystal Hydrated Rochelle Salt

DECOMPOSITION OF ROCHELLE SALTS

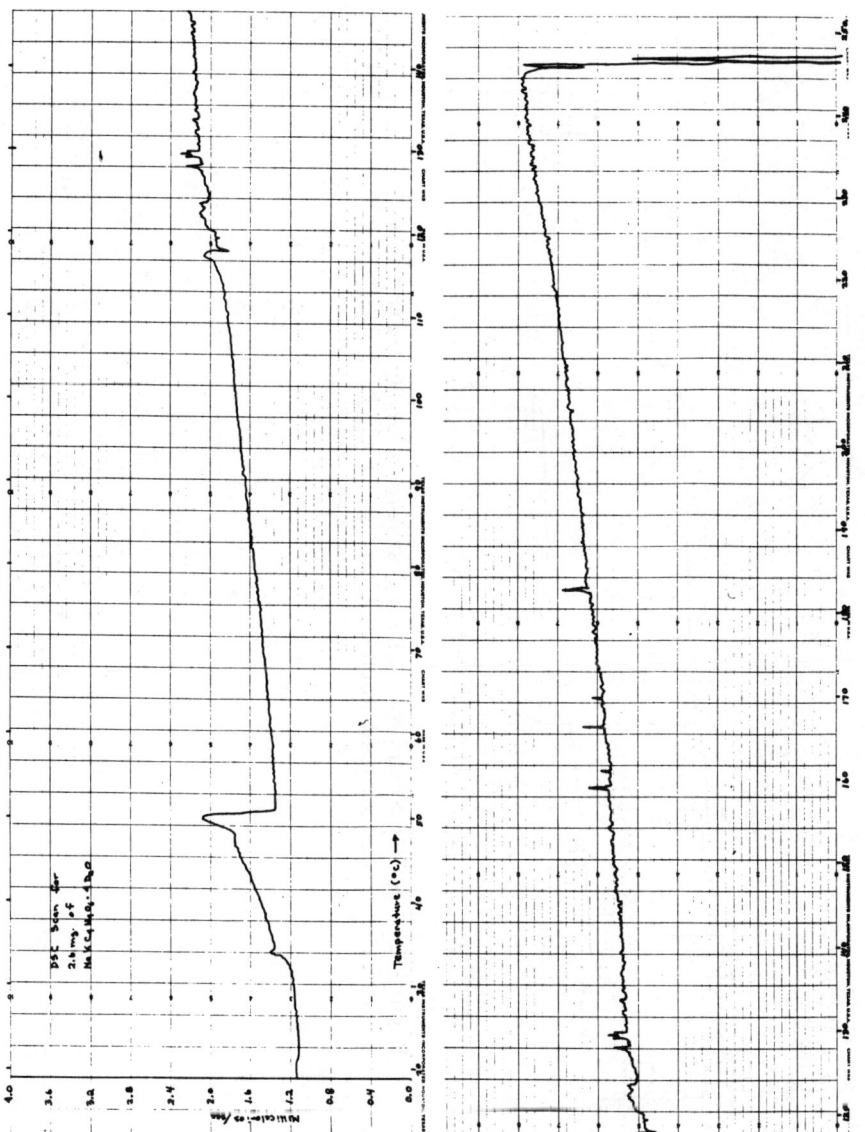

Figure 2-b. - DSC Scan for Single Crystal Deuterated Rochelle Salt

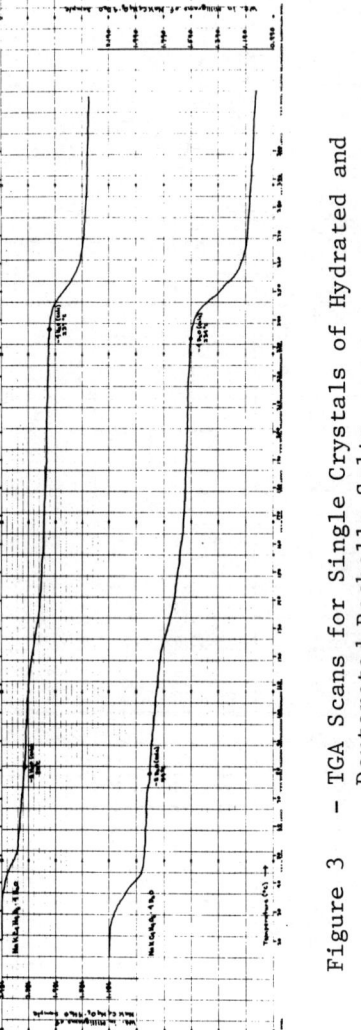

Figure 3 — TGA Scans for Single Crystals of Hydrated and Deuterated Rochelle Salts

peaks observed in the temperature ranges of 108 to 125 °C. for HRS and of 116 to 130 °C. for DRS are probably due to the dehydration of $Na_2Tar \cdot 2H_2O$ to the anhydrous salt. The endothermic peaks observed in the temperature range of 158 to 180 °C. for DRS are probably due to the dehydration of $K_2Tar \cdot 1/2H_2O$ to the anhydrous salt. The TGA curves indicate that at 234 °C. for DRS and 237 °C. for HRS the solid phases consist of a mixture of anhydrous sodium and potassium tartrates.

The TGA curves, shown in Figure 3, show a rapid loss of weight at approximately 250 °C. and the DSC curves, shown in Figure 2, show a large exothermic, decomposition-type reaction at this temperature. After being heated above this temperature, the sample was observed to be a black solid. These observations are consistent with the thermal decomposition of anhydrous sodium and potassium tartrates into the corresponding oxalates with the deposition of carbon, as shown by equation 3.

$$Na_2C_4H_4O_6 + K_2C_4H_4O_6 \xrightarrow[\Delta]{250°C} Na_2C_2O_4 + K_2C_2O_4 + 4C\downarrow + 4H_2O\uparrow \quad (3)$$

Upon further heating the oxalates are slowly decomposed to the corresponding carbonates. It may be observed from the TGA curves (Figure 3) that there is a slow weight loss from 280 to 350 °C., which most likely corresponds to the thermal decomposition of the oxalates to the carbonates. The residue is a black solid, which upon solution in water leaves a black substance (indicating carbon) and the solution has a pH of about 9 (indicating alkaline carbonates). The reaction for the slow decomposition of the oxalates to the carbonates is given by equation 4.

$$Na_2C_2O_4 + K_2C_2O_4 \xrightarrow[\Delta]{280-350°C} Na_2CO_3 + K_2CO_3 + 2CO\uparrow \quad (4)$$

This study has shown that the overall thermal decomposition of HRS and DRS is given by equation 5.

$$2NaKC_4H_4O_6 \cdot 4H_2O \xrightarrow{\Delta} K_2CO_3\downarrow + Na_2CO_3\downarrow + 4C\downarrow + 8H_2O\uparrow + 2CO\uparrow \quad (5)$$

A comparsion of the weights of the residues ($K_2CO_3 + Na_2CO_3 + C$) for both HRS and DRS with those calculated from equation 5 are given in Table I.

TABLE I

Experimental and Theoretical Residue Weights for Hydrated and Deuterated Rochelle Salts

Substance	Initial Wt. (mg)	Residue Wt. (mg) Measured	Calc.	Percent Error
HRS	2.786	1.468	1.383	+ 6.15
DRS	2.190	1.130	1.050	+ 7.62

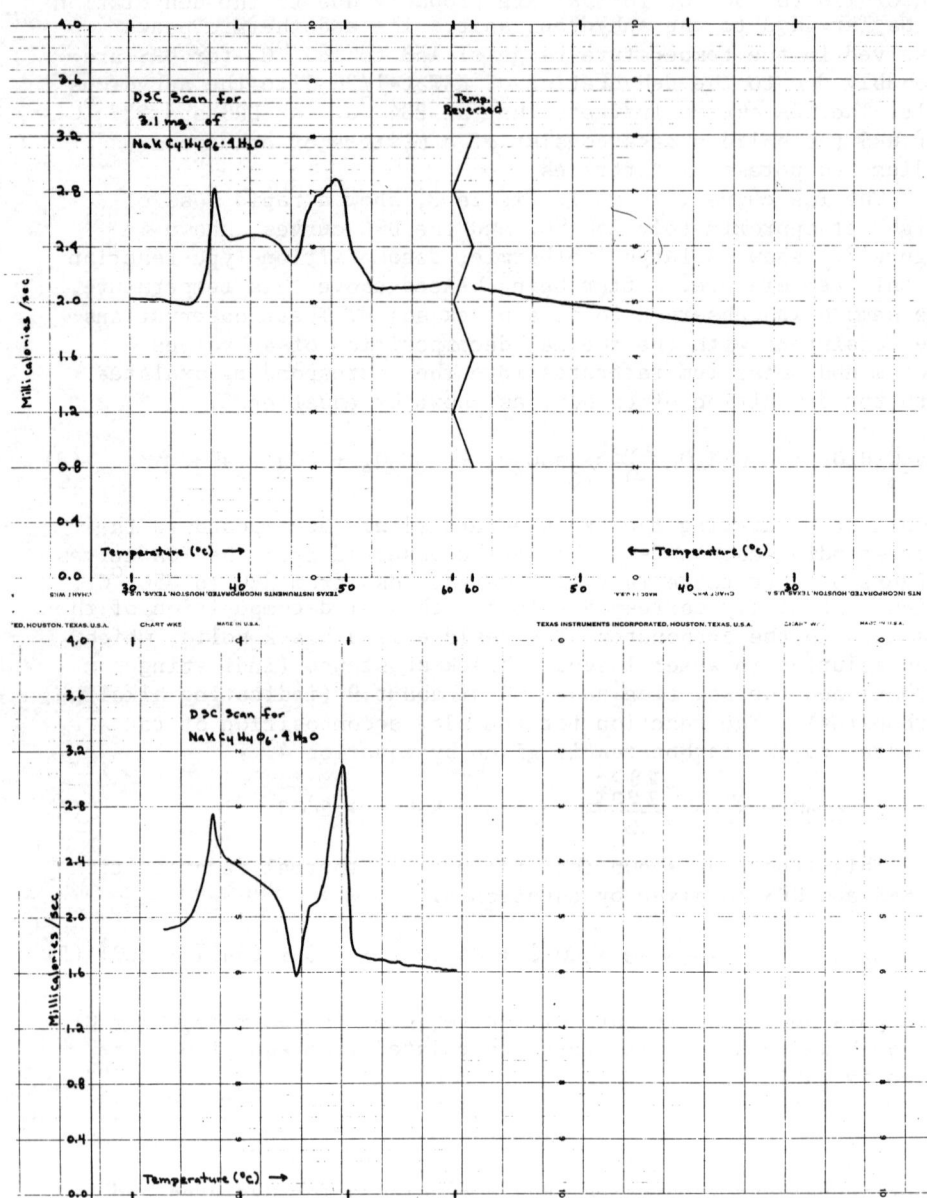

Figure 4 — DSC Scans for Single Crystals of Hydrated Rochelle Salts showing Exothermic Reactions and Irreversible Decomposition

It may be observed from Table I that there is reasonable agreement between the calculated and measured residue weights for both HRS and DRS. This result indicates that the overall reaction for the thermal decomposition of the Rochelle Salts is given by equation 5.

A comparsion between the thermal decompositions of the hydrated and deuterated Rochelle Salts gives the unexpected result that the hydrated salt is more stable than the deuterated salt. Table II gives a summary of the thermal effects that have been observed in this study for these isotopic salts. It may be clearly observed from this Table that HRS appears to be more stable under the same conditions than DRS.

IV. DISCUSSION

The thermal analysis studies reported in this paper indicate that the thermal decomposition of hydrated and deuterated Rochelle Salts occurs via equations 6.

$$NaKTar \cdot 4H_2O \xrightarrow{34°C, \Delta} NaKTar \cdot (4-x)H_2O \text{ (s)} \qquad (6)$$

$$2NaKTar \cdot yH_2O \xrightarrow{36°C, \Delta} Na_2Tar \cdot 2H_2O \text{ (s)} + K_2Tar \cdot 1/2H_2O \text{ (s)}$$

$$Na_2Tar \cdot 2H_2O \text{ (s)} + K_2Tar \cdot 1/2H_2O \text{ (s)} \xrightarrow{50°C, \Delta} Na_2Tar \cdot 2H_2O \text{ (}\ell\text{)} + K_2Tar \cdot 1/2H_2O \text{ (}\ell\text{)}$$

$$Na_2Tar \cdot 2H_2O \text{ (l)} + K_2Tar \cdot 1/2H_2O \xrightarrow{60-180°C, \Delta} Na_2Tar \text{ (s)} + K_2Tar \text{ (s)}$$

$$Na_2Tar \text{ (s)} + K_2Tar \text{ (s)} \xrightarrow{250°C, \Delta} Na_2C_2O_4 \text{ (s)} + K_2C_2O_4 \text{ (s)} + 4C \text{ (s)}$$

$$Na_2C_2O_4 \text{ (s)} + K_2C_2O_4 \text{ (s)} \xrightarrow{280-350°C, \Delta} K_2CO_3 \text{ (s)} + Na_2CO_3 \text{ (s)}$$

These studies have demonstrated that both hydrated and deuterated Rochelle Salts decompose by the same mechanism, as given by equation 6 above.

It has been unexpectedly demonstrated that the hydrated salt is more stable than the deuterated salt. This is a surprising result, because the hydrogens of the water molecules in Rochelle Salt are believed to be hydrogen bonded to oxygens of the - OH and - COOH groups (9) and it is well known that deuterium has a lower zero point energy than hydrogen and that the O-D bond is stronger than the O-H bond (10). Therefore, one would expect that DRS should be more thermally stable than HRS.

A possible explanation for this unexpected behavior is the following; It is believed that the ferroelectric properties of these salts are due to the polarization of specific O-H-O and O-D-O bonds parallel to the <100> crystallographic direction. A diagram of the crystal structure of Rochelle Salt showing the hydrogen bonds (9) is given in Figure 5. It may be observed from this diagram that all of the water molecules are not equivalently

TABLE II

Summary of Thermal Effects Observed
for Hydrated and Deuterated Rochelle Salts

	Observation	HRS	DRS	Remarks
1.	Endothermic Peak at ~ 34°C	34.8°C	34.0°C	Dehydration-DSC(Figure 2)
2.	Beginning of Initial Weight Loss	34°C	26°C	Dehydration TGA(Figure 3)
3.	Endothermic Peak at ~ 50°C	50.5°C	50°C	Decomposition DSC(Figure 2)
4.	Energy of Endothermic Transition (30-52°C)	16.6 $\frac{Kcal}{mole}$	15.2 $\frac{Kcal}{mole}$	Energy for Solution of Salt DSC (Figure 2)
5.	Loss of 2 H_2O/D_2O molecules	82°C	79°C	TGA(Figure 3)
6.	Loss of 4 H_2O/D_2O molecules	237°C	234°C	TGA(Figure 3)
7.	Beginning of weight loss at ~ 240°C	240°C	236°C	Decomposition of Tartrate Molecule-TGA(Figure 3)

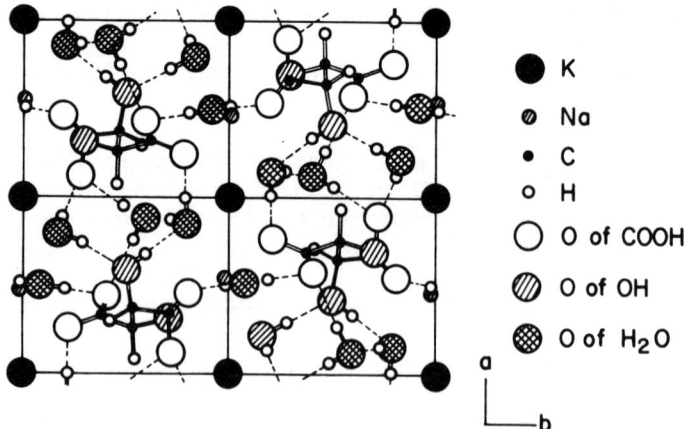

Figure 5 — Hydrogen Bonded Structure of Rochelle Salt According to Frazer et. al. (9)

bonded, and that the material can lose some of its water of crystallization and still maintain its ferroelectric properties. In fact, DRS is ferroelectric up to 35 °C., but we have observed appreciable weight loss prior to attaining this temperature. It is therefore possible, that the D_2O molecules in DRS, that are not contributing to its ferroelectricity, are more weakly bonded than the H_2O molecules in HRS. Recent x-ray studies (11) of the lattice parameters of HRS and DRS have shown that the 'a' parameter for HRS is greater than that for DRS, which is consistent with greater bond strength for the O-D bond in the ferroelectrically important $\langle 100 \rangle$ crystallographic direction: and the 'b' parameter for DRS is greater than that for HRS, which is consistent with other O-D bonds being weaker than the corresponding O-H bonds. In other words, the model being proposed to explain the observed thermal behavior is that the incorporation of deuterium into the crystal lattice, strains the lattice such that some of the O-D bonds are stronger and others are weaker than the corresponding O-H bonds in the hydrated salt.

An alternate explaination for the observed behavior could be that since the vapor pressure of D_2O is less than that of H_2O at the same temperature, the partial pressure of water may prevent the decomposition in HRS but will not inhibit it in DRS. Therefore, a higher temperature would be required for HRS than for DRS to effect decomposition.

Further work is required in order to unequivocally explain this very interesting inverse isotope effect in the Rochelle Salts.

V. REFERENCES

1. W. G. Cady, "Piezoelectricity", McGraw Hill Book Co., New York, N.Y., (1946), pgs. 510-653.

2. W. P. Mason, "Piezoelectric Crystals and their Applications to Ultrasonics", D. Van Nostrand Co. New York, N.Y. (1950), pgs. 114-136.

3. F. Jona and G. Shirane, "Ferroelectric Crystals", Pergamon Press Inc., New York, N.Y. (1962), Pgs. 280-317.

4. H.D. Megaw, "Ferroelectricity in Crystals", Methuen Co. Ltd., London, England, (1957), pgs. 27.

5. A. R. Ubbelohde and I Woodward, Proc. Roy. Soc. (London), A185, 448-465, (1946).

6. Lowry and Morgan, J. Amer. Chem. Soc., 46, 2191, (1924)

7. J.F.G. Hicks and J.G. Hooley, J. Amer. Chem. Soc., 60, 2994-2997, (1938).

8. C.D. Hodgman, editor, "Handbook of Chemistry and Physics", Chem. Rubber Publishing Co., Cleveland, Ohio, (1951), pgs. 1570 and 1574.

9. B. C. Frazer, M. McKeown, and R. Pepinsky, Phy. Rev., 94, 1435, (1954).

10. C.G. Swain and R.F.W. Bader, Tetrahedron, 10, 182-199, (1960)

11. A. Glatz and B. Rubin, Unpublished Data.

STORED ENERGY MEASUREMENTS IN APOLLO 11 LUNAR SAMPLES BY DIFFERENTIAL THERMAL ANALYSIS

John L. Kardos
Dept. of Chemical Engineering
Washington University
St. Louis, Missouri 63130

INTRODUCTION

The moon, unprotected by an atmosphere, is constantly being hit by radiation from the sun and other parts of our galaxy. In general, irradiation produces a wide variety of effects in solids. When these solid-state irradiation effects are examined in lunar samples with a variety of coordinated techniques, one can attempt to unravel the radiation history of the moon.

One such irradiation effect results in energy being stored in the solid. When radiation in the form of nuclear particles from the sun or other sources strikes a rock on the moon's surface, damage results in the form of displaced atoms. Because these atoms are pushed out of their equilibrium positions by the slowing particles, a kinetic energy absorption takes place and the rock in fact stores energy. This energy can be released by heating to a temperature at which the atoms can relax back to their equilibrium positions; when this happens, the stored energy is given off as heat. Prior to their reception it was possible to show theoretically that the moon samples might be heavily radiation-damaged and thus contain large quantities of stored energy.

Differential thermal analysis (DTA) is ideally suited for measuring stored energy in lunar samples because of its ability to measure heat release, both quantitatively and with high sensitivity, from a very small sample over a wide temperature range. In this paper the results of a DTA study on the lunar samples brought back by Apollo 11 are presented and discussed. Three types of samples (fines, rocks, and core tube material) were analyzed over a temperature range from room temperature to 900°C in both air and nitrogen atmospheres. The results are interpreted in terms of possible stored energy and other likely exothermic transitions.

EXPERIMENTAL

Types and Preparation of Samples

Three types of lunar samples were analyzed. In the case of the lunar fines and core-tube samples, the average particle size was small enough to permit direct analysis of the "as-received" material without pregrinding. The third type of sample, the rocks, had to be ground to a coarse powder with a Diamonite mortar and pestle. In all cases the material was carefully placed in either glass or quartz capillaries and weighed by difference.

Selected terrestrial minerals including quartz and hypersthene were also analyzed. Again, when necessary the samples were ground in a Diamonite mortar and pestle prior to placement in capillaries and weighing.

Instrument Calibration and Operating Conditions

A DuPont 900 differential thermal analyzer was used to measure total energy release in both lunar and terrestrial samples. Because of the importance of quantitatively measuring any energy released by the samples, calibration experiments were necessary to determine the effects of sample size, sample packing, and heating rate on measured transition heats in the temperature range of interest. In this study the transition in quartz (ΔH = 1.75 cal/gm., ~580°C) [1] was used to select optimum instrument operating conditions and to calibrate the instrument for quantitative measurements.

The average particle size had little effect on the DTA sensitivity for the size range examined. By increasing the time of grinding, the average quartz particle size was decreased from well over 100 microns to about 10 microns. Since the lunar samples had average particle sizes in this range, this variable was eliminated from further consideration. Sample packing effects were also negligible. There was no significant difference in sensitivity between loosely pouring the sample into the capillary and firmly tamping it down prior to thermocouple insertion.

Both the sample weight and the heating rate significantly affected the sensitivity as expected. These parameters were carefully isolated and their

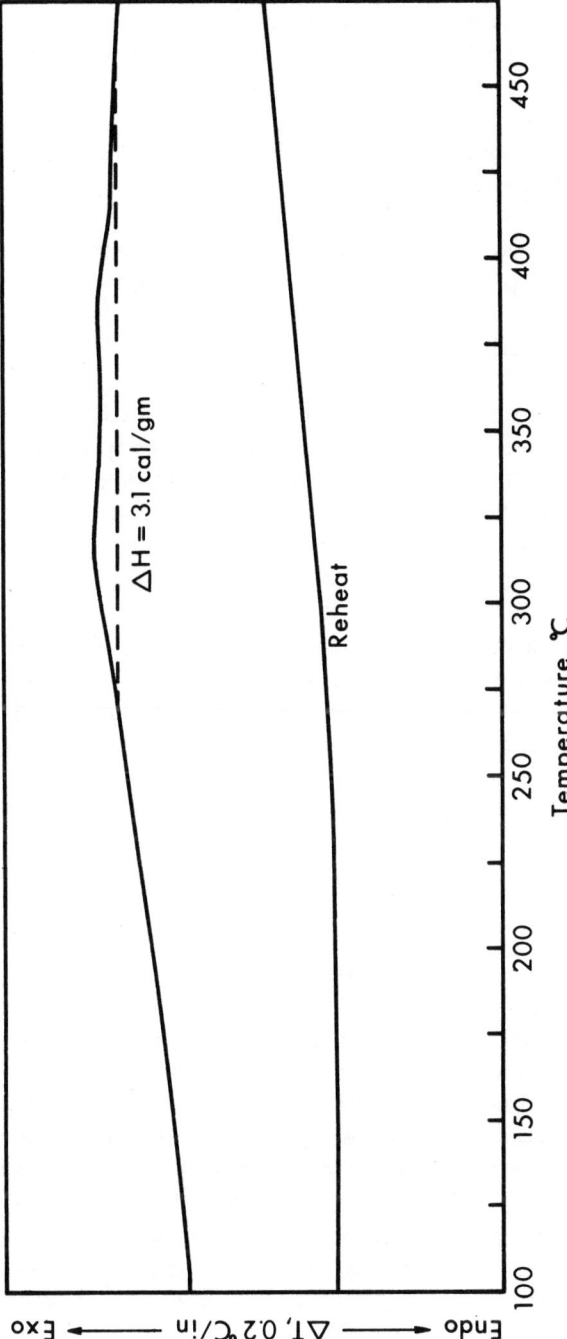

Figure 1. Differential thermal analysis (DTA) scans of lunar surface fines from 100 to 475°C at a heating rate of 20°C/min. in air. The differences between the exothermic peak temperature limits shown here and in subsequent figures and those quoted in the text are due to thermocouple corrections.

individual effects on the $\alpha \to \beta$ transition heat of quartz were determined. Based on these results, an optimum sample size of 12-14 mg. was chosen. For samples smaller than this, the measured transition heat varied sporadically because of incomplete thermocouple coverage, while larger samples did not provide increased intensity and were thus wasteful.

The heating rate was chosen to provide a balance between resolution and sensitivity. The temperature region from 100°C to 500°C was scanned using a standard cell with a heating rate of 20°C/min., while energy transitions in the region from 500°C to 900°C were measured with a less sensitive intermediate temperature cell operating at 30°C/min. Powdered aluminum oxide and glass beads were used as the reference materials in the intermediate temperature and standard cells, respectively.

Transition energies were calculated by comparing them with the known $\alpha \to \beta$ quartz transition at the same sample size, ΔT intensity setting, and heating rate. Specifically, this was accomplished by recording the known and unknown transitions on the same piece of chart paper and then cutting out and weighing the areas under the curves. For a constant chart paper thickness, the weight ratio is equal to the ratio of transition heats.

In general, heating was carried out in an air atmosphere, although several selected samples were heated under nitrogen. Prior to the nitrogen scans, the sample cell was evacuated to less than 1 mm Hg and then purged with nitrogen. This process was repeated five times before a final nitrogen blanket was introduced.

RESULTS

Surface Fines

The energy release pattern of the lunar surface fines in an air atmosphere is characterized by three relatively weak exotherms in the temperature range from 20°C to 900°C. The low temperature peak shown in Figure 1 is quite broad and extends from about 250°C to nearly 450°C. A second scan made under the same conditions without disturbing the sample shows no exotherm. In determining the 3.1 cal/gm of heat released, a base line was drawn as shown (dashed) based on the cooling curve and the slope of the heating curve before and after the exotherm. The energy release value in this and subsequent scans is accurate only to about $\pm$ 20% due to inaccuracies in calibration and in drawing the base line.

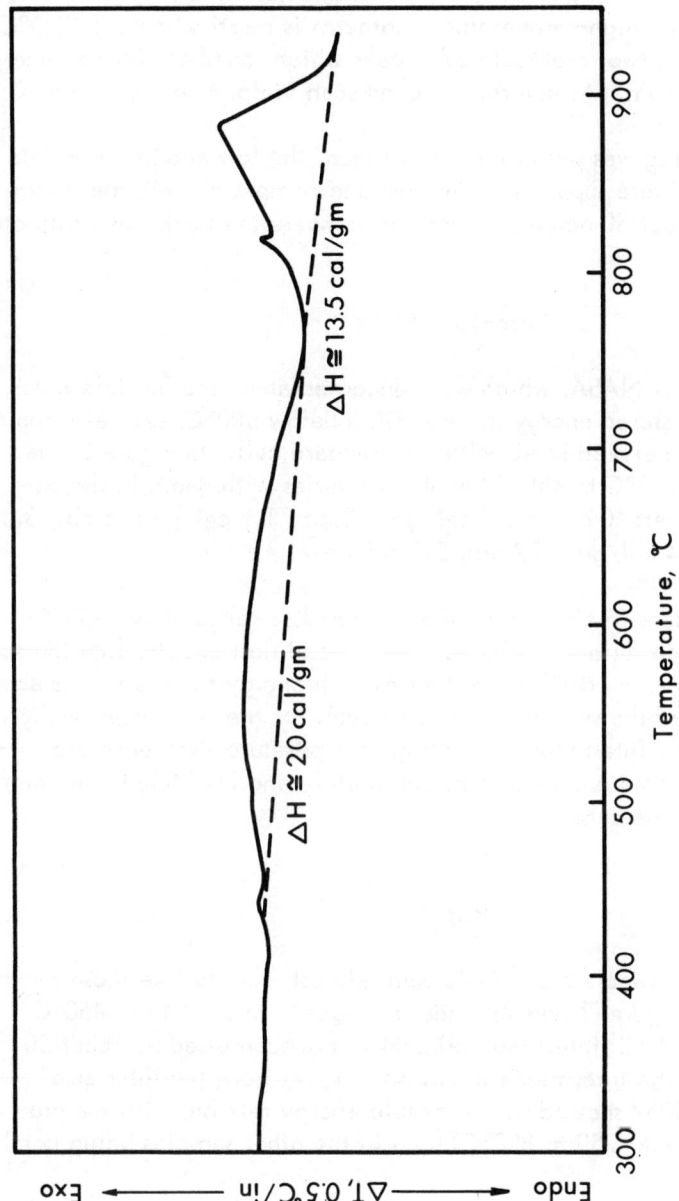

Figure 2. DTA scan of lunar surface fines from 300 to 900°C at a heating rate of 30°C/min. in air. The low-temperature peak seen in Figure 1 is not visible because of reduced sensitivity in this scan.

The intermediate and high temperature exotherms are shown in Figure 2. The low-temperature peak is too weak to be resolved with the less sensitive, intermediate temperature cell used for this scan. The intermediate exotherm covers the temperature range 450°C to 710°C and has an energy release value of ~20 cal/gm. The high-temperature exotherm is relatively sharp (750°C to 850°C) and shows two resolvable sub-peaks which together yield an energy release of 13.5 cal/gm. As before, a second scan yielded no exotherms.

When the heating was performed in nitrogen, the low and intermediate temperature peaks were suppressed, the low one completely and the intermediate one by about 50 percent. The high temperature peak was unaffected.

Core-Tube Fines

At the request of NASA, which was concerned about the possible catastrophic release of stored energy in lunar fines below 500°C, several samples from core 4 were analyzed in air with the standard cell. In Figure 3, the low-temperature (250°C to 450°C) exotherm varies with depth in the core in the following manner: 0 cm, <0.5 cal/gm; 3 cm, 1.9 cal/gm; 6 cm, 3.5 cal/gm; 9 cm, 1.4 cal/gm; 12 cm, 5.7 cal/gm.

Preliminary scans in air through the temperature range above 450°C indicate that the top surface of the core behaves almost exactly like the surface fines; there are two definite exotherms with energy release values about the same as those of the surface fines. However, for the remainder of the core depth, both the intermediate and high-temperature exotherms are suppressed, the intermediate peak almost completely and the high-temperature peak by at least 80 percent.

Rocks

DTA results on breccia rock 10046 were almost exactly like those for the surface fines (see Figures 1 and 2); under nitrogen both the 250 to 450°C (low) and 450 to 710°C (intermediate) peaks were suppressed by about 50 percent. Scans in the intermediate cell on samples from the interior of crystalline rock 10057 showed no noticeable energy release, with the pronounced doublet peak (750 to 850°C) seen in the other samples being notably absent.

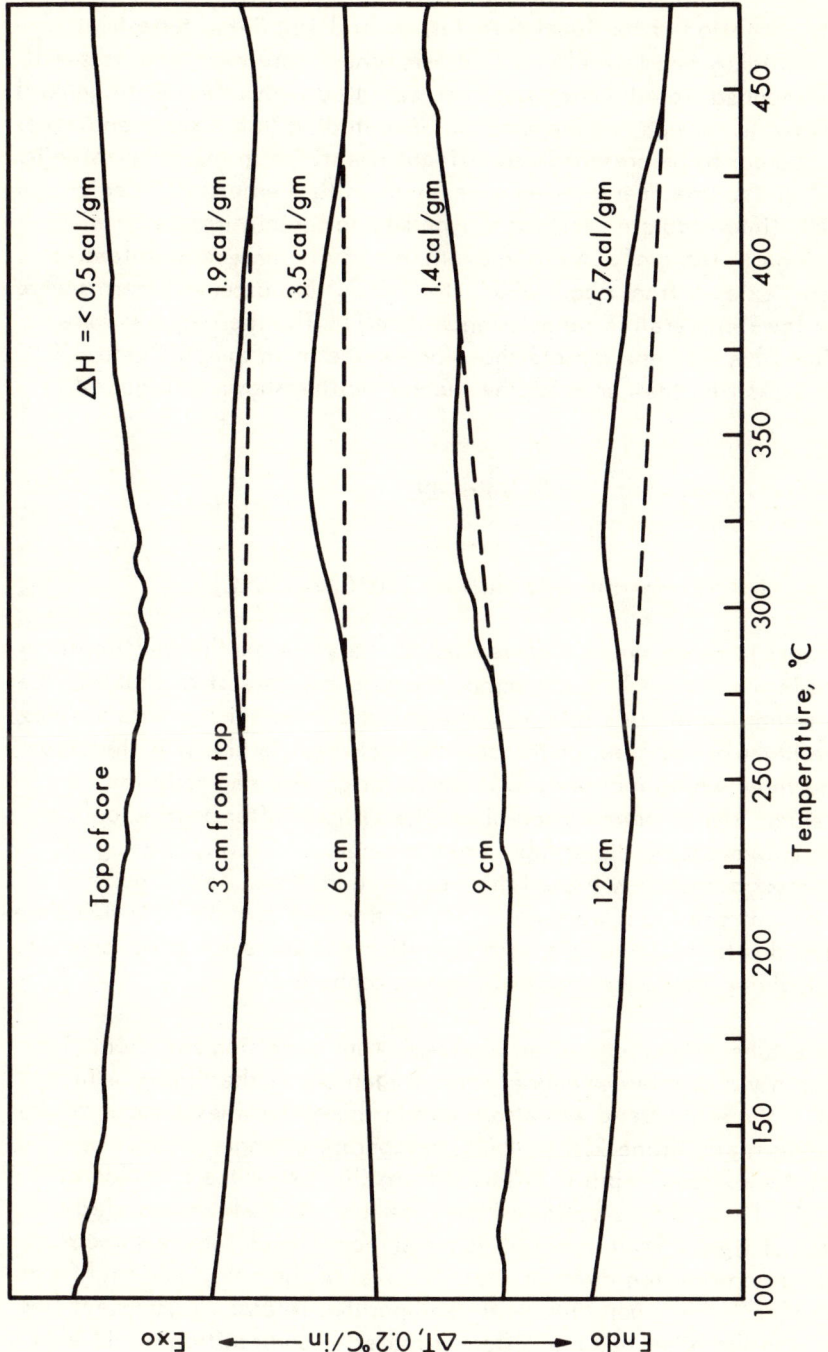

Figure 3. Low-temperature exotherm (250–450°C) as a function of depth for the core-tube material. Scans are at 20°C/min. in air.

Terrestrial Samples

Prior to reception of the lunar samples, several irradiated terrestrial minerals, including assorted felspars and pyroxenes, were examined to see if artificially induced stored energy could be detected with DTA. Both general types of minerals, though not the specific ones studied (labradorite and hypersthene) turned out to be present in significant quantities in the lunar samples. Figure 4 illustrates the energy release behavior in the temperature region above 300°C (intermediate cell) for a hypersthene sample irradiated with 3.7×10^{15} α particles/cm^2. A broad exotherm having an energy release of 15.5 cal/gm. extends from about 500°C to 800°C. No other exotherms were seen in the low-temperature range (standard cell). The thermogram for a non-irradiated hypersthene sample shows no exotherm in the 500°C to 800°C region and looks like the curve for the second heating shown in Figure 4.

DISCUSSION

Low-Temperature Transition (250°C to 450°C)

The lowest in temperature and weakest of the three exothermal transitions found in the Apollo 11 lunar samples is also the hardest to characterize. Its energy release value (~3 cal/gm.) is about the same for the surface fines, the surface layer of the core, or the breccia rock. Curiously, it is the only peak of the three which increases in intensity at greater depths in the core (see Figure 3). Visual examination of all the samples after heating only to 500°C in air revealed no discernable color change. However, heating in nitrogen almost completely erased this peak in both the surface fines and breccia rock (nitrogen runs were not made on the core samples). It therefore appears that this transition arises from oxidation or from some other chemical reaction involving the oxygen-containing atmosphere.

The core tube results present an interesting but confusing situation. The six-cm. depth yields a larger energy release than either the three- or nine-cm. depths. The same trend was also found in thermoluminescence intensity measurements made in the 250 to 450°C temperature range [2, 3]. Verification that the six-cm. depth is different from the rest of the core comes from the initial visual observation of the core tube [4], which revealed a narrow band of lighter gray material at the six-cm. level. If the six-cm. stratum has a composition different from the rest of the core, one might expect it to react with oxygen differently and perhaps release larger quantities of energy. On the other hand, a different material composition might also contain a larger quantity of stored energy, although this is not very likely. In

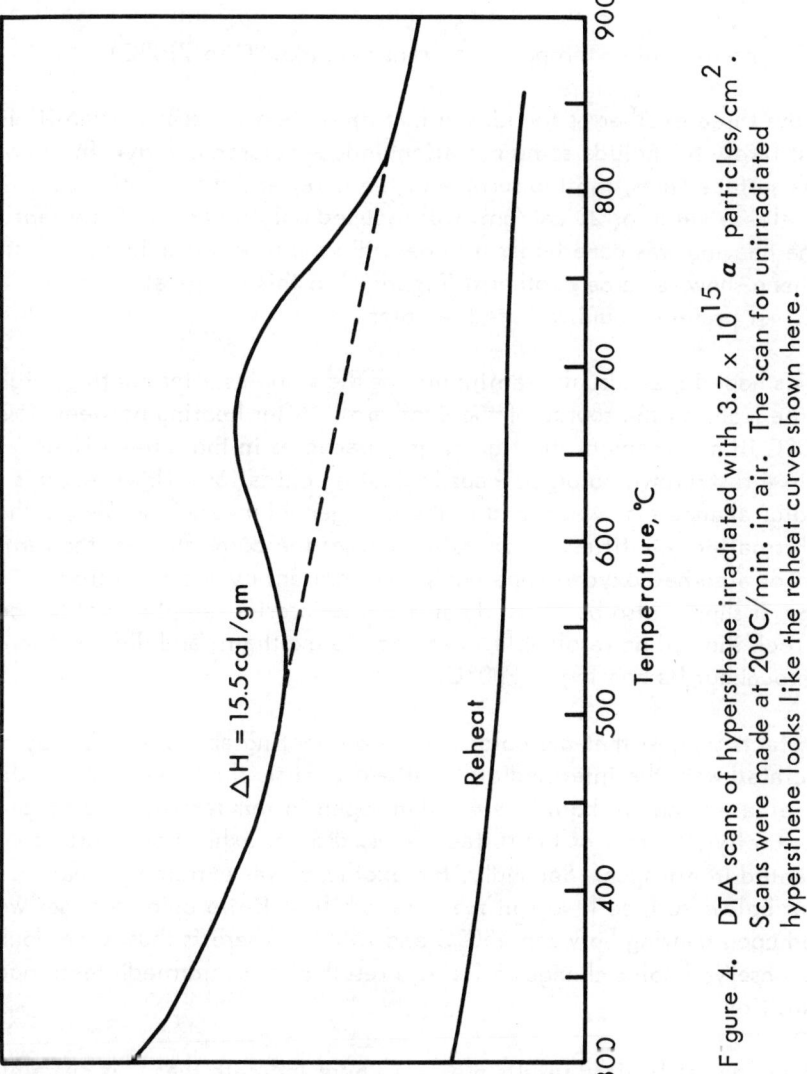

Figure 4. DTA scans of hypersthene irradiated with 3.7×10^{15} α particles/cm^2. Scans were made at 20°C/min. in air. The scan for unirradiated hypersthene looks like the reheat curve shown here.

the absence of nitrogen DTA results, it is impossible to distinguish between oxidation and radiation-induced stored energy as possible causes for the low-temperature exotherm in the core samples, although extrapolation of results from the surface fines points toward oxidation.

Intermediate-Temperature Transition (450°C to 710°C)

Of the three exotherms found, the intermediate-temperature transition is the most likely to include some radiation-induced stored energy. In samples from the surface fines, the top surface of the core, and the breccia rock, the energy release value of 20 cal/gm. was reduced only by about 50 percent when the heating was done under nitrogen. Furthermore, irradiated terrestrial hypersthene shows a large exotherm (Figure 4) in this temperature region, which is not present in unirradiated samples.

It was hoped that visual examination of the samples after heating might shed some light on the source of the exotherm. After heating between 450°C and 750°C in air, many of the lighter gray particles in the surface fines turned to a rust-brown color, perhaps indicating oxidation. However, this same color change also was found in the nitrogen-blanketed samples, although in much smaller quantities. If oxidation causes the color change, then small amounts of adsorbed oxygen apparently were present even after nitrogen flushing. It should also be pointed out that the interior sample from the crystalline rock showed no resolvable intermediate exotherm and did not change color throughout its heating to 900°C.

Two facts suggest that the color change on heating above 450°C may not be associated with the intermediate exotherm. First, the breccia rock, whose energy release behavior both in air and nitrogen in this temperature range is almost exactly like that of the surface fines, did not exhibit any color change when heated in nitrogen. Secondly, the exotherm was extremely small or absent at below-surface levels in the core, while definite color changes were observed upon heating between 450°C and 750°C. There is thus some doubt that the observed color changes arise as a result of the intermediate-temperature transition.

The somewhat limited results presented here indicate that it is certainly possible that the intermediate-temperature exotherm could be partially due to release of stored energy, although there is no compelling evidence that this is the case.

High-Temperature Transition (750°C to 850°C)

The high-temperature doublet peak between 750°C and 850°C is unaffected by heating in nitrogen and probably represents an intrinsic process. The absence of this peak in the crystalline rock suggests that it may be associated with the glass in the fines and the breccia rock. This is consistent with three other observations: (1) There are noticeably fewer glass spherules present in samples heated to 900°C than in those which were unheated. (2) At approximately 800°C several euhedrally shaped glassy grains were observed to burst explosively [2]. (3) X-ray analysis showed the transition of a glassy spherule to a felspathic crystalline material after annealing at 900°C [2]. In view of this evidence it appears possible that the high-temperature peak is due to crystallization of the glass. The doublet nature of the peak remains unexplained at this writing.

Stored Energy Considerations

The results presented here show that the stored energy release, at least above 200°C, is very small. As will now be outlined, the absence of a large stored energy peak is consistent with our current knowledge of the radiation history of the lunar samples. By measuring the density of deep damage tracks caused by high-energy cosmic rays and solar flare particles in the lunar samples and by applying other realistic assumptions, it has been estimated [3] that a maximum energy release of 0.5 cal/gm. can be expected from recovery of these tracks. Assuming that the lunar material is completely saturated with damage to a depth of 200 Å, a maximum of 5 cal/gm. was calculated [3] for the energy release due to recovery of solar wind-induced damage. Recovery of defects produced by spallation reactions probably also contributes less than 5 cal/gm. Thus, the sum of the major contributing causes of stored energy in the lunar samples yields a value small enough to be consistent with the interpretation of the intermediate exotherm, yet large enough to be easily detected by DTA.

ACKNOWLEDGMENTS

I am grateful to R. E. Lavengood for obtaining much of the calibration data on artificially irradiated terrestrial minerals and to Dennis Britton for the optical microscopy characterizations. I am also indebted to Prof. Robert Walker for our many discussions and his critical review of the manuscript.

This work was supported in part by contracts from NASA and the McDonnell-Douglas Corporation.

REFERENCES

1. P. Lakodey, C. Eyraud and M. Prettre, Compt. Rend., 242, 3071 (1956).

2. G. Crozaz, U. Haack, M. Hair, H. Hoyt, J. Kardos, M. Maurette, M. Miyajima, M. Seitz, S. Sun, R. Walker, M. Wittels, and D. Woolum, Science, 167, 563 (1970).

3. H. P. Hoyt, J. L. Kardos, M. Miyajima, M. G. Seitz, S. S. Sun, R. M. Walker, and M. C. Wittels, Geochim. Cosmochim. Acta, in press.

4. Lunar Preliminary Examination Team, Science, 165, 1211 (1969).

DETERMINATION OF THE HEAT OF TRANSITION OF SODIUM, RUBIDIUM AND CESIUM TETRAFLUOROBORATE *

R. T. Marano and E. R. Shuster

Nuclear Materials and Equipment Corporation

INTRODUCTION

This work is part of a continuing study of the thermal properties of the alkali metal tetrafluoroborates by the authors[1,2].

This investigation reports the heat of transition for Na, Rb & Cs tetrafluoroborate determined under two different sets of conditions and calculated by two methods for each. Using the reported heat of transition[3] for KBF_4 of 10 cal/g as the standard value, the compounds were evaluated using a method similar to one described by Yagfarov[4] which is based on the internal standard method for a heating rate of 10°C min^{-1}. The heats of transition were also determined at a heating rate of 20°C min^{-1}, using external KBF_4 standardization.

EXPERIMENTAL PROCEDURE

The study was performed using a standard commercial thermal analyzer manufactured by Fisher Scientific Company consisting of the Fisher Model 360 Linear Temperature Programmer and the Model 260 Furnace and Specimen Holder. The curves were recorded on a Texas Instrument two pen strip chart recorder with 1 mv full scale response. The specimen holder was an Inconel cylinder with eight symetrically arranged cylindrical holes.

*Work supported by the U.S. Atomic Energy Commission under Contract AT-(40-1)-3292

The sample and reference material were placed in borosilicate glass or graphite crucibles which in turn were placed in the specimen holder. The differential thermocouple was inserted into the sample and reference, and the holder, crucibles, thermocouples, sample, and reference placed in the furnace. The lead cable from the furnace was connected through the programmer to the recorder.

Argon was introduced into the furnace through the handle of the specimen holder and dispersed at four points just above the tops of the crucibles. The atmosphere filled the furnace and then vented at the furnace top. All data was collected using chromel-alumel thermocouples for the differential temperature, a Platinel I thermocouple to measure the alumina reference temperature, and Argon furnace atmosphere at 2SCF/H.

The $NaBF_4$ and $RbBF_4$ used for this investigation were purchased from ALFA Inorganics and are 99%+ pure while the $CsBF_4$ was prepared in our laboratory and is also 99%+ pure.

In the first set of analyses made at a heating rate 10°C min^{-1} equal weights of sample and KBF_4 were mixed and packed into borosilicate glass crucibles. A crucible filled with platinum foil was positioned in the specimen holder between the sample and the alumina reference positions. This was done to thermally isolate the sample from the reference material. Table I contains a summary of the data obtained on sodium tetrafluoroborate by integrating the total peak area using a plainmeter. Table 2 contains a summary of the data obtained on sodium tetrafluoroborate by integrating the front half of the peak.

From the data it can be seen that there is a difference in the ΔH_T calculated by these methods. At this time the reason for this difference is undetermined but it is believed to be a function of the heat capacity and heat flow characteristics of the system.

The second compound investigated was $RbBF_4$. Tables 3 and 4 contain a summary of the data collected for this material. From these tables the values calculated by the two methods again appear to be somewhat in disagreement. The standard deviation of the measurement is greater than that calculated for the $NaBF_4$, but the values are in better agreement. Further investigation is needed to ascertain why the standard deviation is larger.

The third compound investigated was $CsBF_4$. Tables 5 and 6 contain a summary of the data collected for this material. These values are in much better agreement than the values for the two previous materials.

TABLE I

Heat of Transition
Sodium Tetrafluoroborate

Heating Rate 10° Min^{-1}
Full Peak Integration

Run	NaBF$_4$ Area (in^2)	KBF$_4$ Area (in^2)	$\frac{A, NaBF_4}{A, KBF_4}$	ΔH_T(Cal/g)
1	0.54	0.82	0.66	6.6
2	0.49	0.89	0.55	5.5
3	0.35	0.65	0.54	5.4
4	0.35	0.64	0.55	5.5
5	0.35	0.50	0.70	7.0
				Avg = 6.0 ± 0.7

TABLE II

Heat of Transition
Sodium Tetrafluoroborate

Heating Rate 10° Min^{-1}
Half Peak Integration

Run	NaBF$_4$ Area (in^2)	KBF$_4$ Area (in^2)	$\frac{A, NaBF_4}{A, KBF_4}$	ΔH_T(Cal/g)
1	0.24	0.54	0.44	4.4
2	0.25	0.55	0.45	4.5
3	0.16	0.35	0.46	4.6
4	0.18	0.35	0.51	5.1
5	0.15	0.35	0.43	4.3
				Avg = 4.6 ± 0.3

TABLE III

Heat of Transition
Rubidium Tetrafluoroborate
Heating Rate 10° Min^{-1}
Full Peak Integration

Run	RbBF$_4$ Area (in^2)	KBF$_4$ Area (in^2)	$\frac{A, RbBF_4}{A, KBF_4}$	ΔH_T(Cal/g)
1	0.53	1.06	0.50	5.0
2	0.46	0.92	0.50	5.0
3	0.48	0.58	0.83	8.3
4	0.36	0.42	0.85	8.5
5	0.36	0.53	0.70	7.0

Avg = 6.8 ± 1.5

TABLE IV

Heat of Transition
Rubidium Tetrafluoroborate
Heating Rate 10° Min^{-1}
Half Peak Integration

Run	RbBF$_4$ Area (in^2)	KBF$_4$ Area (in^2)	$\frac{A, RbBF_4}{A, KBF_4}$	ΔH_T(Cal/g)
1	0.27	0.70	0.38	3.8
2	0.26	0.63	0.41	4.1
3	0.27	0.35	0.77	7.7
4	0.19	0.28	0.68	6.8
5	0.23	0.32	0.72	7.2

Avg = 5.9 ± 1.6

TABLE V

Heat of Transition
Cesium Tetrafluoroborate

Heating Rate 10° Min^{-1}
Full Peak Integration

Run	$CsBF_4$ Area (in^2)	KBF_4 Area (in^2)	$\frac{A, CsBF_4}{A, KBF_4}$	ΔH_T(Cal/g)
1	0.18	0.84	0.21	2.1
2	0.20	1.00	0.20	2.0
3	0.07	0.61	0.11	1.1
4	0.11	0.59	0.19	1.9
5	0.07	0.63	0.11	1.1

Avg. = 1.6 ± 0.4

TABLE VI

Heat of Transition
Cesium Tetrafluoroborate

Heating Rate 10° Min^{-1}
Half Peak Integration

Run	$CsBF_4$ Area (in^2)	KBF_4 Area (in^2)	$\frac{A, CsBF_4}{A, KBF_4}$	ΔH_T(Cal/g)
1	0.10	0.51	0.20	2.0
2	0.08	0.64	0.13	1.3
3	0.03	0.40	0.08	0.8
4	0.05	0.38	0.13	1.3
5	0.03	0.39	0.08	0.8

Avg. = 1.2 ± 0.4

A second series of determinations were made at 20°C Min^{-1} using graphite crucibles. In this set the samples were run independently from the KBF$_4$ standard. The platinum foil filled crucible was not used. Based on the average peak area obtained using KBF$_4$, as reported in Table 7, a constant K has been calculated to calibrate the system using both the front half peak area and the total peak area as follows:

$$K = \frac{\text{Peak area (in}^2\text{)}}{(\Delta H_T)(\text{Sample wt. g})}$$

For the full peak area integration, K = 0.54 and for the half peak integration K = 0.28. This calculation is based on the KBF$_4$ sample weight of 0.3423 g and the reported ΔH_T of KBF$_4$ of 10 Cal/g.

Based upon the calculated constant K for the system and the equation rewritten as $[\Delta H_T = \frac{\text{Peak area (in}^2\text{)}}{(K) \times (\text{sample weight})}]$ the heat of transition for the three tetrafluoroborates were calculated.

Table 8 contains a summary of the data collected for NaBF$_4$ at 20°C Min^{-1}. The value of ΔH_T calculated by integrating the whole peak and the value calculated by integrating the half peak are in very good agreement. It appears that the increased heating rate does beneficially affect the results.

Next RbBF$_4$ was analyzed by this same method. Table 9 shows a summary of this data. Both the agreement in values between integration methods and the standard deviation of the measurement are significantly improved over values obtained using Yagfarov's technique.

The final compound investigated was CsBF$_4$. Table 10 contains a summary of the data collected for this material. The data again is in better agreement than by the previous method with good measurement precision.

DISCUSSION

From the heat of transition data determined for the Rhombic-Cubic transition of the three alkali metal tetrafluoroborates reported, it can be noted that the method described by Yagfarov and based on the method of internal standards is less desirable than the simpler method of separate determinations on standard and sample. For this reason it is believed that the values obtained by the Yagfarov method are less valid than the values calculated based on separate determinations. This increased inaccuracy is probably due to the interaction of the heat capacity of the different systems and the probable solid state reaction of the mixed salts.

TABLE VII

Heat of Transition
Potassium Tetrafluoroborate

Area Determinations at 20°C Min^{-1}

Run	Total Area (in^2)	Front Area (in^2)
1	1.78	0.96
2	1.70	0.87
3	1.92	0.96
4	1.93	1.02
5	1.85	0.97
Avg.	1.84 ± 0.09	0.96 ± 0.05

TABLE VIII

Heat of Transition
Sodium Tetrafluoroborate

Heating Rate 20°C Min^{-1}

Run	Total Area Integration (in^2)	ΔH_T(Cal/g)	Half Peak Integration (in^2)	ΔH_T(Cal/g)
1	0.92	5.5	0.44	5.0
2	0.83	4.9	0.39	4.5
3	0.85	5.0	0.39	4.5
4	0.88	5.2	0.40	4.6
5	0.89	5.3	0.40	4.6
Avg.	0.87	5.2 ± 0.2	0.40	4.6 ± 0.2

TABLE IX

Heat of Transition
Rubidium Tetrafluoroborate

Heating Rate 20°C Min^{-1}

Run	Total Area Peak (in^2)	ΔH_T(Cal/g)	Half Area Peak (in^2)	ΔH_T(Cal/g)
1	0.85	5.2	0.55	6.5
2	1.17	7.1	0.61	7.2
3	1.00	6.1	0.60	7.1
4	1.05	6.4	0.56	6.6
5	1.07	6.5	0.54	6.4
Avg.	1.03	6.3 ± 0.6	0.57	6.8 ± 0.3

TABLE X

Heat of Transition
Cesium Tetrafluoroborate

Heating Rate 20°C Min^{-1}

Run	Total Area Peak (in^2)	ΔH_T(Cal/g)	Half Area Peak (in^2)	ΔH_T(Cal/g)
1	0.60	3.9	0.28	3.5
2	0.42	2.8	0.24	3.0
3	0.52	3.4	0.26	3.3
4	0.44	2.9	0.24	3.0
5	0.38	2.5	0.21	2.7
Avg.	0.47	3.1 ± 0.5	0.25	3.2 ± 0.2

For this reason the values assigned as first approximation values of the ΔH_T of these salts are the average values calculated from separate run determinations. Thus ΔH_T for $NaBF_4$ equals 4.9 ± 0.3 Cal/g, ΔH_T for $RbBF_4$ equals 6.6 ± 0.6 Cal/g, ΔH_T for $CsBF_4$ equals 3.2 ± 0.4 Cal/g.

It is hoped that other investigators will carry this effort further. A great amount of fundamental data on these compounds is unknown and unreported. At this time, programmatic considerations preclude the authors from further work in this area.

REFERENCES

1. Marano, R. T. and Shuster, E. R., "Thermal Analysis". Ed. by R. F. Schwenker & P. D. Garn, Vol. 2, Academic Press, New York, 1969, p. 709.

2. Marano, R. T. and Shuster, E. R., "Determination of the Rhombic to Cubic Transition Temperatures of the Alkali Metal Tetrafluoroborates Using Differential Thermal Analysis" (in publication).

3. Ryss, I. G., "The Chemistry of Fluorine and Its Inorganic Compounds", Part 2, State Publishing House for Scientific Technical and Chemical Literature, Moscow, 1956, p. 529.

4. Yagfarov, M. S., Russ J. Inorg. Chem. (English Transl.), 6, 1236 (1961)

THERMAL ANALYSIS OF HYDROXYLAMMONIUM PERCHLORATE

J. N. Maycock and V. R. Pai Verneker

Research Institute for Advanced Studies

Martin Marietta Corporation, Baltimore, Md. 21227

ABSTRACT

Hydroxylammonium perchlorate (HAP) is a hygroscopic energetic perchlorate which is known to exhibit two endothermic processes, one attributable to its melting and one attributable to a crystal phase change or a dehydration. The present study has been a re-examination of these endothermic processes by the use of differential thermal analysis, thermogravimetric analysis, mass spectroscopy and thermal microscopy techniques. By a combination of these different techniques it has been established that the endotherm at 55-60°C is due to a crystal phase change. For samples of HAP, using different manufacturing techniques the enthalpy of the phase change varies from 2.5 to 3.0 calories per gm. These same techniques have been used to show that the presently available HAP contains occluded pockets of a liquid phase. The identity of this liquid phase is still undetermined but several possible explanations are presented as a guide for further study.

INTRODUCTION

Hydroxylammonium perchlorate (HAP) is a white, granular, hygroscopic solid whose thermogram indicates an endothermic event at 55-60°C, a melting endotherm at 85-90°C and a decomposition exotherm at 240-250°C. Three crystalline phases of HAP have been determined by Dickens (1) using X-ray diffraction techniques. The crystal structure of the most stable phase was determined as orthorhombic having a a_0 space group of $P2_1$ cn and cell dimensions of 7.52, 7.14 and 15.99 A at 25°C. The general physical structure is considered to consist of chains of perchlorate ion tetrahedra held together by hydrogen bonding from parallel chains of NH_3OH^+ ions.

A complete presentation of the various crystal phases of HAP is given in Figure 1.

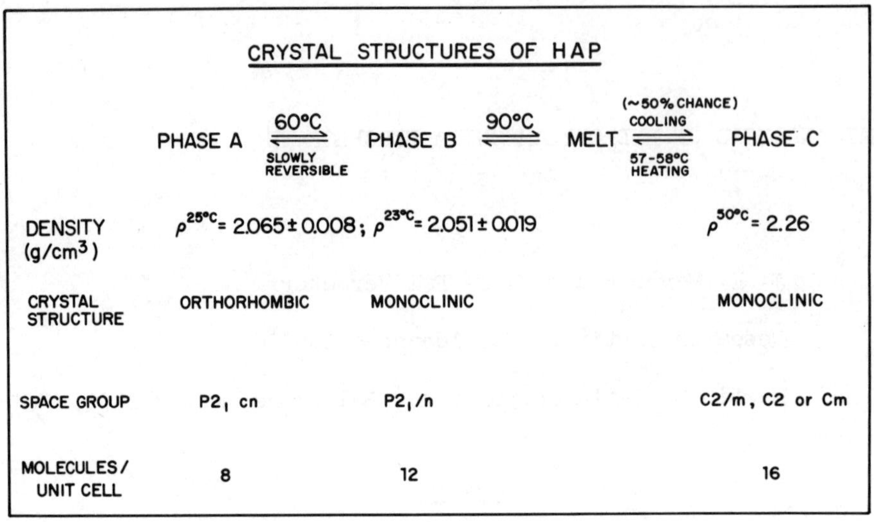

CRYSTAL STRUCTURES OF HAP

	PHASE A $\underset{\text{SLOWLY REVERSIBLE}}{\overset{60°C}{\rightleftarrows}}$ PHASE B	$\overset{90°C}{\rightleftarrows}$ MELT	(~50% CHANCE) $\underset{\text{57-58°C HEATING}}{\overset{\text{COOLING}}{\rightleftarrows}}$ PHASE C
DENSITY (g/cm³)	$\rho^{25°C}= 2.065 \pm 0.008$; $\rho^{23°C}= 2.051 \pm 0.019$		$\rho^{50°C}= 2.26$
CRYSTAL STRUCTURE	ORTHORHOMBIC MONOCLINIC		MONOCLINIC
SPACE GROUP	$P2_1 cn$ $P2_1/n$		$C2/m$, $C2$ or Cm
MOLECULES/ UNIT CELL	8 12		16

FIG. 1 – Description of the various crystal phases of HAP.

The thermal decomposition of HAP using small heating rates (2) indicates that the initial step in the decomposition involves dissociation of the salt to free amine and perchloric acid. This reaction is then followed by the oxidation and decomposition of hydroxylamine and perchloric acid to give a multitude of side products including NH_4ClO_4 (AP) and O_2. This dissociation step has been verified by direct inlet mass spectrometric studies at the U.S. Air Force Rocket Propulsion Laboratory. Using the relative intensity peak height technique (for NH_2OH) an activation energy of 20 kcal mole^{-1} has been calculated for the formation of NH_2OH and $HClO_4$. However, a break in the activation energy, E_a, slope was observed at 60°C with a lower E_a being observed above this temperature. The conclusion to be drawn from these data is that the 55°C endotherm is somehow responsible for an increase in the chemical reactivity at this temperature.

The objective of the present program is to define the nature of the first endotherm, phase change or dehydration, and also to examine any relationship between the postulated dissociation mechanism and crystal behavior.

EXPERIMENTAL TECHNIQUES

Materials and Thermal Analysis Testing

Two different batches of material were used. One batch was obtained from the U.S. Naval Propellant plant and is designated as Navy HAP. The other batch from the Thiokol Chemical Corporation, Elkton, Md. is designated as Thiokol HAP. Thiokol HAP was received in dessicated bottles and specified as being 98% minimum assay with a particle size range of 44 to 840 microns. The Navy HAP was received as a carbon tetrachloride slurry. Prior to testing, the Navy material was filtered through a Buchner funnel and vacuum dried for two days with continuous pumping. An additional sample of Navy HAP was also received which had been packed dry rather than as a slurry with carbon tetrachloride.

In tests involving differential thermal analysis (DTA) measurements, 15 mg of calcined Al_2O_3 were weighed into a platinum cup (3mm dia. $\times$ 5mm ht.) and placed in the reference holder of the Mettler thermoanalyzer. The thermoanalyzer was then pumped down to approximately 8×10^{-5} Torr with diffusion pumps and filled with dry helium. The HAP sample was then loaded into a similar cup inside a dry box with the amount of sample being determined volumetrically to give a weight of approximately 30 mg. Precise weight determination was made gravimetrically after testing. The sample was transferred from the dry box to the thermoanalyzer in a closed container having a dry nitrogen atmosphere. Loading the sample into the thermoanalyzer was performed inside a large bell jar through which a high flow rate of helium was being maintained (3). A similar procedure was followed for tests involving only thermogravimetric analysis,TGA, measurements except that elimination of the need for reference material and differential thermocouple permitted the use of a 16mm diameter platinum macro cup capable of holding up to 700 mg of sample. All thermal testing was performed using the most sensitive weight scale, permitting detection of weight changes as small as one microgram. DTA sensitivity was either 2 or 10 microvolts per inch using a Pt vs. Pt-10% Rh thermocouple for which a differential temperature of 1° corresponds to about 7 microvolts over the temperature range of 50-100°C. A helium flow rate of 5 liters per hour was maintained during testing, using helium which had been passed through H_2SO_4 and P_2O_5 drying towers and deoxygenated over hot copper. Estimates of enthalpy values are based upon measurements of DTA peak areas using a calibration factor derived from potassium nitrate melting and freezing peaks. No adjustment has been attempted for factors such as thermal conduction, specific heat, etc. which can influence peak size but are very difficult to evaluate.

THERMAL MICROSCOPY

Due to the nature of the problem being investigated, i.e. phase change or dehydration, a thermal microscopy study of HAP upto its melting point was made. The equipment used was a Mettler FP2 programmed thermal microscope* completely enclosed in a dry bag containing dry helium. The attractive feature of the Mettler hot stage for this study is that the sample under investigation is placed between two identical heating plates. This allows temperature measurement accurately without imbedding a thermistor in the sample or calibrating a thermometer.

A typical program involved preparing a uniform sample between a microscope slide and a cover glass, inserting the sample into the hot stage and selecting the appropriate thermal program for the study. With this instrument it was possible to select heating and cooling rates between $10°$ and $0.2°C$ min^{-1} and also to maintain isothermal conditions. To photographically record phenomena seen through the Nikon Model S-KE microscope a Leica 35mm camera was used loaded with Kodachrome X film. To improve the background illumination the camera was synchronized with a Nikon electronic flash, power unit model MS-1.

RESULTS

Thermal Analysis (DTA-TGA)

Upon initial heating at $6°C/min$ HAP shows two distinct low temperature endotherms occurring at about $66°$ and $95°C$, Figure 2. The $95°$ peak is roughly ten times as large as the lower temperature peak and visually corresponds with sample melting.

The $66°$ endotherm has an associated small but sharp weight loss. Upon cooling no corresponding exotherm is observed and neither the sharp weight loss nor the endotherm is observed if the sample is immediately reheated. Upon retesting the same sample 2 or 3 days later an endotherm is again noticed. This endotherm temperature is lower by about 4 to $10°$, has no sharp weight loss associated with it and is reproducible through at least two successive heating cycles.

The sharp weight loss associated with the endothermic peak is a function of particle size. Table 1 shows weight loss values obtained with the different particle size ranges for both Navy HAP and Thiokol HAP. Navy HAP loses up to 20 times more weight than the Thiokol material with a maximum loss of 0.2% of sample weight. For Navy HAP the weight loss at the largest particle size (> 500 microns) is less than that of particles in the 354-500 micron range. This could result from formation of the largest particles by coalescence or fusing together of smaller particles but, more likely, suggests occurrence of a maximum point on the weight loss curve.

---

* This equipment was kindly loaned to us by Mr. H. Vaughan, Mettler Instrument Corporation, Princeton, New Jersey.

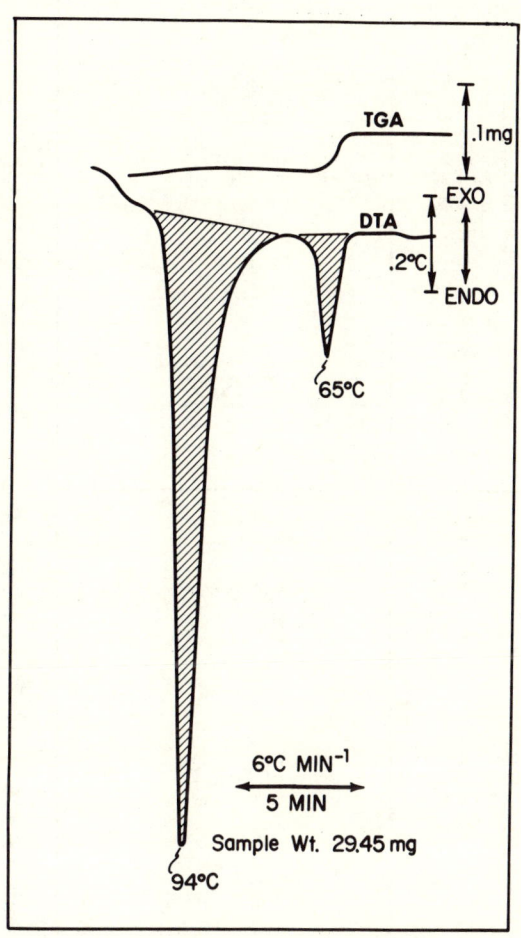

FIG. 2 - Simultaneous DTA-TGA of Thiokol HAP (29.45 mg). The sample was held in a flowing (5 liters hour^{-1}) atmosphere of H_2O and O_2 free He and heated at 6°C min^{-1}. The phase change is recorded at 65°C and the melting of the sample at 94°C.

TABLE 1
HAP Wt. Loss for Various Particle Size Ranges*

Particle Size (μ)	Navy HAP			Thiokol HAP		
	Sample Wt. (mg)	Wt. Loss (mg)	Wt. Loss ppm	Sample Wt. (mg)	Wt. Loss (mg)	Wt. Loss ppm
< 53 μ	31.7	1	31.5	539.3	0	0
53–149	30.0	11	367	611.1	7	11
149–210	28.5	33	1156	587.1	11	19
210–354	29.5	59	1999	695.9	45	65
354–500	34.4	53	1541	468.6	45	96
> 500	29.5	40	1536			

*These data are relevant to a 6°C min^{-1} heating rate.

TABLE 2

Enthalpy and Peak Temperature Values for Various Particle Sizes

	AP Phase Change (15 mg, 6°/min in He) Peak Temp. (°C)	Cal/g	Thiokol HAP (6°/min in Helium) Peak Temp.	Cal/g	Navy HAP (2°/min in Helium) Peak Temp.	Cal/g
< 53 μ	249	24.9	70	2.84, 2.61		
53-149	248	24.0	67	3.13	63	3.72
149-210	247	22.5	67	2.49	63	3.33
210-354	247	23.7	66	2.47	62	2.97
354-500	247	23.1	66	2.67	62	3.28
> 500	246	22.9			62	2.73, 2.35

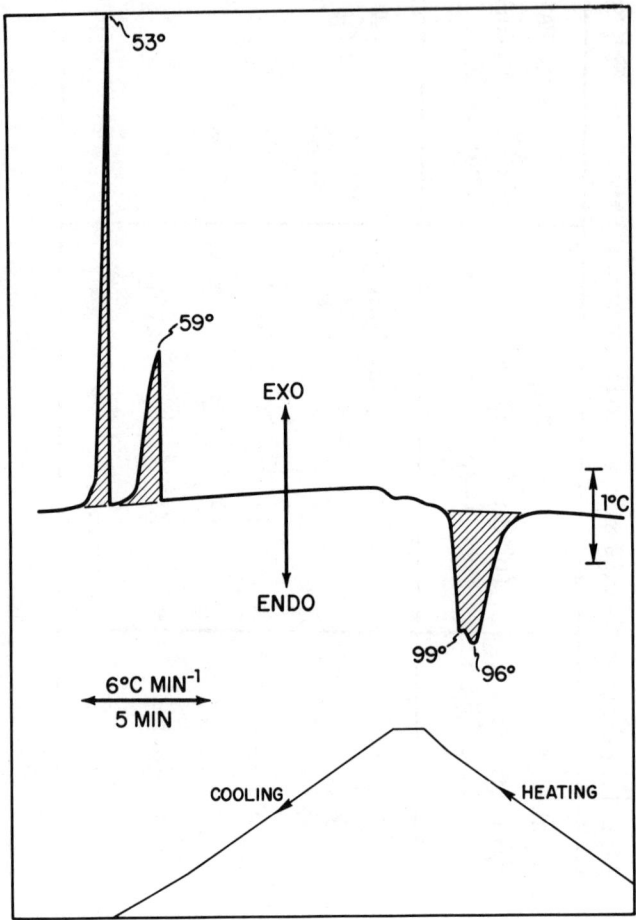

FIG. 3 - A DTA trace of Thiokol HAP which is heated ($6°C$ min^{-1}) up to its melting point at $96°C$ and then cooled at the same rate. The two freezing points, $59°$ and $53°C$ are clearly shown. The sample was in a flowing atmosphere (5 liters $hour^{-1}$) of H_2O and O_2 free He.

Other than the apparent maximum shown by the Navy HAP it is abundantly clear that larger the particle size, larger the weight loss. These data are all normalized to the same mass. If this weight loss were strictly a surface desorption then the largest effect would be seen with the smallest particle size. The data shows that the reverse is true, which can tentatively be interpreted as an evolution from within the particles.

Twenty four hours additional vacuum drying had little or no effect toward reducing the sharp weight loss of the Navy HAP and conversely soaking the Thiokol HAP in carbon tetrachloride for 6 days did not noticeably increase its weight loss. Testing an additional sample of Navy HAP which was received in a dessicated container rather than as a carbon tetrachloride slurry gave the same type of weight loss as the main sample of Navy HAP.

Enthalpy and peak temperature values for the low temperature endotherm observed with various particle size ranges of HAP are shown in Table 2. As a comparison with another perchlorate system similar data obtained for the phase transition of ammonium perchlorate from orthorhombic to cubic are also presented in Table 2.

With both HAP and AP a decrease in particle size results in increased DTA peak temperatures and also an increase in the estimated calories per gram values. This behavior is probably due to reduced heat conduction with the smaller particles rather than to any significant enthalpy variations. Poor heat conduction causes a higher temperature DTA peak because the center portion of the sample tends to lag behind the temperature of the sensing thermocouple. Reduced thermal conduction also results in higher thermal gradients within the sample causing the DTA peak to be broader but of smaller amplitude. Peak height values for AP shown in Table 2 indicate this behavior. Poor conduction also increases the time required for the sample to return to ambient test temperature resulting in a larger DTA area for a given enthalpic occurrence.

Values of estimated enthalpy for the low temperature endotherm of HAP average about 3.06 cal/g (409 cal/mole) for Navy material and 2.71 cal/g (362 cal/mole) for the Thiokol material. The difference in values between the two batches amounts to about 12% and is of questionable significance.

The melting endotherm reappears upon successive heatings and during cool down of the molten material a sharp exothermic indication of freezing is indicated by a large single peak about 30% of the time and by two smaller exothermic peaks about 70%, Figure 3, of the time. In every instance the Navy Material gives a single peak in freezing. Typical temperatures and estimated enthalpies for melting and freezing of the Thiokol material are listed in

TABLE 3

Temperature of Endotherm Peak and Estimated Enthalpy Melting and Freezing of Thiokol HAP

Sample Wt. (mg)	Heating/ Cooling Rate °/Min	1st Heat °C	1st Heat cal/g	2nd Heat °C	2nd Heat cal/g	3rd Heat °C	3rd Heat cal/g	1st Cool °C	1st Cool cal/g	2nd Cool °C	2nd Cool cal/g	3rd Cool °C	3rd Cool cal/g
28.4	2	94	33.7	94	29.1	94	31.4	63	36.0	59	13.4	59	13.4
24.3	2	94	32.5	93	30.5	92	29.1	64	25.8	55	19.2 / 32.6	35	5.2 / 18.6
25.4	6	97	39.5	96	33.7	96	35.0	63	31.1	56	40.7	59 / 40	12.9 / 18.3 / 31.2
8.6	6	95	25.3	94	23.5			56	12.5	62	35.0	60 / 44	14.9 / 20.8 / 35.7
								49	16.1	58	8.2		
									28.6	44	15.1		
25.9	6/2	94	44.4	96	37.2	96	35.4	61	29.3	60	23.3	61	13.8
										59	15.9 / 18.5 / 34.4	52	17.3 / 31.1

Table 3. The following observations apply to Thiokol HAP.

1 - The same sample can exhibit both single and double freezing exotherms when it is cycled through the melting point.
2 - The combined area of double peaks is about the same as the area of a single peak.
3 - The initial freezing peak usually occurs between 56 and 63° irrespective of whether it is the only peak or not.
4 - Where two peaks occur, the second (i.e. lower temperature) peak is usually larger and may occur anywhere over a fairly large temperature range (35-60°).
5 - Occurrence of a double freezing does not appear related to either sample size or heating rate.
6 - The likelihood of a double peak seems to increase on successive thermal cycles.
7 - Often a small irregularity is noted in the shape of the melting endotherm suggestive of a possible double melting peak, Figure 3.

The higher apparent heat content (larger DTA area) obtained upon initial melting is attributed to the poorer thermal conduction of particulate HAP as opposed to the conduction with a large fused slug of HAP.

Heating of Navy Material used for the low temperature endotherm study was continued up to 120° at 2°/min. Samples were then cooled and reheated at 6°/min. The melting endotherm peak was observed at the following temperatures, with a heat of fusion of 27 cal/gm.

Melting Point of HAP for Different Particle Sizes

Particle Size (Microns)	Initial Melt (2°/min)	2nd Melt (6°/min)
53-149	90°C	93°C
149-120	91	93
210-354	91	93
354-500	92	95
> 500	91	94
> 500	92	95

The temperature at which the initial melting endotherm peak occurs tends to increase with particle size contrary to behavior at the lower temperature endotherm. Furthermore, this trend is repeated during the second melt where one would not expect any effect due to particle size because of previous melting and freezing. Two obvious possibilities exist:

1 - Contamination or reduced purity of the smaller particles.

2 - Some form of memory effect associated with initial freezing (e.g. more fissures or stratification occurring during freezing at the larger particles).

MASS SPECTROMETER STUDIES

Samples of Thiokol HAP have been heated up to $80°C$ in the direct inlet probe to a Bendix Time-of-Flight mass spectrometer. All samples were heated at $6°C/min$ in order to compare the data with the DTA data. In preliminary analyses it was apparent that the only unusual phenomena was a rapid increase in the H_2O^+ peak at about $65°C$. This has been studied further by cyclic scanning, as a function of temperature, the m/e ranges 14-18, 14-46, 28-32, 38-46, and 0-100. Only the ranges 14-18 and 0-100 showed any unusual activity in the temperature range studied. A typical spectrum between 14 and 18 at $40°C$ is shown in Figure 4. From these data it is apparent that there is no unusual activity in this temperature region. Figure 5, however, recorded at $65°C$, shows the rapid increase in the H_2O^+ and OH^+ peaks associated with a fairly sudden liberation of H_2O from the HAP. The agreement between the mass spectrometrically observed evolution of H_2O at $65°C$ and the abrupt weight losses seen by TGA at $65°C$ implies that these two observations are probably of the same phenomenon.

Water does not reappear at $65°C$ if the material is thermally cycled. However, a small H_2O^+ peak can be observed if the material is stored for several days (~ 4 days). This undoubtedly is due to the extreme hygroscopicity of HAP thereby allowing the stored sample to pick up some water.

All Navy samples tested showed an evolution of carbon tetrachloride in the region of $65°C$. No water evolution was recorded since the Navy preparation is done in nonaqueous organic solvents whereas the Thiokol preparation is based on an aqueous system.

THERMAL MICROSCOPY STUDIES

Upon heating at $2°C\ min^{-1}$ samples (multicolored under polarized light) changed appearance at $55°-60°C$ to give a reasonably uniformly dark image. This color transformation is best interpreted as due to a crystalline phase change. If the sample were then melted and cooled to room temperature the original multicolored state was produced. On a second heating cycle this color

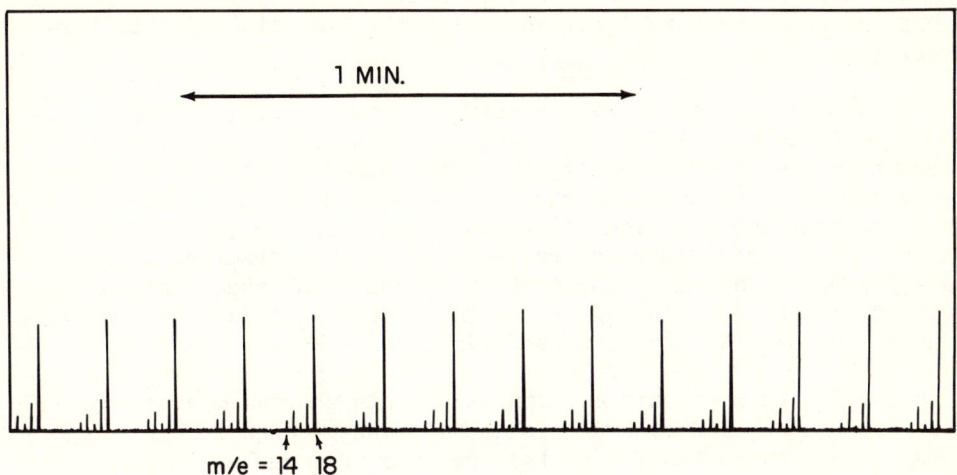

FIG. 4 - Mass spectrometric data for cyclic scans between m/e = 14 to 18 for Thiohol HAP held at $40^\circ C$.

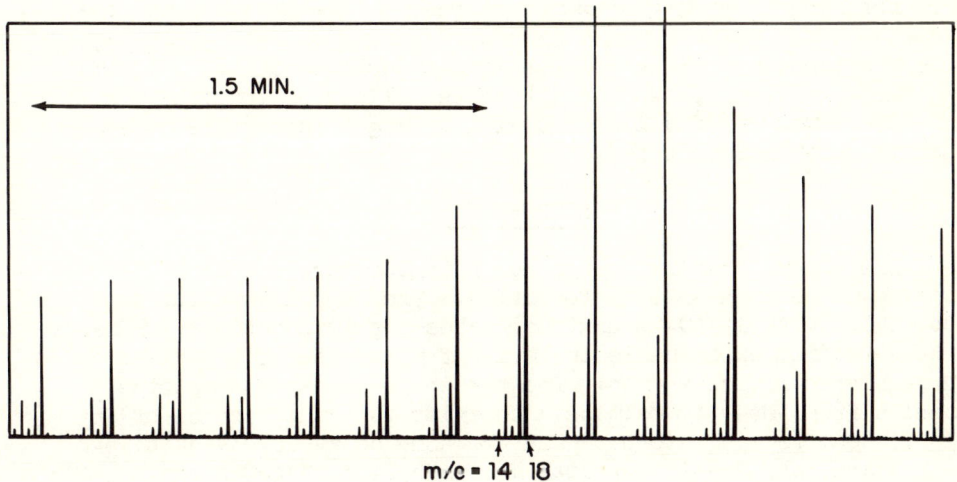

FIG. 5 - Mass spectrometric data for cyclic scan between m/e = 14 to 18 for Thiokol HAP being heated at $60^\circ C$ min^{-1}. The region of intense m/e = 18 peaks is at $65^\circ C$.

remained as the sample temperature passed through the range 50-70°C indicating that no phase change occurred. These visual observations are completely in accordance with the DTA data reported earlier.

The melting process was invariably gradual, spread over a temperature range of ~ 3°C, indicating the possibility of impurities present in the HAP. Freezing, at a controlled cooling rate, also appeared over a temperature range of several degrees. With the Navy HAP only one freezing point was noted. However, with the Thiokol HAP double separate freezing points were observed approximately 70% of the time. The instant at which the second solid phase forms is shown in Figure 6. These data again are in accordance with the DTA data discussed earlier.

One of the most unique features of this thermal microscopy study is that both Navy and Thiokol HAP presented an inhomogeneous appearance when molten due to the presence of a number of small dark circles which invariably appeared to have a point source of light at their center. It is unlikely that the black circles are not trapped gas since they can also be seen at 20°C. The circles were usually noticeable prior to melting and tended to remain visible during cool down even after freezing. The greatest number of circles was present in liquid HAP. The circles themselves were fluid and capable of uniting to form larger circles or dividing into smaller ones. On several occasions one or more of these circles were observed to spontaneously disappear.

The identity of the circles is difficult to define although the fact that they show a bright center allows us to state that they are composed of a liquid having a larger refractive index than the host HAP melt (4).

DISCUSSION

The reversible low energy lower temperature endotherm is considered to be a phase transformation. The very low weight loss associated with it (0.2% maximum) makes chemical activity unlikely but is readily explainable in terms of a release of volatiles through crystal relaxation. The possibility that the endotherm results from impurity melting also exists but the absence of any associated freezing exotherm and the time delay involved in recovery are more consistent with a crystalline phase change.

The origin of the carbon tetrachloride released upon an initial heating of Navy HAP is unlikely to be caused by diffusion of carbon tetrachloride into the material since the Thiokol material soaked in carbon tetrachloride did not exhibit a similar weight loss. One would, in fact, not expect the large carbon tetrachloride

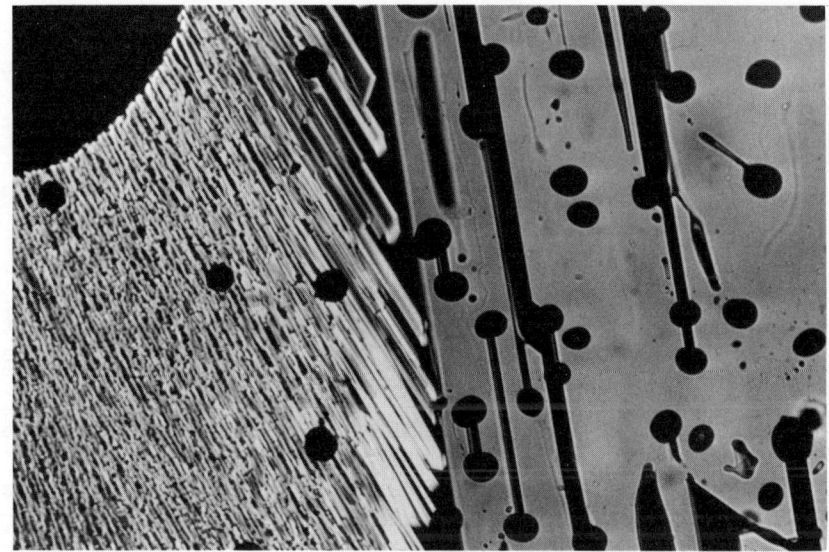

FIG. 6 - Thermal microscopic photograph of HAP at the instant of the second phase crystallizing during a controlled cool of the sample.

molecules to diffuse very readily. It must be concluded therefore that the carbon tetrachloride present in the Navy HAP was occluded internally at the time of crystallization. The water which is observed in the mass spectra of both batches of material is likely due to the hydroscopic nature of HAP. Unlike carbon tetrachloride, water could be expected to diffuse readily into the HAP crystals. The appearance of water in bursts during mass spectrometer measurements suggest that water is present internally. The fact that the weight loss in parts per million for Thiokol HAP is larger with the larger particle sizes suggests that the water pick up is not a function of total surface area as much as it is a function of average volume and leads to the suspicion that some water was introduced into the Thiokol sample at the time of crystallization. A somewhat contrary indication is given by the Navy HAP where the melting point peak temperatures suggest reduced purity for the smaller particle sizes. For a unit weight, smaller particles have larger surface areas and the evidence suggests that Navy HAP acquires some water through surface pick up. In this connection it will be noted that Baker analyzed spectrophotometer quality carbon tetrachloride contains up to 0.05% water so that the contamination might well result from prolonged exposure to carbon tetrachloride rather than to any atmospheric exposure.

In the absence of evidence to the contrary both the water evolution at the assumed phase change and the more gradual weight loss sometimes observed below the melting point are attributed to a drying out of the material at elevated temperatures.

The microscopic observation reinforces the conclusion that the low temperature endotherm of HAP is due to a phase change. The dark circles observed in HAP are presumed to be an effect of water although no completely satisfactory explanation is possible at present. If the dark spots are water or a water-HAP solution the question arises as to why such a phenomenon should be localized rather than uniform. The weak attractive forces between water and HAP would be affected by HAP's changing from an ionic crystal to a liquid. The assumed phase change might influence the mutual solubility of water and HAP. The circular shapes suggest that the material has a high surface tension with respect to the host HAP. The observed disappearance of these circles undoubtedly contributes to the low weight loss observed with HAP at temperatures above the phase change. The greater incidence of dark circles in molted HAP is probably due to the greater optical clarity of molten HAP which permits one to see what previously was hidden. A less satisfactory explanation of the inhomogeneity could be based on a contaminant other than water, e.g. perchloric acid or hydroxylamine so that the observed circles represent HAP-contaminant or water-contaminant solutions. Such a contaminant could result from processing, from hydrolysis reactions or from auto decomposition.

The confirmed presence of water naturally invites speculation regarding its possible effects on HAP. The drying method suggested by Thiokol of vacuum heating up to 70° suggests a weak interaction between water and HAP. A HAP sample which was allowed to dissolve itself through atmospheric moisture pick up still looked wet after 1 hour at 120° at 760mm pressure. Should any hydrolysis reaction occur, however, it might result in the formation of perchloric acid monohydrate by a reaction such as:

$$NH_3OHClO_4 + 2H_2O \rightleftharpoons NH_4OH + H_3OClO_4 + \frac{1}{2} O_2.$$

In the case of ammonium perchlorate increased reactivity has been attributed to the presence of small quantities of perchloric acid monohydrate in an earlier investigation (5). The reversibility of the specific reaction indicated would depend on the extent to which oxygen can diffuse out of the crystal. When oxygen is unavailable for recombination, the resulting ammonium hydroxide and perchloric acid monohydrate would combine upon melting to give AP and water.

In conclusion it has been clearly shown that HAP undergoes a crystalline phase change at about 60°C which is reversible after about three days. All samples of tested HAP contain occluded pockets of some high surface tension liquid, possibly a saturated aqueous solution of HAP, perchloric acid or hydroxylamine. Some volatiles are released during the phase change presumably due to the crystal relaxation. Presently all indications are that H_2O is the major volatile released.

Thiokol HAP also very clearly exhibits two freezing points which have been clearly identified as the solidification of two separate solid phases. Navy HAP does not exhibit this phenomenon.

ACKNOWLEDGEMENTS

This work was supported by the U. S. Air Force, Rocket Propulsion Laboratory, Contract No. F04611-68-C-0068. We would also like to thank the Thiokol Chemical Corporation and the U. S. Navy Propellant Plant for making samples of HAP available to us.

REFERENCES

1. Dickens, B., N.B.S. Washington, D. C.
2. R. Foscante, U.S.A.F. R.P.L., private communication.
3. Maycock, J. N., Pai Verneker, V. R., Anal. Chem., 40, 1938, (1968).
4. Van de Hulst, M. C., "Light Scattering by Small Particles," Wiley and Sons, New York, (1957).
5. Maycock, J. N., Pai Verneker, V. R. and Rouch, L. L., Inorg. Nucl. Chem. Letters, 4, 119, (1968).

GLASS TRANSITION PHENOMENA IN VAPOR QUENCHED Bi-Se ALLOYS

Mark B. Myers, John C. Schottmiller, William J. Hillegas

Xerox Corporation, Rochester Research Laboratory,

Xerox Square - 114, Rochester, New York 14603

INTRODUCTION

Concepts of the nature of the glassy state of the chalcogenide glasses are primarily developed from the viewpoint of supercooling a melt from above the liquidus temperature through the glass transition range. It is tacitly assumed that the molecular configurations in the glassy state are only slightly perturbed from those of the liquid state during this quenching process. For many materials such as the As-S and As-Se glasses this appears to be the case. Thermal and optical studies on these systems have been able to discern only small differences due to quenching effects.

This assumption is difficult to apply to the chalcogenide system Bi-Se. As indicated in Fig. 1, a liquid or melt in the Bi-Se system is immiscible in the composition range of 72 to near 100% Se (1). A single phase liquid cooled through this region will separate into two phases, i.e. a selenium rich liquid and a Bi_2Se_3 rich liquid. The liquid immiscibility coupled with the rapid crystallization of the Bi_2Se_3 rich phase prohibits homogenous glass formation from supercooled melts. The vitreous state can only be realized by a very rapid quench such as occurs during a vapor deposition (2). Further, the consideration of metastable extensions of the liquid immiscibility boundries on the phase diagram indicates that it homogenous single phase Bi-Se glasses are realized by vapor quenching, they will not only be thermodynamically unstable with respect to crystallization, but they will also be unstable with respect to subliquidus two-liquid phase separation (3).

A useful characterization of the nature of the glassy state in chalcogenide materials has been the study of the glass transition

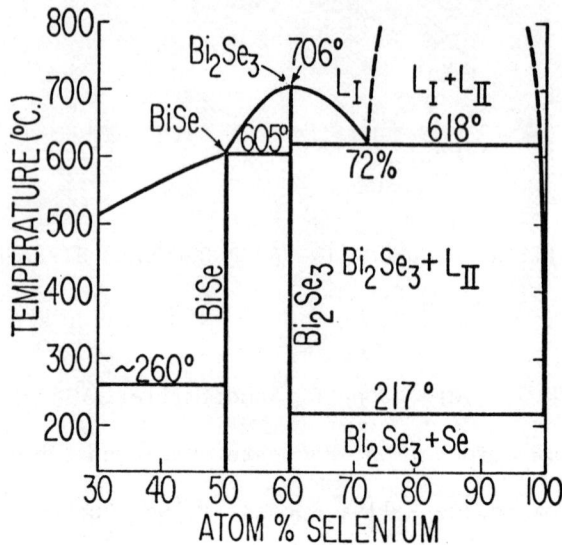

Figure 1   Phase diagram for the Bi-Se system.

phenomena by techniques such as differential thermal analysis (4). The variations of the glass transition temperatures with respect to changes in composition have been found to reflect changes in molecular constitution; hence, their study provides very useful insights concerning the state of the materials. In this study, the glass transition phenomenon of Bi-Se vapor quenched glasses is examined as a function of composition, and a comparison is made to the glass transition behavior of the structurally similar As-S and As-Se glasses which can be prepared by both melt and vapor quenching techniques.

EXPERIMENTAL

The glass transition temperatures were observed by the change of Cp as indicated by a Perkin-Elmer DSC-1B differential scanning calorimeter. The transition temperatures were defined by an extrapolated onset technique and corrected for the nonlinearity of the DSC-1B temperature read-out by calibration with metals with melting points appropriate for the temperature range of interest. Heats of crystallization of the Bi-Se films were determined by comparing the areas of the exothermic peaks to those of In standards (5); and the specific heat measurements were made by comparison to a sapphire standard as prescribed by the Perkin-Elmer Corporation (6). A heating rate of 10°C per minute was used for all of the calorimetric determinations.

The vapor quenched Bi-Se glasses were prepared by coevaporation of ASARCO 99.999 percent pure elemental components. This technique involves the simultaneous evaporation of bismuth and selenium from separate heating sources and a subsequent reaction and quenching of their vapor phases at a constant substrate temperature. The substrate temperature was held at $50 \pm 2^{\circ}C$, and the evaporations were done in a bell jar vacuum system at 10^{-6} torr. The As-Se and As-S glasses were prepared by reaction of Gallard Schlesinger, 99.9999 percent pure elemental components in silica glass ampuls at $600^{\circ}C$ for 24 hours and air cooling to room temperature.

To define a specific thermal history for the melt quenched glasses, the glass transition temperatures were measured 24 hours after the samples had been heated above the liquidus temperature and then rapidly cooled to room temperature. Similar thermal histories could not be imposed on the vapor deposited films because the analyses would no longer be representative of vapor quenched materials. The time interval between preparation and thermal analysis of the Bi-Se films was several weeks, thus they displayed rather prominent exothermic effects at Tg characteristic of relaxed samples. Although the combination of the thermal history and heating rate factors produced fluctuations and uncertainities in the measurements, these were small compared to the systematic changes found by varying the composition of the materials.

Finally, the compositions of all the vapor deposited materials were determined by x-ray fluorescence analyses. The broad halo of an electron diffraction pattern coupled with the absence of detectable diffraction rings established the noncrystalline nature of the films.

RESULTS & DISCUSSION

The thermal transitions observed in Bi-Se vapor-quenched films are demonstrated in the DSC-1B trace shown in Fig. 2. The thermogram indicates all the thermal phenomena characteristic of a glassy material. A glass transition occurs at $81^{\circ}C$, followed by a metastable liquid region, crystallization starting at $125^{\circ}C$, and two melting peaks at 212 and $219^{\circ}C$. The aspects which differ from that expected from a melt-quenched glass are the slight dip in the curve above Tg and the existence of two melting peaks where there should only be the one corresponding to the eutectic temperature of $217^{\circ}C$.

The variation of Tg with composition in the Bi-Se system is shown in Fig. 3. These data are compared to the Tg's for the melt quenched As-S and As-Se glasses. The similarity in the variation of Tg with composition in certain ranges is believed to be due to the similarity in the way the molecular configurations are modified.

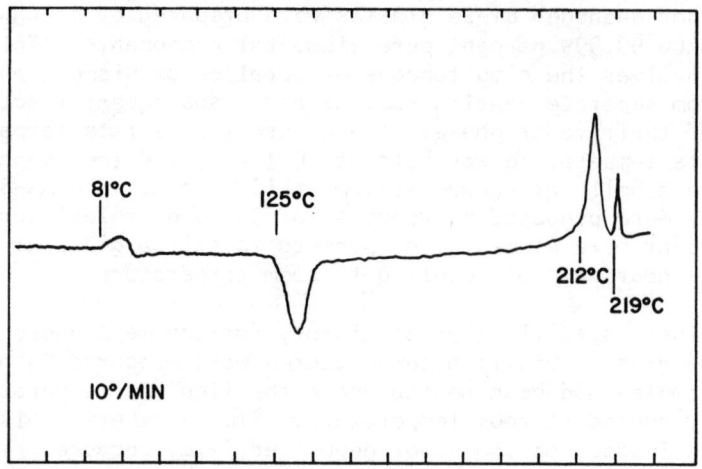

Figure 2 DSC-1B trace of vapor quenched vitreous $Bi_{.09}Se_{.91}$

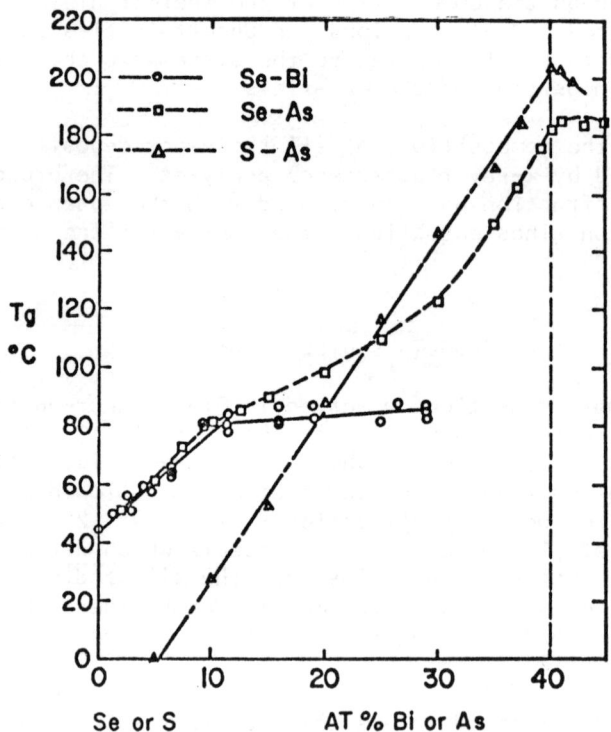

Figure 3 Glass transition temperatures for the vapor quenched Bi-Se and melt quenched As-S and As-Se glasses.

For example, the compositional dependence of Tg in the As-S and As-Se systems should be compared in the region of 30 to 40% As and the As-Se and Bi-Se systems in the region of 0 to 10% As or Bi.

The compositional variation of the glass transitions in the As-S and As-Se systems have been explained (4) by variations in the molecular configurations such as illustrated in Fig. 4. As the arsenic content increases, the Tg increases in both cases due to branching of the sulfur or selenium chains by trivalent arsenic to form $AsS_{3/2}$ or $AsSe_{3/2}$ linkages. In the intermediate composition range they differ, however, because the linkages occur randomly in the As-S system and non-randomly in the As-Se system.

The initial increase in the Tg's for the vapor-quenched Bi-Se materials is similarily attributed to the bismuth branching of selenium chains. The near invariance of Tg with composition as the bismuth content increases is due to a more pronounced tendency for non random association of like molecular types in this system than in that of the As-Se and As-S systems. This is compatible with the observation of liquid immiscibility in the system.

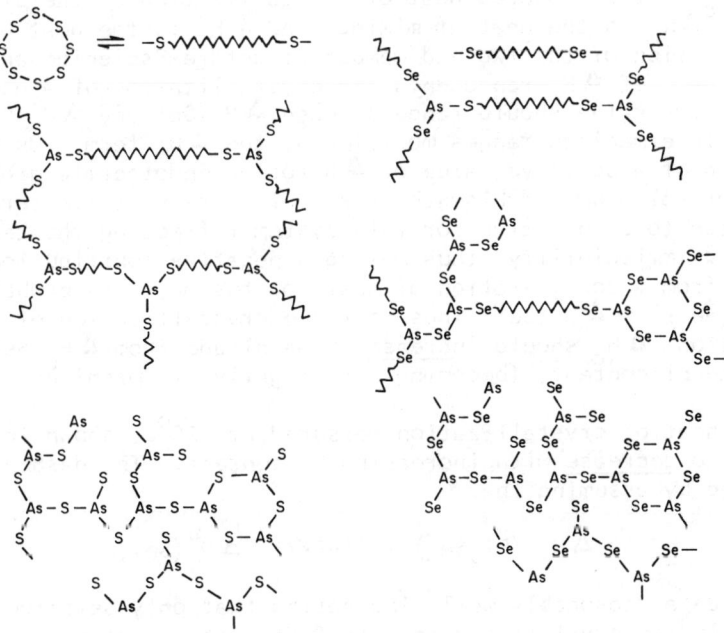

Figure 4 Variations in molecular configurations with increasing As content in the $S-As_{.4}Se_{.6}$ and $Se-As_{.4}Se_{.6}$ systems.

The observations of the variation of Tg with composition indicate little difference in thermal properties between these vapor-quenched vitreous Bi-Se materials and the melt quenched As-Se and As-S glasses. However, rather subtle but important differences become apparent by considering in detail the Bi-Se thermal transitions shown in Fig. 2.

Assuming that the Bi-Se film is a homogenous vitreous solution up to $125°C$, the crystallization at that temperature can be represented by the reaction

$$Bi_x Se_y \rightarrow (1-5x/2) \, Se_{(crystal)} + (5x/2) \, Bi_{.4}Se_{.6 \,(crystal)}$$

where $x + y = 1$.

The equation is written to denote that the solution crystallizes to the end members Se and $Bi_{.4}Se_{.6}$, which are a subbinary of the system Bi-Se. The heat evolved during the crystallization would be given by

$$\Delta H_c (Bi_x Se_y) = -\Delta H_m (Bi_x Se_y) + (1-5x/2) \, \Delta H°(Se)$$
$$+ (5x/2) \, \Delta H°(Bi_{.4}Se_{.6})$$

where ΔH_c is the measured heat of crystallization of the $Bi_x Se_y$ solution, ΔH_m is the heat of mixing, and $\Delta H°$ is the heat of crystallization of the two end member structures selenium and bismuth triselenide. If ΔH_c represents the crystallization of a homogenous solution, its value should range between $\Delta H°(Se)$ and $\Delta H°(Bi_{.4}Se_{.6})$ with the intermediate ranges modified by the ΔH_m term. Using the convention of a positive value of ΔH for an endothermic process, the $\Delta H°$'s for selenium and bismuth triselenide are negative terms. ΔH_m is expected to be positive for this system reflecting the tendency for liquid immiscibility, thus making a negative contribution to ΔH_c. Further, from a consideration of heats of fusion it is probable that $\Delta H°(Bi_{.4}Se_{.6}) > \Delta H°(Se)$. Thus, for the crystallization of a homogenous solution, ΔH_c should <u>increase</u> in magnitude from $\Delta H_c(Se)$ with increasing Bi content, (becoming more negative in magnitude).

The heat of crystallization measured at $125°C$, shown in Fig. 5, is found to <u>decrease</u> with increasing Bi content. The dashed line, calculated by assuming that

$$\Delta H_c (Bi_x Se_y) = (1-5x/2) \, \Delta H°(Se),$$

fits the data reasonably well, indicating that only selenium is involved in the crystallization. In fact, the deviation of the data below the line at higher concentrations of bismuth suggests that part of the selenium was previously crystallized. Hence it appears that the $Bi_{.09}Se_{.91}$ solution was decomposed and partially crystallized

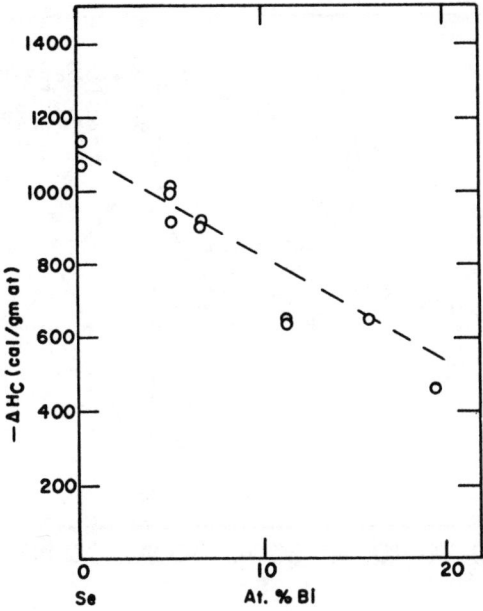

Figure 5 Heats of crystallization for Bi-Se films as a function of composition.

before the thermal event at $125°C$.

This observation is clarified by examining Cp in the region above and below Tg for both vapor quenched (Se, Bi-Se and As-Se) and melt quenched (Se and As-Se) glasses. The data are shown in Fig. 6. The large endotherms for the selenium and Bi-Se films are due to room ambient relaxation effects on Tg and are not of concern in this discussion. The important consideration is the behavior of Cp approaching Tg and in the metastable liquid region above Tg. Normal behavior is observed for both the melt and vapor-quenched selenium and As-Se glasses. The approach to Tg from the glass phase is normal for the Bi-Se film; however, in the metastable liquid region just above Tg the film becomes unstable and undergoes a very rapid relaxation with a corresponding exothermic heat effect.

This behavior can be further elucidated by consecutive thermal analyses of the films. That is, heating the samples through their Tg's into the metastable liquid region, cooling back into the glass regime, then reheating the samples through Tg. The results are shown in Fig. 7.

The $Bi_{.09}Se_{.91}$ sample heated through Tg has undergone an

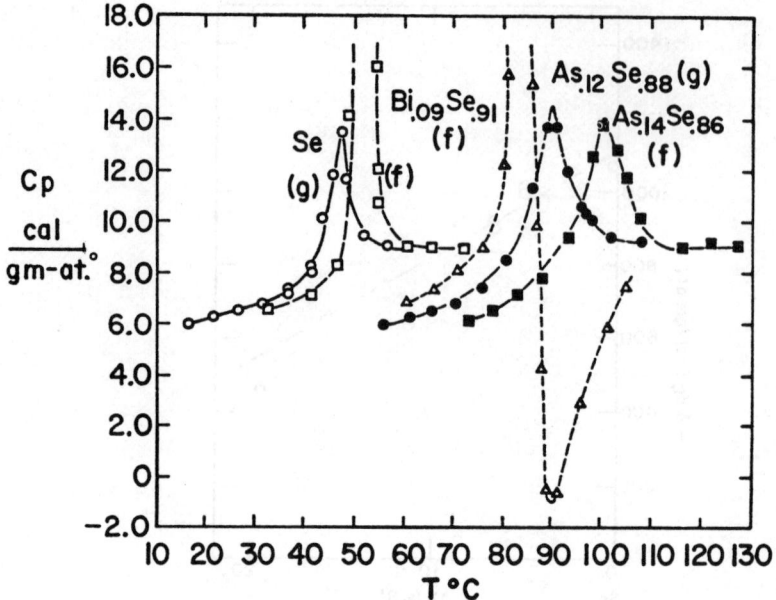

Figure 6 Specific heat data for vitreous Se, Bi-Se, and As-Se in the vicinity of Tg. Vapor quenched-(f), melt quenched-(g).

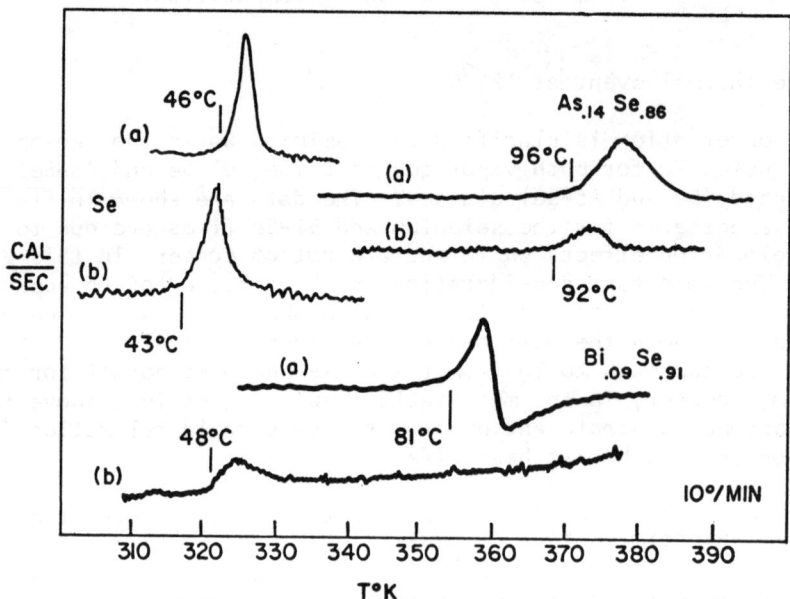

Figure 7 Consecutive thermal analyses on vapor quenched films of Se, Bi-Se and As-Se.

irreversible change in the nature of Tg indicating an irreversible separation in the glass structure. The Tg on the subsequent determination is characteristic of a glass containing less bismuth, i.e. $Bi_{.02}Se_{.98}$. In contrast to the Bi-Se sample, the Tg's of the vapor-quenched selenium and As-Se samples remain unchanged with thermal cycling, except for ambient relaxation effects. It is concluded that the different thermal behavior of the vitreous Bi-Se materials is not directly attributable to vapor quenching but to the specific thermodynamic nature of their binary solutions.

Although the Bi-Se samples heated through Tg still visually appear quite glassy, electron diffraction analyses indicate the presence of very small and poorly developed crystallites. The exact mechanism of the decomposition and crystallization has not been established; however, considering the nature of the system, it would appear to involve crystallization assisted by a precursor liquid-liquid phase separation.

Finally, the two melting peaks at $212°C$ and $219°C$ in Fig. 2, are due to the nonequilibrium melting of crystalline selenium in two different compositional environments. The part of the selenium which separated and crystallized with the bismuth triselenide at Tg is not in thermodynamic equilibrium with the selenium which crystallized at $125°C$. These two melting peaks blend into one equilibrium eutectic peak by thermal cycling the sample.

CONCLUSIONS

It is seen that the thermal properties of these vapor quenched Bi-Se glasses are basically the same as melt quenched glasses. Subtle but significant differences in thermal properties arise due to the unusual nature of the glassy states of these materials. The rapid quenching of the vapor deposition process allows glassy states to be realized in a system such as Bi-Se which could not be realized by a melt quenching technique because of kinetic limitations.

ACKNOWLEDGEMENTS

The authors wish to acknowledge the assistance of J. O'Neill, F. Ryan and T. Taylor with the experimental work.

REFERENCES

1. M. Hansen, <u>Constitution of Binary Alloys</u>, McGraw-Hill Book Company, p. 335 (1958).
2. J. C. Schottmiller, C. C. Bowman, and C. Wood, J. Appl. Phys. <u>39</u>, 1663 (1968).

3. J. Cahn, J. Amer. Ceram. Soc. $\underline{52}$, 118 (1969).
4. M. B. Myers and E. J. Felty, Mater. Res. Bull. $\underline{2}$, 535 (1967).
5. American Institute of Physics Handbook, McGraw-Hill Publishing Co., Inc., New York (1963).
6. Perkin-Elmer Corporation, Thermal Analysis Newsletter, $\underline{3}$, Norwich, Conneticut.

KINETICS OF DISSOLUTION OF MAGNESIUM OXIDE IN A SODIUM SILICATE MELT

Samuel F. Hulbert and Furman H. Brown

Division of Interdisciplinary Studies, College of

Engineering, Clemson University, Clemson, S.C.

ABSTRACT

A continuous measurement process is needed for studying the kinetics of refractory corrosion. Dissolution kinetics of magnesium oxide in a sodium silicate melt in the temperature range of 1100°C to 1220°C were studied. The rate of reaction was observed by using a continuous measurement process which was designed especially for this investigation but can be used for other refractory corrosion reactions. With this process, the weight loss can be measured at any time during the experiment, thus providing many advantages over previous techniques in which only the starting and ending points can be measured.

A typical graph of weight loss versus time can be divided into two parts: transient and steady-state. Diffusion through the boundary layer is the controlling process after the corrosion reaction has reached steady-state. The thickness of the boundary layer is controlled by the viscosity of the melt, the difference in the density of the melt and refractory, and the velocity of the melt. Porosity does not have an effect on the rate of dissolution as long as the pores are small so that the melt cannot enter the structure of the refractory. The mechanism of the dissolution of magnesium oxide in a sodium silicate melt can be described on the microscopic level using the screening concept.

INTRODUCTION

Each year millions of dollars and thousands of man-hours are lost because of refractory corrosion. The results of refractory

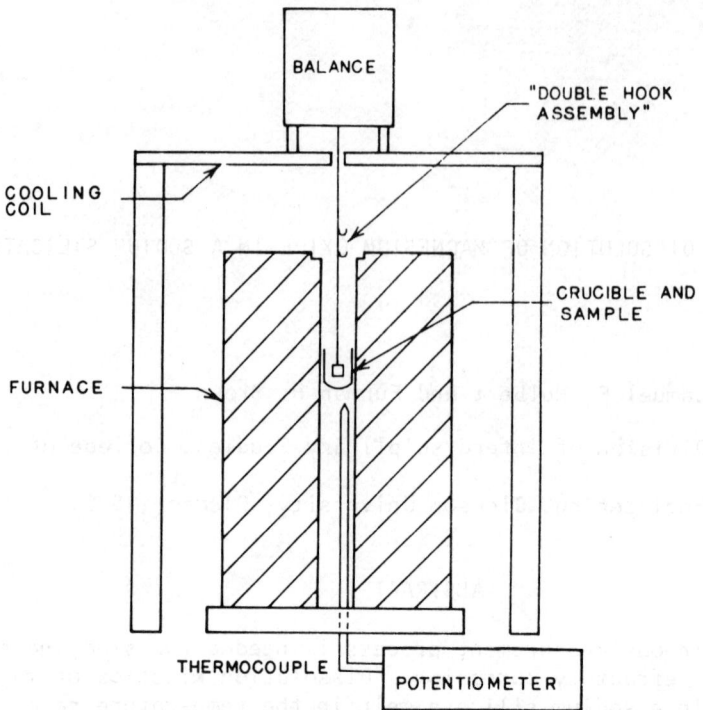

Fig. 1. Schematic of test equipment.

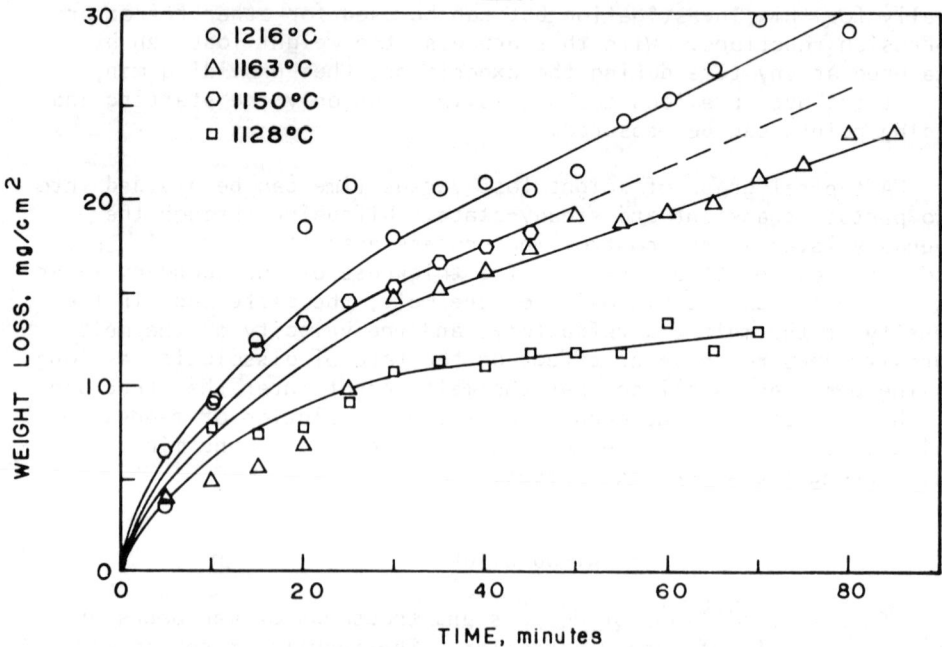

Fig. 2. Weight loss of single crystal MgO in a sodium-silicate melt.

corrosion have two major effects on the campaign of an industrial furnace: 1) it causes hours of "downtime" and the expense of repairs, 2) it lowers the quality of the product through the introduction of impurities. Most of the work reported to date on refractory corrosion has been directed toward the selection; by testing, of more suitable refractories for specific applications.[1-3] The general mechanism of refractory corrosion has been discussed by Moore and Heeley,[4] Scott,[5] Sosman,[6] and Endell, Fehling, and Kley.[7] Porosity and permeability effects on refractory corrosion have been reported by Busby,[8] Schurecht,[9] and Massengale, Mong, and Heindl.[10] The effect of the properties of the slag on corrosion rate has been investigated by Comeforo and Hursh,[11] and Laing, et al.[12]

Cooper and Kingeny[13] have discussed the fundamental features of dissolution of refractories in liquid slags and glasses. In the past, refractory corrosion experiments have not been designed to test the validity of the assumption that refractory corrosion is a diffusion-controlled process and to shed light on the significance of various parameters. Barham and Barrett[14] have recently reported a technique which is applicable for studying the fundamentals of dissolution of refractories in glasses and slags.

The rate of dissolution of magnesium oxide in a $Na_2O \cdot SiO_2$ melt were measured in an attempt to gain a better understanding of refractory corrosion and to obtain reaction data suitable for testing present theories of refractory corrosion. A modified Barham-Barrett[14] technique in which the MgO specimen is continuously weighed while immersed in the $Na_2O \cdot SiO_2$ melt was used to determine the rate of corrosion. The rate of dissolution was observed as a function of reaction temperature and microstructure of magnesium oxide.

EXPERIMENTAL PROCEDURE
Description of Equipment

A schematic of the refractory corrosion apparatus specifically designed and constructed for this investigation is shown in Figure 1.

A mullite-lined tube furnace (silicon carbide heating elements) served as the heating chamber. Placed within 3/4 of one inch of the corroding sample was a platinum versus platinum-13 percent rhodium thermocouple. The end of the combustion tube was sealed to reduce air currents which might disturb the sample.

The furnace, with its axis perpendicular to the floor, was placed in a metal support. A wooden balance table was built around the furnace in order to reduce the vibrations which might disturb the experiment. A cooling coil soldered to a brass plate bolted to the bottom of the table kept the top of the plate within 2°C of room temperature.

The analytical balance used is sensitive to ±0.1 milligram. A double hook assembly with the top hook 3/4 inch above the bottom was attached to the weighing unit of the balance. (The purpose of this assembly will be explained later.) Inside the furnace was a platinum crucible, suspended by two chromel or platinum wires depending on the temperature of the furnace. A mirror of chrome plated steel with a hole in the center was placed above the furnace in order that the experiment might be observed at any time.

Preparation of Samples and Glass

Single crystal MgO samples were cleaved from larger crystals, each being approximately uniform in shape and size. The average surface area of the rectangular crystals was 1.6 square centimeters, and the average weight was 225 milligrams. When single crystal magnesium oxide is cleaved, it does so along the (100) plane. Since these samples were single crystals, the porosity and permeability are assumed to be zero.

Reagent grade magnesium oxide powder with a particle size of -100 mesh was used to make the cylindrical polycrystalline samples. Samples were dry pressed in a tool steel die with a pressure of 3060 pounds per square inch. They were placed on platinum foil, sintered at 1450°C for either twelve or twenty hours, and then stored in a desiccator to prevent contamination prior to use in the dissolution tests. A mercury intrusion porosimeter was employed to determine the pore size distribution for representative polycrystalline samples. A group of seven samples from each of the polycrystalline groups was examined using ASTM[15] Test Designation: C20-46 to determine bulk density and apparent porosity.

The glass composition was 1.0 mole Na_2O to 1.0 mole SiO_2, (by weight 50.8% Na_2O and 49.2% SiO_2.) Both reactants were Reagent grade. The glass was processed in a platinum crucible to minimize contamination. The glass was premelted and crushed into cullet.

Test Procedure

Each magnesium oxide sample was heated to remove moisture, weighed, and measured for geometric surface area. The sample was wrapped with platinum wire. One end of the wire was looped and hung from the top hook of the double hook assembly previously mentioned. At this point, the magnesium oxide sample was approximately 1/4 of one inch above the glass (cullet) which has been placed in the platinum crucible. The temperature of the furnace was gradually raised to the desired temperature over a period of five hours. After the desired temperature was reached, a period of two hours was allowed for the temperature of the melt to become constant and for the melt to become homogeneous. The sample was then lowered into the melt by placing the wire loop on the bottom hook of the double hook assembly. The weight of the sample was then recorded at 2 minute intervals until the experiment was terminated. Immediately after the last reading, the sample was removed from the melt and allowed to cool. Warm water was used to dissolve the glass which remained on the surface and in the pores. After a period of twenty-four hours, the sample was dried by heating to 200°C and weighed. The percentage of weight lost was calculated.

Other Tests

A series of representative samples was prepared similar to those for the weight-loss versus time experiment. These were placed in a melt at predetermined temperatures. After a specific time -- ten, twenty-five, or fifty minutes -- the entire crucible was removed from the furnace and placed in a tempering furnace and slowly cooled to room temperature. An alumina crucible was used instead of platinum for this experiment.

A crucible was cut with a diamond saw at the approximate place where the sample had been frozen in the glass. The section was ground (without water) until the magnesium oxide appeared, and polishing was continued until a 600-grit finish was achieved. These samples were observed microscopically to determine the boundary layer thickness.

Spectrographic analysis was used to check the type of impurities in both the single crystal and polycrystalline samples, and an x-ray diffraction analysis verified the chemical nature of compounds formed during this reaction.

A contact angle experiment was performed by placing a small particle of glass on both single crystals and polycrystalline samples. The samples were raised to a predetermined temperature and left for a specific time, removed, placed in a tempering furnace, and cooled slowly to room temperature. The angle that the glass made with the magnesium oxide sample was measured. This experiment was performed to determine the effect that porosity had on the dissolution rate.

PRESENTATION AND DISCUSSION OF RESULTS

The dissolution rate curves for the single crystal MgO are

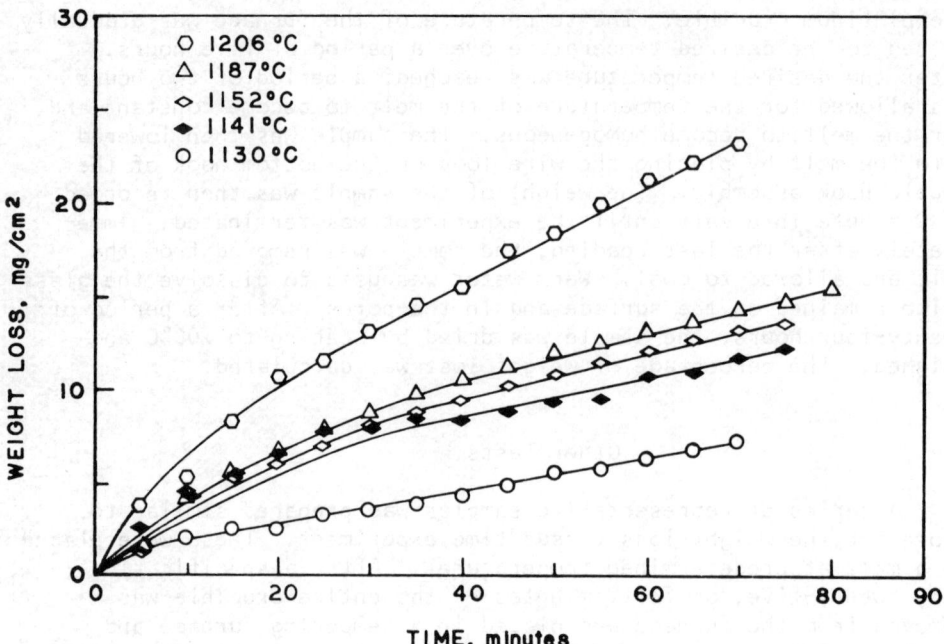

Fig. 3. Weight loss of polycrystal 1 MgO in a sodium-silicate melt.

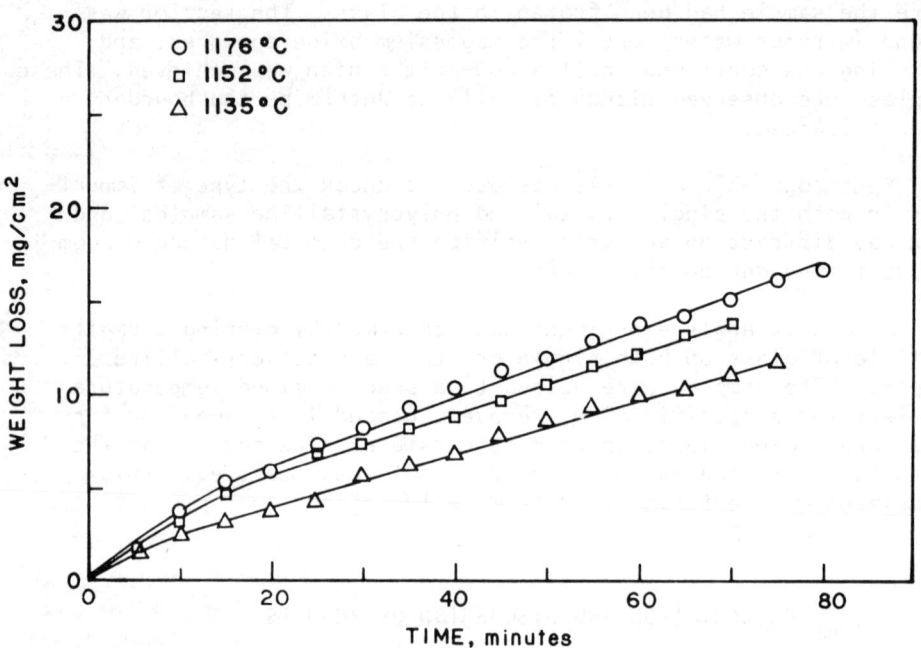

Fig. 4. Weight loss of polycrystal 2 MgO in a sodium-silicate melt.

shown in Figure 2 and for the polycrystalline samples in Figures 3 and 4. Table 1 summarizes the properties of each test set. The following points pertain to each set of dissolution rate curves. 1) the amount of dissolution increases with an increase in temperature, 2) the curves can be divided into two parts, 3) the slope of the linear portion of each curve increases with temperature, 4) the transient or non-steady state period lasts longer at higher temperature.

TABLE I

Properties of Test Samples

	Single Crystal	Poly-Crystal #1	Poly-Crystal #2
Density (GM/CM^3)	3.58	2.92	2.82
Apparent Porosity	---	21.0%	19.0%
Pore Size (Microns)	---	1.25-0.60	1.5-0.50
Largest Pores (Microns)	---	1.5	2.5
Boundary Layer Thickness at 1150°C			
10 minutes	---	.015 cm	.017 cm
25 minutes	.017 cm	.017 cm	.017 cm
50 minutes	.017 cm	.015 cm	---
Contact Angle at 1140°C			
5 minutes	---	10°	
20 minutes	10°	10°	
35 minutes	10°	10°	
Firing Schedule			
Temperature	---	1450°C	1450°C
Time	---	12 hours	20 hours

The first three observations can be explained on the basis that the reaction is controlled by diffusion through a product phase boundary layer. The fourth observation is more difficult to explain, since several events must take place before steady-state conditions are established: 1) the boundary layer approaches constant thickness, 2) the system recovers from the initial disturbances caused by the magnesium oxide sample being placed in the melt, and 3) all other irregularities, which may have entered the system at the start of an experiment, are damped out. The duration of the transient period may be determined by any of these events.

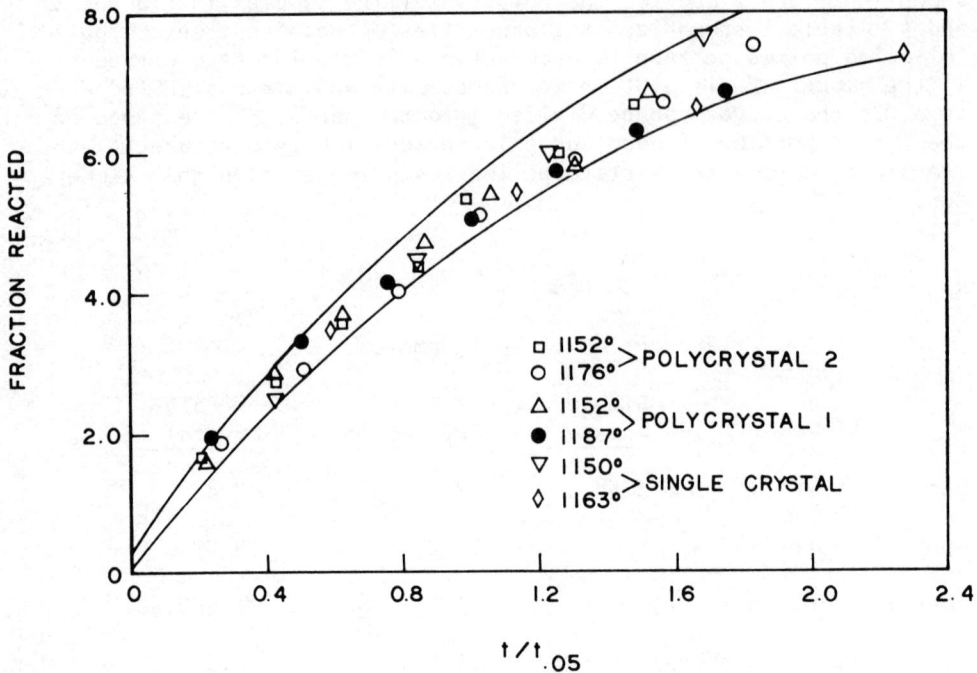

Fig. 5. Fraction of reaction completed with respect to the time for 5 percent reaction.

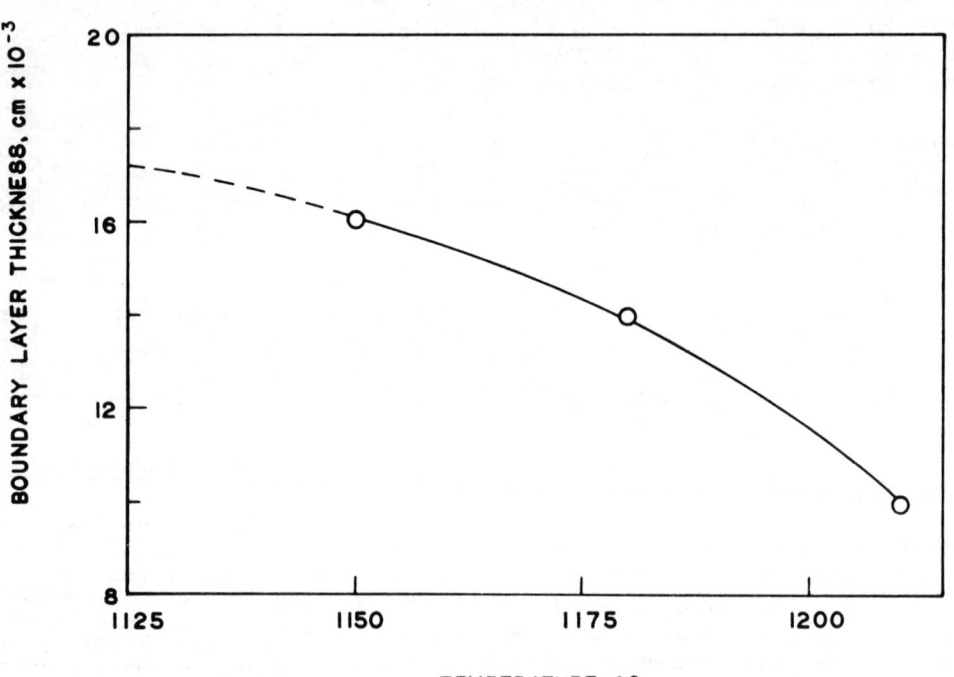

Fig. 6. Boundary layer thickness as a function of temperature

A van't Hoff[16] differential analysis performed on each set of data indicated that reaction equations based on the concept of an order of reaction are mathematically inappropriate for expressing and correlating the data. For solid-liquid, phase-boundary-controlled, contracting-cylinder geometry reactions[17] (the geometry of the polycrystalline MgO samples) the order of the reaction should be 1/2. This suggests that the dissolution of MgO in $Na_2O \cdot SiO_2$ glass is not phase boundary rate controlled over the range of boundary conditions employed in this investigation.

A plot of percent reaction versus a reduced time scale, t/t_x, where t_x is the time for five percent of the reaction to occur, is presented in Figure 5. Brindley[18] has concluded that if the curves closely approximate each other a similar set of reaction mechanism is operative in all cases. From this analysis it was concluded that the same set of mechanisms was occurring with each sample set tested.

An Arrhenius analysis was performed and apparent activation energies ranging from 37 kcal $mole^{-1}$ to 58 kcal $mole^{-1}$ were observed. These values are similar to that observed by Reed and Barrett[19] for the dissolution of alumina in a melt of 89% $CaO \cdot SiO_2$ - 11% Al_2O_3.

Several factors determine the effect of porosity on dissolution rate: 1) the total amount of void space in the sample, 2) the pore size and pore size distribution, 3) the pores which are open to the surface, and 4) the connected pores and the size of the necks which join them. In this investigation, the pores which were open to the surface were so small (1.0 - 2.5 microns) that the corroding medium could not enter the structure of the magnesium oxide. This is supported by the fact that the contact angle was the same for both the polycrystalline and single crystal samples over the boundary conditions investigated. Thus, porosity has little or no effect in this experiment.

The screening demand[20,21] and the free energy associated with it are driving forces for the dissolution of magnesium oxide in a sodium silicate melt. As MgO contacts the melt, the Mg^{+2} ion "sees" a structure in which it can be better screened and leaves the MgO structure. The product formed is a magnesium silicate which has a melting point of approximately 1500°C. As this product is formed, it precipitates out on the surface of the MgO. With increased time, this precipitate forms a layer around the MgO, and its thickness increases until a steady-state condition is achieved in which the amount of reaction product being formed on the surface of the MgO equals the amount of dissolution product (magnesium silicate) going into the corroding reservoir (sodium silicate melt).

As can be seen by the curves in Figures 2, 3, and 4, a typical weight loss versus time plot at constant temperature can be broken

Fig. 7. Photograph of boundary layer of polycrystalline #1 sample which remained in the melt for 10 minutes before it was cooled. Upper portion is $Na_2O \cdot SiO_2$ and the lower MgO. (Scale: 1" = 0.005")

into two parts. The first is a transient portion in which the thickness of the boundary layer is being established. When the sample is first placed in the melt, the chemical reaction process is rate controlling. As the boundary layer thickness increases, the rate of diffusion of reactants to the reaction sites begins to have a controlling effect on the dissolution rate. After a constant thickness of the boundary layer has been established, the reaction rate reaches a constant value. The thickness of the boundary layer is inversely related to reaction temperature. Figure 6 is a plot of boundary layer thickness versus temperature. Figure 7 is a photomicrograph demonstrating the presence of the boundary layer.

The steady state thickness of the boundary layer is influenced by 1) the viscosity of the melt, 2) the difference in density between the MgO and sodium silicate melt, 3) the flow velocity of the melt. These three factors are temperature dependent, but they have interacting effects on the boundary layer. As the viscosity decreases with increased temperature, the rates of diffusion of reactants to and products from the reaction site increases, however, a decrease in viscosity also decreases the friction between the solid boundary layer and the melt. The boundary layer thickness in the absence of a velocity field is proportional to viscosity raised to the 1/4 power.[13] An increase in temperature decreases the density of the melt and, therefore, increases the density gradient across the boundary. This tends to decrease the boundary layer thickness. An increase in the thermal gradient at high temperature causes an increase in the flow velocity of the melt which also decreases the boundary layer thickness. Diffusion through the boundary layer increases exponentially with temperature.

The amount of dissolution per unit is greater with the single crystal samples than it is with the polycrystalline samples. The single crystal samples were cleaved from a larger crystal exposing fresh surfaces; while the polycrystalline samples were dry pressed and sintered. During the sintering process the samples were raised to a temperature of 1450°C and then cooled slowly to room temperature. At this time, if the Mg^{+2} ion is not sufficiently screened, it can adsorb components from the furnace atmosphere. Thus, at this elevated temperature, which provides the thermal energy needed for the reaction, the polycrystalline samples have had an opportunity for better screening of Mg^+ ions.

The shape of the sample can also affect the rate of dissolution. Single crystals were rectangular in cross-section and the polycrystals were cylindrical. The sharp corners and edges of the rectangular samples can result in a faster rate of dissolution because the corners have more surface energy and are more apt to be broken off or eroded.

Another possibility which should be considered is the amount of impurities that may be present. A spectrographic analysis was run to check on impurities. Comparison of spectra revealed that the two types of samples used in this investigation were identical except for a small trace of sodium in the polycrystalline material. If there had been a significant difference in the amount and type of foreign ions, one would conclude that the less pure sample would be more susceptible to attack by the melt because these ions would enter the lattice as either interstitial or substitutional ions, causing the lattice to be stretched and distorted. This distortion would leave the Mg^{+2} ions less screened, and the energy needed to enter a magnesium silicate structure would be decreased. The impurities would also lead to an increase in the diffusion rate of the Mg^{+2} ions.

If both the single crystals and the polycrystalline samples were made from the same raw materials, and if the single crystals were grown so that they were the exact size needed and required no cleaving, it is postulated that the single crystals would have a greater resistance to attack, since the absence of grain boundaries would leave the Mg^{+2} ions better screened.

It should be noted that laminar flow does exist in the sodium silicate melt even though no external agitation is applied. This movement was observed visually, and in several instances caused prematured termination of an experimental run. Movement can be caused by either a concentration gradient or a temperature gradient. As the melt first reaches a temperature at which it is completely molten, there will be great movement as the system tries to minimize concentration differences. This convection can easily be reduced if the melt is allowed to remain at a constant temperature for a significant length of time. For this reason, the furnace was maintained at temperature two hours before the dissolution experiment was begun.

A second source of error cannot be controlled as easily as the first. The platinum crucible is heated by radient heat from the mullite tube, and the melt is heated by conduction from the walls of the crucible. Therefore, the melt closest to the crucible wall will be several degrees hotter than that in the middle. Fluid flow was observed as shown in Figure 8. The buoyancy resulting from the motion caused the magnesium oxide sample to "bounce." Three readings were taken and averaged to compensate for this effect.

In the two hour period while the furnace was at temperature before the actual dissolution experiment was begun, it is possible that corrosion could be taking place as a result of vapors being given off from the melt. However, the maximum weight change observed during this period was a gain of 0.1 milligram which is less than .05 percent of the total weight. It was concluded that

KINETICS OF MAGNESIUM OXIDE DISSOLUTION

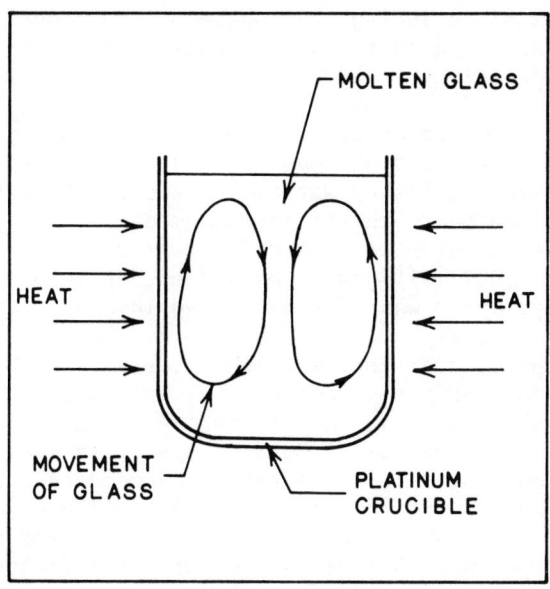

Fig. 8. Schematic showing the movement of molten glass caused by thermal gradients.

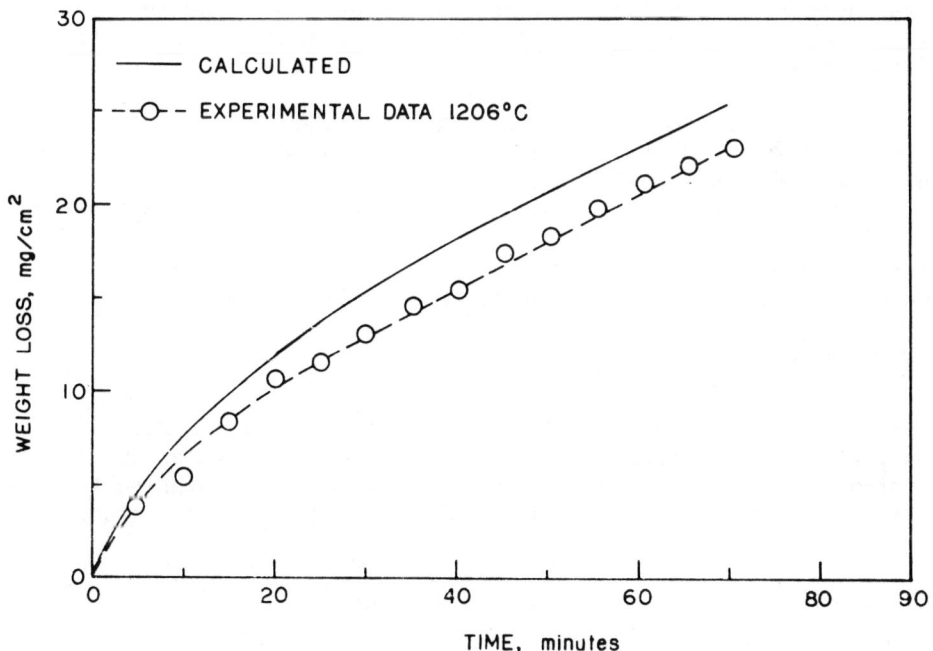

Fig. 9. Comparison between experimental data and mathematical calculations.

no corrosion took place during this phase of the experiment. It should not be overlooked that such a reaction may occur with different refractory corrosion systems.

Another source of error could have occurred during the twenty-four hour period when the glass was being dissolved from the surface of the magnesium oxide by water. It was possible that the water could have reacted with the magnesium oxide. To check this possibility, seven samples were placed in water for three days. Only an average of 0.2 milligrams was gained per sample; therefore, any error from hydration was considered negligible.

A contact angle experiment was run to determine whether a difference existed between the single crystal and polycrystalline samples, and whether there was a difference in the affinity of the samples for the corrosive melt. Both groups of samples were run similarly and simultaneously. The results showed that all samples (the exposure time varying from five to twenty-five minutes) had a contact angle of 10°. This suggests that porosity and surface roughness of the polycrystalline samples were not factors in this investigation.

In order to derive a mathematical expression for the dissolution of magnesium oxide in a corrosive melt, each step of the reaction must be considered. It was noted previsouly that the boundary layer determines the length of the diffusion path; thus the reaction rate at the MgO surface is inversely related to the thickness of the boundary layer.

From the reaction at the surface of the MgO, $MgO \cdot SiO_2$ is being formed causing the boundary layer to thicken in the melt. At the same time, $MgO \cdot SiO_2$ is being removed from the boundary layer and is entering the melt. Hence, both constructive and destructive reactions occur simultaneously. An equation, relating the change in the thickness of the boundary layer (x), in centimeters, to these two reactions, may be expressed as follows:

$$\frac{dx}{dt} = \frac{k_1}{x} - k_2 \quad \text{for} \quad t \leq t_{ss} \tag{1}$$

where k_1 (cm^2/min) is a constant related to the rate of reaction at the surface of the MgO, and k_2 (cm/min) is a velocity term which is related to the speed at which the boundary layer is entering the melt. The time, in minutes, at which steady-state conditions begin, is t_{ss}.

Separating the variables t and x in equation (1) yields

$$dt = \frac{x \, dx}{k_1 - xk_2}$$

and subsequent integration of the above gives

$$t = \frac{1}{k_2^2} [k_1 - k_2 x - k_1 \ln(k_1 - k_2 x)] + C$$

where C is the constant of integration.

To evaluate C, the initial conditions $t = 0$ and $x = 0$ are applied to produce

$$t = \frac{1}{k_2^2} [k_1 - k_2 x - k_1 \ln(k_1 - k_2 x)] - \frac{k_1}{k_2^2} (1 - \ln k) \quad (2)$$

Now consider a representative unit volume which is moving through the boundary layer. The volume (V) of this unit can be expressed by

$$V = \bar{v} t A \quad (3)$$

where $\bar{v}$ = velocity at which the unit travels, t = time, and A = area of the surface of the unit perpendicular to the direction of movement.

Differentiating (3) gives

$$\frac{dV}{dt} = \bar{v} A \quad (4)$$

But

$$V = m/\rho$$

where m = mass of the unit and ρ = density of the boundary layer. Here

$$\frac{dV}{dt} = \frac{d}{dt}(m/\rho) \quad (5)$$

Equating equations (4) and (5) yields

$$\bar{v} A = \frac{d}{dt}(m/\rho);$$

$$\bar{v} = \frac{d}{dt}(m/\rho A) = \frac{1}{\rho}\frac{d}{dt}(m/A)$$

Since k_2 is the rate at which the boundary layer is going into solution, k_2 equals the velocity $\bar{v}$. Thus,

$$k_2 = \frac{1}{\rho}\frac{d}{dt}(m/A)$$

where $\frac{d}{dt}(m/A)$ is the slope of the weight loss curve.

For steady-state conditions $\frac{dx}{dt}$, in equation (1), is equal to zero; therefore,

$$k_1 = x\,k_2$$

The polycrystalline Curve #1 at 1206°C was chosen for comparing the preceding derivation with experimental work. The slope of this curve was 0.25 mg/cm-min under steady-state conditions. The actual value for the density of the boundary layer was not known, but a good approximation was 2.5 gm/cm^3 (this was approximately the mean of the densities of the melt and the MgO). The steady-state thickness of the boundary layer at 1206°C was observed to be 0.011 cm. Therefore,

$$k_2 = \frac{1}{\rho}\frac{d}{dt}(m/A) = 1 \times 10^{-4}\ \text{cm/min},$$

and, hence,

$$k_1 = xk_2 = 1.1 \times 10^{-6}\ \text{cm}^2/\text{min}.$$

Substituting these values into equation (2), values for t and x can be found. See Table II. Equation (2) only applies when $t \leq t_{ss}$. For this experiment t_{ss} equaled 25 minutes.

The weight lost (w) will be the product of the velocity term associated with the reaction rate at the surface of the MgO, the density, and time (t); i.e.,

$$w = \frac{k_1}{x}\,\rho t \qquad \text{for } t \leq t_{ss} \qquad (6)$$

To be more precise, equation (6) could be written as

$$w_i = w_{i-1} + \int_{t_{i-1}}^{t_i} \frac{k_1}{x_i}\,\rho\,dt,\ \text{for } t \leq t_{ss}$$

For $t > t_{ss}$, the weight lost (2) is expressed as

$$w = w_t + k_2\rho\,(t - t_{ss})$$

where w_t is the weight lost during the transient period. (See Table II for values of w.)

TABLE II

Calculated Values for the Time Dependence
of the Boundary Layer Thickness
and Weight Loss

t (minutes)	x (centimeters)	w (mg/cm^2)
0	0	0
6	.0034	4.8
12	.0047	8.3
18	.0054	11.3
24	.0060	14.0
30*		15.5
40		18.0
50		20.5
60		23.0
70		25.5

* for $t \geq t_{ss}$; x = constant

It must first be pointed out that the calculated value of x differs from the observed value when the constants (k_1 and k_2) are evaluated from the steady-state portion of the observed weight loss curve. Note that the calculated values are comparable to the "Effect on boundary thickness" discussed by Cooper and Kingeny[13]. The derivation is based on the assumption that the density across the boundary layer is constant; however, in actuality, this is not the case. The density probably varies as Cooper and Kingeny predicted which complicates this analysis.

Secondly, it should be noted that the model predicts more weight loss than was experimentally observed. This behavior may be rationalized because the derivation was based on a specialized simple model in which factors such as density gradients, changes in the corroding medium, and other physical factors mentioned in the discussion were not taken into account. Considering the assumptions, the equations do well in fitting the experimental data. (See Figure 9).

CONCLUSION

For the dissolution of MgO in a $Na_2O \cdot SiO_2$ melt, a typical weight loss versus time curve can be divided into two parts: transient and steady-state. Diffusion through the boundary layer becomes the controlling process after the corrosion reaction has been established. The steady-state thickness of the boundary layer is controlled by the viscosity of the melt, the difference in density of the melt and refractory, and the flow velocity of the melt. Porosity does not have an effect on the rate of dissolution as long as the pores are small and the melt cannot enter the structure of the refractory. The mechanism of the dissolution of magnesium oxide in a sodium silicate melt can be described on the microscopic level using the screening concept.

REFERENCES

1. Vago, E. and Griffith, F., "The Corrosion of Zircon Refractories by Molten Glasses. Part 1," Glass Tech., 2: 218-223 (1961)
2. Refractories Committee of the Society, "Summary of Test Results from Cooperative Corrosion Tests," Glass Tech., 5: 207-211 (1964)
3. Norton, F.H., Refractories, 3rd Edition, 453-478, McGraw-Hill Book Company, Inc., New York (1949)
4. Moore, H and Heeley, R., "An Experimental Investigation of Alumina-Silicate Refractories of High Purity, for Use in Glass-Melting," J. Soc. of Glass Tech., 34 (12): 274-304 (1950)
5. Scott, A., "Corrosion of Steel Furnace Refractories," J. Brit. Ceramic Soc., 25: 339-350 (1926)
6. Sosman, R.B., "Some Fundamental Principles Governing the Corrosion of a Fire Clay Refractory by a Glass," J. Amer. Ceramic Soc., 8 (4): 191-204 (1925)
7. Endell, K., Fehling, R., and Kley, R., "Influence of Fluidity, Hydrodynamic Characteristics, and Solvent Action of Slag on the Destruction of Refractories at High Temperatures," J. Amer. Ceramic Soc., 22 (4): 105-116 (1939)
8. Busby, T.S., "Porosity and Refractory Corrosion," J. Soc. of Glass Tech. 41: 318T-329T, (1957)
9. Schurecht, H.G., "Reactions of Slag with Refractories: Part I, Surface Reactions," J. Amer. Ceramic Soc., 22 (5): 116-129 (1939)
10. Massengale, G.B., Mong, L.E., and Heindl, R.A., "Permeability and Some Other Properties of a Variety of Refractory Materials: 11" J. Amer. Ceramic Soc., 36 (8): 273-278 (1953)
11. Comeforo, J.E., and Hursh, R.K., "Wetting of Al_2O_3-SiO_2 Refractories by Molten Glass: II, Effect of Wetting on Penetration of Glass into Refractory," J. Amer. Ceramic Soc., 35 (6): 142-148 (1952)

12. Laing, K.M., Emhiser, P.E., Fitzgerald, J.V., and Jones, R.E., "The Location of Sodium by Nuclear Activation in Glass-Corroded Tank Block Refractories," J. Amer. Ceramic Soc, 34 (12): 380-383 (1951)
13. Cooper, A.R., and Kingery, W.D., "Corrosion of Refractories by Liquid Slags and Glasses," Kinetics of High Temperature Processes, 85-99, Wiley and Sons, Inc., New York (1959)
14. Barham, D. and Barrett, L.R., "The Dissolution of Magnesium Aluminate Spinel in Sodium Silicate Melts," Trans. Brit. Ceramic Soc., 67, 49-56 (1968)
15. "Standard Methods of Test for Apparent Porosity, Water Absorption, Apparent Specific Gravity, and Bulk Density of Burned Refractory Brick," ASTM Standards, Designation: C 20-46 (Reapproved 1967), 13: 8-10 (1968)
16. Benson, S.W., The Foundation of Chemical Kinetics, p. 82, McGraw-Hill Book Company, New York (1960)
17. Hulbert, S.F., "Models for Solid State Reactions in Powder Compacts," J. Brit. Ceramic Soc., 6: 11-20 (1969)
18. Brindley, G.W., Sharp, J.H., Patterson, J.H., and Harahari Achar, B.N., "Kinetics and Mechanisms of Dehydration Processes, Part I," Am. Minerologists, 52: 201-211 (1967)
19. Reed, L. and Barrett, L.R., "The Slagging of Refractories. Part II. The Kinetics of Corrosion," Trans. Brit. Ceramic Soc., 63: 509-534 (1964)
20. Weyl, W.A. and Marboe, E.C., The Constitution of Glasses, John Wiley and Sons, Inc., New York (1962)
21. Fajans, K., "Electronic Structure of Some Molecules and Crystals," Ceramic Age, 53: 288-292 (1949)

ADDITIONAL INFORMATION

This investigation was sponsored by the Kaiser Aluminum and Chemical Corporation. The writers are: Samuel F. Hulbert, Associate Professor of Materials Engineering and Biomedical Engineering and Associate Dean for Engineering Research and Interdisciplinary Programs, Clemson University, Clemson, S.C.; and Furman H. Brown, Research Associate in Materials Engineering, Clemson University, Clemson, S.C.

IMPROVEMENT OF CALORIMETRIC ACCURACY OF DTA BY PRECISION SAMPLE PACKING*

Edgar M. Bollin and A. J. Bauman

Jet Propulsion Laboratory

California Institute of Technology

Pasadena, California

ABSTRACT

A precision sample loading device for differential thermal analysis is developed that readily allows reduction in variations in technique and previously unaccounted heat losses. The device forms cylindrical samples of uniform size with closely controlled packing densities in which the thermocouple occupies a central position. An envelope of loosely packed porous granular aluminum oxide supports the sample in the center of a metal thermal well. Transmission of heat from the thermal head to the sample is adequate to provide a low base line deviation, while dissipation of the heat of reaction to the thermal head is restricted through a medium which is constant in its thermal properties. An exchange column for reactions involving a gaseous phase is provided which allows easy entrance or exit of reaction products.

Dissipation of the heat of reaction is more rigorously controlled than in normal techniques by confining the sample to the immediate area of the thermocouple and by the elimination of direct contact of the sample with the thermal head.

---

*This paper presents the results of one phase of research carried out at the Jet Propulsion Laboratory, California Institute of Technology, under Contract No. NAS 7-100, sponsored by the National Aeronautics and Space Administration.

Preliminary tests show that variation in packing density produces a family of reaction temperature curves that are a function of the resulting linear increase of internal partial pressure of the reaction products with an increase in packing density.

INTRODUCTION

One of the most serious variables affecting reproducibility and quantitative accuracy of DTA has been the problem of sample packing. The lack of a satisfactory quantitative method to study this effect has relegated this variable to a technique status. The ability to control the packing density factor (PDF) adds a significant parameter to the study of gas phase reactions.

The problem of sample packing is universally recognized, and nearly all early DTA investigations directed considerable effort to this factor and its effect upon reproducibility. Recent investigators in the field of thermal analysis have become aware that it is the reaction kinetics of various compounds under dynamic conditions of analysis that are highly susceptible to many variables rather than the DTA technique itself. The supposed quantitative inaccuracy of DTA is not a basic lack of reproducibility, but rather, a high resolution capability that readily allows the study of these real variations in reaction kinetics produced by variable environmental conditions.

The objections to the use of DTA in the quantitative determination of thermal reactions have been based to some extent on the lack of recognition of the magnitude and the complexity of the environmental dependence of many thermal reactions. As complimentary techniques of investigation, such as thermogravimetric analysis and dynamic thermal X-ray diffraction, have become more sophisticated, it is increasingly apparent that in many cases it has been the reaction which is highly complex and nonreproducible under variations in technique rather than DTA.

Studies previous to dynamically swept gas techniques (Stone, 1960) have generally failed to distinguish the differences between reactions involving gaseous phase transfer and solid state reactions. The effects of variations in the PDF may be divided into two distinct categories: (1) those concerned with gaseous phase interchange and (2) those concerned with concomittant changes in thermal diffusivity due to increased heat transfer with increased particle to particle contact. The latter effect is more generalized in that it is operative throughout the temperature range of analysis, affects base line drift and the magnitude of the reactions, but does not directly affect the temperature at which reactions take place.

Variations in PDF on the grain to grain contact may to some extent affect the magnitude and temperature range of exothermic reactions such as encountered in the calcination of ceramic materials. These effects are generally rather minor and are beyond the scope of the present investigation.

The die extrusion technique produces an extremely reproducible packing density and precision placement of a DTA sample within a thermal head. The reproducibility of the technique and the greatly reduced heat of reaction losses in the thermal head allow quantitative control of heat losses neglected in normal sample packing techniques.

Such heat losses, often neglected, include: the variable heat loss of the thermal head caused by different thermal conductivities of the samples, the insulation of the thermocouple from the total reaction by the low conductivity of the sample, and the highly variable loss of the heat of reaction of the sample in direct contact with the head.

Previous DTA packing techniques have yielded reasonably reproducible results when a specific technique has been rigidly followed, but often reproducibility has been a function of uniformity in the technique of a particular investigator rather than reproducibility of a specific thermal head design. Discrepancies result when the method of packing, the configuration of the thermal head, or the personnel factor is varied. A die-extrusion method of packing yields improved overall reproducibility, since variations in technique which depend upon manipulation are minimized. Samples produced have improved geometry and thermal behavior.

THEORETICAL CONSIDERATIONS

The extrusion die molds a cylindrical sample of uniform size and compaction immediately around the thermocouple junction. The theoretical ideals of uniform dimensions, thermocouple placement, and precision control of sample density are largely realized. A reduction in the loss of the heat of reaction is achieved by introducing a low conductivity barrier between the sample and the thermal head. In effect, a composite thermal head is produced that has the stability of a metallic thermal head and the sensitivity of a ceramic thermal head.

A similar use of aluminum oxide previously has been shown to be an effective medium of heat transfer from the thermal head to the sample (Bollin and Kerr, 1961). The envelope of aluminum oxide serves as a medium for the transfer of heat from the head to the sample, while reducing the transfer of heat of reaction to the head.

Such a selective heat transfer is effective because of the relative time constants of the rate of heating of the furnace and the rate of the reaction produced in the sample. The relatively slow increase in the temperature of the head is effectively transferred through the aluminum oxide to the sample because the heating rate is constant. However, the relatively rapid reaction of the sample is prevented from dissipation in the thermal mass of the head by the low thermal diffusivity of the aluminum oxide. The effect of the thermal barrier becomes more pronounced as the ratio of the path length of the heat flow through the aluminum oxide increases in relation to the path length of the heat flow through the sample.

In normal DTA packing techniques, the dissipation of the heat of reaction in the head is dependent upon the variations in the thermal diffusivity of each differing sample and the variations in the thermal diffusivity caused by differing packing densities, as well as the thermal contact of the sample with the head. As these parameters are highly dependent upon the grain size of the individual crystallites of the sample, as well as its previous thermal history, they contribute to a range of results even with different samples of the same material. In the present technique, the flow of the heat of reaction takes place through a constant medium for the major portion of its travel to the sample, and the slow time constant of the aluminum oxide allows the major portion of the reaction to be recorded by the thermocouple before dissipation occurs in the system.

The major factor that is often neglected in DTA instrument design is the nearly complete loss of the heat of reaction of the sample that is in direct contact with the thermal well of the head. The thermal diffusivity of the sample and its position in the thermal well determines in a large measure, the relative amount of reaction that is dissipated in the system before it can activate the thermocouple. Thus, equal weights of samples with differing thermal diffusivities result in differing calorimetric values. In reality, many of the materials such as fine powders which are usually studied by DTA have very poor thermal diffusivities and the heat of reaction is insulated from dissipation in the thermal mass of the head. However, for the same reasons, the portions of the sample removed any appreciable distance from the thermocouple are insulated from it and the heat of reaction is not recorded. Thus, the outer portion of the sample in contact with the thermal well (Fig. 1, position a) is dissipated in the head, the middle portion of the sample (Fig. 1, position b) is thermally insulated from the thermocouple by the poor thermal conductivity of the sample, and only the portion of the sample immediately surrounding the thermocouple (Fig. 1, position c) is effectively recorded.

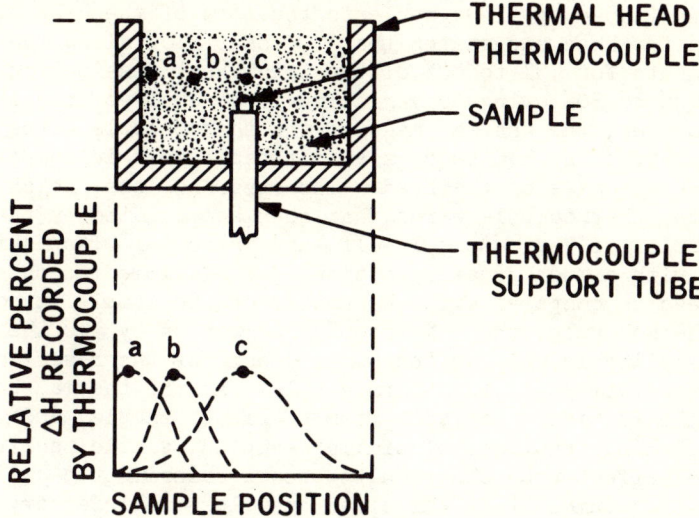

Fig. 1. Normal distribution of heat of reaction

The degree to which the above relations are effective in distorting the thermal record of the sample is not only dependent upon the thermal diffusivity of each sample and its packing density, but is also dependent upon the configuration of the thermal well. Deep thermal wells with a small diameter have a relatively greater proportion of the sample in direct contact with the thermal well than shallow thermal wells with a large diameter. This effect cannot be corrected by the use of extremely wide, shallow thermal wells because of excessive distortion of the isothermal contours at the ends of the thermal wells and insulation effects mentioned above. Thus, the proportion of the reaction that is dissipated in the head varies for each design of a thermal head and the manner in which it is used.

The combination of the previously mentioned effects of variable dissipation of the heat of reaction before it can be recorded by the thermocouple has in many instances been the reason for a lack of correlation between differing types of thermal apparatus and packing procedures. Because of the interrelationship of these variable parameters a systematic evaluation of the loss of the heat of reaction to the head is difficult, which explains to a great degree why this loss has been largely neglected by many investigators. The loss to the head is many times more serious than the loss of heat of reaction that is carried away by the thermocouple wires, which has received much attention.

Kingery (1959a, p. 259) neglects the loss of the heat of reaction in the head and states that "...heat flow along the thermocouple accounts for 75% to 80% of the total heat dissipation and varies by 20% to 30% owing to variation in assembly of the same setup." However, the thermal mass of the thermocouple wires in a normal block-type thermal head is almost negligible in comparison with the thermal mass of the head. The apparent loss, which is caused by the thermocouple wires, has been accentuated by the thermal barrier of the sample itself, and in normal DTA packing procedures with a much greater portion of the sample in contact with the head as compared with the amount of the sample in contact with the thermocouple, the 75% to 80% dissipation by the thermocouple is applicable only to the small amount of the reaction that is capable of activating the thermocouple. As the thermal barrier of the sample is highly variable from sample to sample, and with differing packing densities of similar samples, sample packing has a pronounced effect upon DTA results, hence calorimetric calibration of DTA instrumentation has often been less satisfactory than is desirable.

METHOD OF DIE-LOADING THE SAMPLE

The steps in the extrusion of uniformly compacted sample are shown in Fig. 2. A precisely centered thermocouple is first fixed in the thermal well (Fig. 2a). The thermocouple insulator tube forms the bottom of the sample and furnishes a solid base upon which the sample is compacted. The machined loading die, which closely fits in the top of the thermal well, accurately centers the sample loading tube around the thermocouple. The sample is poured inside the sample tube, and 60 mesh aluminum oxide is poured around the exterior (Fig. 2b). The sample is compacted to a uniform depth by the introduction of a closely fitting piston in the tube (Fig. 2c). The sample tube is moved upwards to clear the sample while the compacting piston is held fixed to prevent any distortion or removal of the sample (Fig. 2d). Finally, the entire assembly is removed and the sample well is completely filled with aluminum oxide (Fig. 2f).

The result of the die-loading is an equidimensioned sample placed immediately around the thermocouple. The porosity of the aluminum oxide surrounding the sample serves as an exchange column for reactions involving the vapor state.

CONSTRUCTION OF LOADING-DIE

Several factors govern the design of a loading-die which will yield a uniformly dense sample around a thermocouple. A major

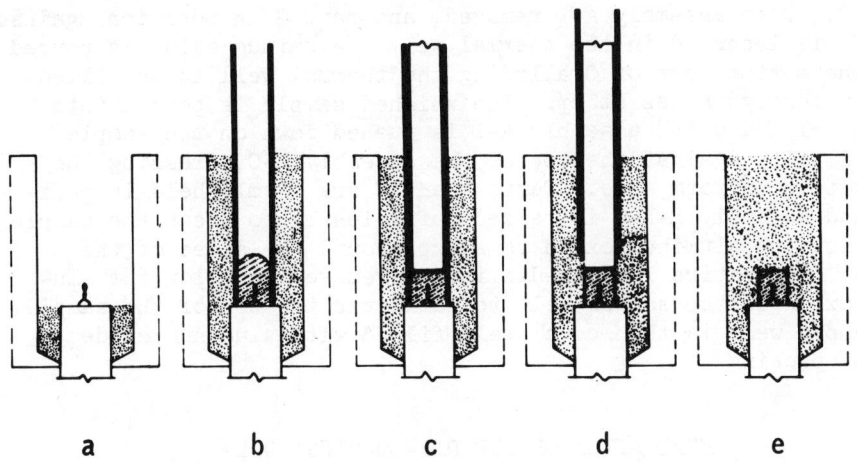

Fig. 2. Method of die-loading sample

requirement is the accurate placement of the thermocouple element in the thermal well. Another is the use of a rigid support for the thermocouple. Further, since the rigid ceramic insulator constitutes a source of dissipation of the heat of reaction, as well as the thermocouple wires, the thermal head should be designed to minimize this effect (Bollin and Kerr, 1961).

A brass plug P (Fig. 3) is used as a location jig to cement the thermocouple and the thermocouple insulator at the center of and parallel to the bore of the thermal well. The thermocouple junction is welded in the form of a small sphere (Bollin and Kerr, 1961) and projects exactly one-half of the diameter of the extruded sample above the thermocouple insulator tubing.

The construction of the die-loading tool (Fig. 3) is simplified by the use of telescoping brass tubing, although precision stainless steel tubing would be a more satisfactory choice in the construction of a standardized loading tool. The remainder of the machined parts of the tool may also be constructed of stainless steel in place of the brass. However, brass was used for a preliminary model because of its immediate availability and ease of working.

LOADING PROCEDURE

The thermal well is filled with aluminum oxide to approximately the top of the thermocouple insulator tube. The loading-die

(Fig. 3), with assembly A-D removed, and part G in position against part H, is inserted in the thermal well. Aluminum oxide is poured into the bottom part of C allowing the thermal well to be filled by flow through holes at H_1. The weighed sample is poured into the funnel E and the assembly A-D is pushed down on the sample until it fits against the top of the assembly B-C, pressing the sample to a uniform size. Parts A and C are firmly held in position and assembly E-F-G is raised sufficiently to clear the sample and allow the aluminum oxide to slump around the sides of the sample. The entire loading-die is now removed to allow the aluminum oxide in the sample well to fall over the top of the sample. The sample well is then completely filled with aluminum oxide without packing.

EVALUATION OF THE DIE-LOADING TOOL

Actual calorimetric calibration of the present thermal head and die-loading tool has not been attempted since the design is preliminary and described as a method intended to increase the accuracy of calorimetric measurements obtainable by DTA.

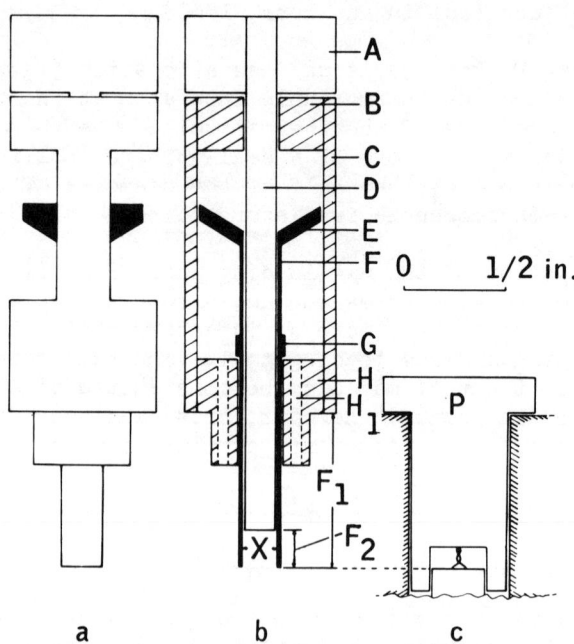

Fig. 3. Die-loading tool and locating fixture

Kaolinite (A. P. I. No. 1a, Murfreesboro, Arkansas) was used for evaluation as a mineral which has been studied extensively with reactions that are relatively well known. Two reactions are produced in the range of 25 to 1000°C, dehydroxylation (endothermic, 500-600°C) and recrystallization (exothermic, 950-1000°C).

Only the dehydroxylation reaction of kaolinite is considered because the recrystallization reaction does not involve the exchange of a vapor state reaction and is relatively unaffected by packing density in the range under study. The dehydroxylation of kaolinite provides a definitive test of the die-loading tool as "...the reaction is very sensitive to low water vapor pressures, not exceeding about 1 atm" (Brindly, 1959, p. 312). The decomposition of the mineral has been studied extensively and the decomposition rate has been shown to be controlled by the rate at which the partial pressure of the water is able to disperse from the reaction zone, thus becoming lower than the dissociation pressure of the hydroxyl (Kingery, 1959b, pp. 294-313).

A die-loading tool that produced a compacted sample 5/32 in. diameter and length in a thermal well 3/8 in. in diameter and 1 in. deep has been used for the following evaluation. The total diameter of the thermal head is 2 1/2 in. in diameter, and number 22 gauge chromel-alumel thermocouples are supported by a 4-bore, 3/8 in. diameter thermocouple insulator tube. The tool will hold 20 mg of kaolinite without compaction, and 70 mg is nearly the maximum that will allow operation of the tool without difficulty.

The peak temperatures of the dehydroxylation reaction of kaolinite for various sample weights are shown in Fig. 4. Sample weights less than 20 mg do not completely fill the sample chamber of the die-loading tool in the compacted position. The volume of the sample is maintained constant by adding sufficient 120-mesh alumina to fill the sample chamber with these small sample weights.

Sample dilution techniques have not been employed in the range of sample weights, which would more than fill the sample chamber when uncompacted as it is desirable to present an undistorted evaluation of the packing itself. The reasons advanced by various investigators for sample dilution are as valid for the die-loading tool as in normal packing procedures. However, if calorimetric values are to be determined on small samples, sample dilution must be practiced with care. Dilution material of high thermal conductivity should be used to prevent insulation of the heat of reaction from the thermocouple. Sample dilution techniques do not need to be as extreme with the die-loading tool as in normal packing procedures because of the vapor exchange capabilities of the alumina envelope and the relatively small size of the sample.

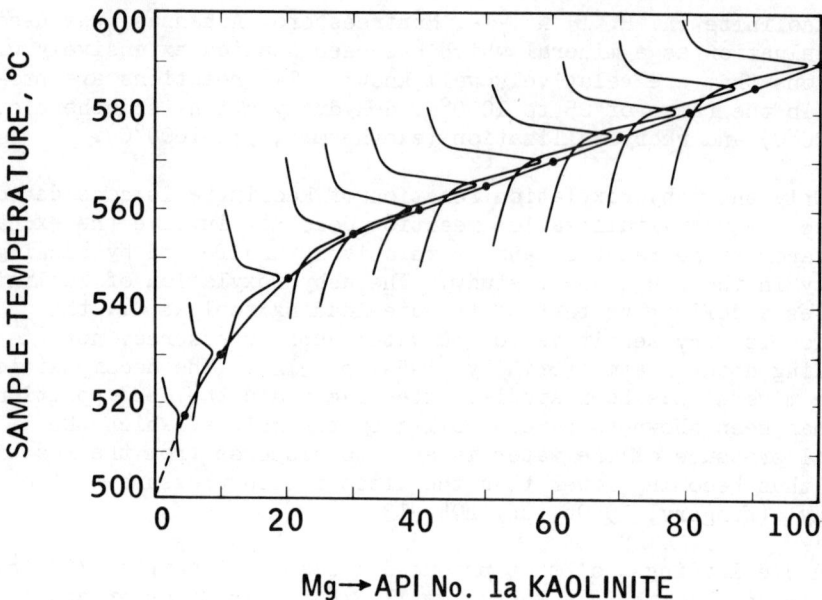

Fig. 4. Change in reaction temperatures for various sample weights of kaolinite

The linearity of the compacted sample weights is evident in Fig. 4. The nonlinearity of the curve for samples less than 30 mg in weight is caused by the lack of compaction and the increasing dilution of the sample allowing the dissociated water vapor to diffuse through the alumina exchange column at an accelerated rate.

The extrapolation of the curve to a vanishing sample weight is tentative as the dissipation of the heat of reaction by the relatively large thermocouple wires does not allow observation of sample weights below 5 mg. The dispersion of the sample into the relatively large volume of the sample chamber for sample weights in this range similarly does not allow efficient transfer of the heat of reaction to the thermocouple. The fact that such an extrapolation is at all possible indicates the relative consistency of results that the die-loading tool produces.

It should be emphasized that the results shown in Fig. 4 are derived from a single preliminary analysis of each sample weight, and averaging of values about a mean have not produced the linearity of the curve. The pronounced linearity of the shift in the peak temperature of the dehydroxylation of kaolinite with differing packing densities is believed to be somewhat noteworthy in view of the variability commonly experienced in DTA. The linearity of the curve is a more sensitive test of the reproducibility of the curve to possible variations in technique than a series of identical samples.

DIFFUSION EXCHANGE COLUMN

There appears to be a certain amount of fortunate coincidence in the overall design parameters of the present experiment which contribute to the linearity of the family of curves obtained. One of the important parameters is the provision of a competitive adsorption and/or storage site for the reaction products. The granular aluminum oxide surrounding the sample provides this function in that it allows the reaction products to diffuse away from the sample while retaining these products in close enough proximity to allow the buildup of the local partial pressure of these products. The surface area of this exchange column as well as its volume contributes to the effective diffusion exchange between sample and column. Thus the partition between the sample and the exchange column will exceed the partition between the sample and the external environment. These conditions will pertain as long as sufficient path length is provided so that a dynamic saturation value is obtained that is constant within reasonable limits and is dominantly a function of the surface area and chemical nature of the column itself. These conditions are readily controlled by sizing, chemical, or thermal pretreatment. The exchange column may be considered as a buffer between the sample and the environment which allows the reverse partial pressure sensitivity of the sample to be accentuated sufficiently as to be readily observed as a function of controlled packing density.

The total range of the sensitivity to the reverse partial pressure effect is obtained from a minimum under hard vacuum conditions to a maximum under superpressure of the reaction products. An approach to each end of the range is provided by the dynamically swept gas techniques of Stone (1960) representing a pseudominimum and the inhibited diffusion technique of Garn (1965) representing a pseudomaximum. Neither of these techniques provides a method of continuous variation of parameters, and the hard vacuum and superpressure end points require special equipment not readily available to the majority of workers in the field. In contrast, the controlled packing density technique provides a relatively constant reference column against which packing parameters can be varied in a continuous and linear manner which is within the range that can be affected by the fugacity of the sample.

DESIGN LIMITS OF LOADING TOOL

A second die-loading tool was designed to fit the R. L. Stone Company's "liquids" thermal head No. SH-11BP2. Sample size (3/32 in. dia. x 3/32 in. long) was much smaller than the original tool because of the smaller thermocouple support tube and extended from 20 mg downwards. Results (Fig. 5) were similar to the original tool but linear only in the upper end of sample weights. Apparently sample size was too small to allow a sufficiently wide range in packing density and the thermal well was too shallow to allow dynamic equilibrium of local partial pressure.

The areas under the curve of the larger sample weights of the large tool show a relatively constant value for increasing sample weights. This indicates that insulation of the heat of reaction by poor thermal diffusivity of the sample is effective in this range. Therefore, it appears that a loading tool producing a sample diameter between 5/32 and 3/32 in. would be optimum.

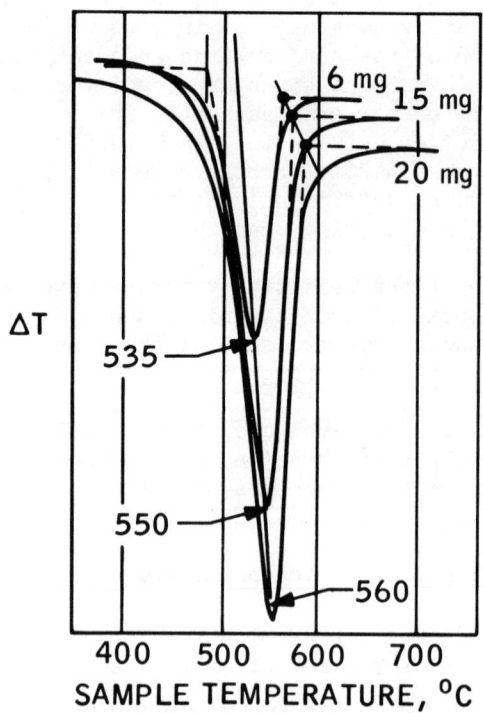

Fig. 5. Change in reaction using small die-loading tool

CONCLUSIONS

The technique described provides the capability of a continuous variation of sample packing parameters allowing an assessment of the sensitivity to reverse partial pressure of reversible vapor phase reactions.

In practice, a great advantage of the die-loading tool is the ease and speed of placing a sample into a thermal well. The use of small samples or "sandwich techniques" are greatly facilitated even if a weighed sample or the controlled density function capability is not used.

The cylindrical "sandwich" produced with the die-loading tool is a more efficient geometry than the planar "sandwich" (Barshad, 1952) in which a layer at the plane of the thermocouple extends to the wall of the thermal well. The vertical losses of the heat of reaction which are reduced by the planar "sandwich" technique are further reduced in the radial direction by the cylindrical sample configuration.

REFERENCES

BARSHAD, I., (1952) Temperature and Heat of Reaction Calibration of the Differential Thermal Analysis Apparatus: Amer. Mineral., 37, 667-694.

BOLLIN, E. M., and KERR, P. F., (1961) Differential Thermal Pyrosynthesis: Amer. Mineral., 46, 823-858.

BRINDLY, G. W., (1959) in Kingery, W. D., (Ed.) Kinetics of High-Temperature Processes, John Wiley and Sons, Inc., New York, 326 p.

GARN, P. D., (1965) Differential Thermal Analysis Using Self-generated Atmospheres at Sub- and Supra-Atmospheric Pressures: Anal. Chem. 37, 77-78.

KINGERY, W. D., (1959a) Property Measurements at High Temperatures: John Wiley and Sons, Inc., New York, 416 p.

KINGERY, W. D., (Ed.), (1959b) Kenetics of High-Temperature Processes: John Wiley and Sons, Inc., New York, 326 p.

STONE, R. L., (1960) Differential Thermal Analysis by the Dynamic Gas Technique: Anal. Chem., 32, 1582-1588.

THE APPLICATION OF HIGH PRESSURE DSC TO CATALYTIC REDUCTION STUDIES

William E. Collins

E. I. du Pont de Nemours & Company(Inc.)

Photo Products Dept., I&E Division,

Wilmington, Delaware   19898

INTRODUCTION

The efficient reduction of aromatic nitro compounds for use in the synthesis of dyes, pharmaceuticals and numerous intermediates depends to a large extent on the activity of precious metal catalysts such as Platinum and Palladium. The measure of this activity dictates the ratio of catalyst to starting material and in turn the economics of the batch reduction process. Catalysts are presently evaluated by duplicating the reaction in a laboratory autoclave and measuring either the rate of hydrogen consumption or product yield per unit time. These reactions can be time consuming, expensive and hazardous.

This work was undertaken to demonstrate the applicability of high pressure differential scanning calorimetry (DSC) to the evaluation of catalyst activities and explore the possibility of following the reduction of organic compounds in the DSC cell.

DSC records the differential heat flow between a sample and reference material as a function of temperature or time. The integrated area, therefore, is directly proportional to the total heat either evolved or absorbed by the sample.

The Du Pont Pressure DSC Module attached to Du Pont's 900 Thermal Analyzer was used for this study. The recent introduction of this pressure cell makes possible the direct quantitative measure of the heat evolved or absorbed by a sample in either active or inert atmospheres from 10 microns of reduced pressure up to 1000 psig.

Emmett, Sabatier and Reid[1] discussed several techniques that correlate the activity of a catalyst to the results attained by actually running small scale test reactions.

Komers et. al.[2] studied the thermally induced desorption of various gases and reaction products from Platinum on silica catalysts. A gas chromatograph, connected directly to the reaction vessel, was used to identify and measure the products.

Stone and Rase[3] showed that the exothermic adsorption of various gases such as steam and ammonia on silica-alumina cracking catalysts could be detected by DTA and the peak heights correlated with catalyst activities as measured by other tests.

David[4] demonstrated the reduction of Dinitrotoluene with Raney nickel in methanol in a DTA cell under 250 psig of hydrogen.

Brusie, Keely and Mathes[5] showed that "good" and "bad" hydrogenation catalysts (returned from the plant) could be distinguished by the size of their exothermic reaction when small amounts of carbon monoxide were injected into a dynamic hydrogen atmosphere at 260°C and atmospheric pressure.

May and Bsharah[6] ranked the catalytic effect of a series of metals on the oxidation temperature of rubber using a DTA cell in a 300 psig oxygen atmosphere.

This work was initiated to duplicate the evaluation of a series of commercial and experimental catalysts which had been tested by the standard hydrogen up-take method in a Parr autoclave. The DSC procedure developed for this evaluation is about 25 times faster and uses about 10 times less catalyst than the autoclave technique. Furthermore, since most of the catalytic reductions studied are highly exothermic, these micro scale reactions are potentially safer to personnel and equipment.

APPARATUS

The instrument used in this study consists of a standard Du Pont DSC cell[7] enclosed in a specially designed pressure housing. The pressure module is shown in FIGURE 1.

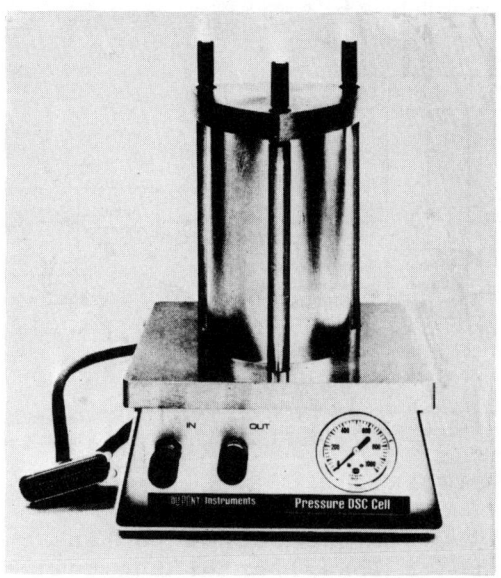

FIGURE 1

Both the top of the cell and the cylinder are removable by simply unscrewing the three bolts with the fingers. Normally only the top is removed for loading and unloading samples. "O"-ring seals, set in grooves in the top and bottom portions of the enclosure, seat against the side wall of the cylinder and increase their contact as either pressure or vacuum is applied.

The assembly is designed to operate at a maximum working pressure of 1000 psig and to attain a vacuum of under 10 microns with a good pumping system attached. The cell, valving and connections are diagrammed in FIGURE 2.

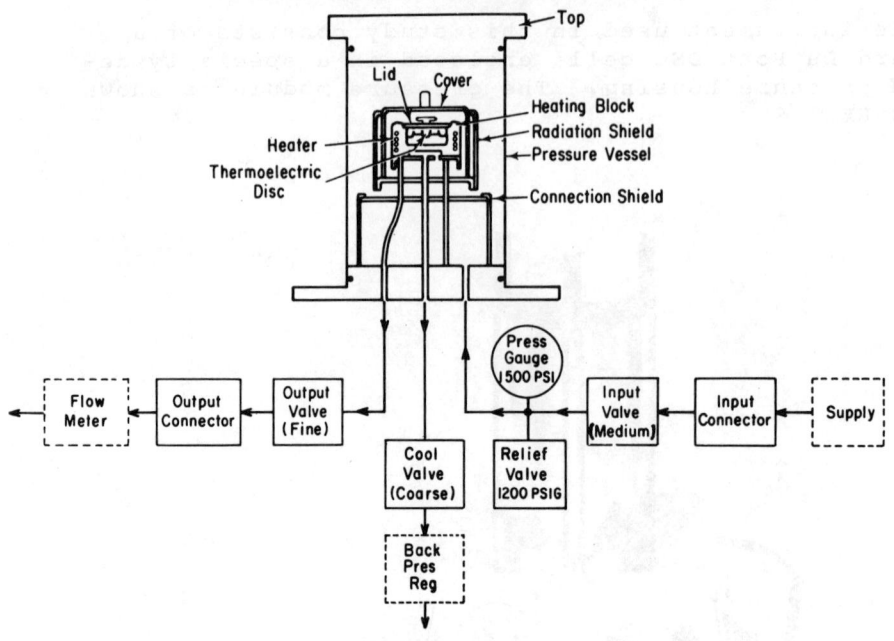

FIGURE 2

One or more supply gases (connected to tank regulators set at 1000 psig or lower) enter the cell through the input connector and are controlled by the input control valve. The incoming gas enters the cell through a port in the baseplate and floods the entire volume of the cylinder. A spring loaded safety valve in the supply line between the input valve and the baseplate is set up open at 1200 psig. A pressure gauge with a maximum range of 1500 psig (red lined above 1000 psig) is also connected at this point.

Gas exiting the cell during purging enters the sample chamber through a hole in the silver lid, passes over the sample, through an orfice in the side wall of the sample chamber and out through a tube connected to the baseplate. Gas flow rate during purging is controlled by a fine thread (14 turn) high pressure needle valve. A flow meter may be connected to a line from the exhaust valve.

A third valve, labeled "cool control valve" in FIGURE 2, is connected internally in the cell to a point under the heating block outside of the sample chamber.

This is a coarse (3 turn) stop valve which is used to rapidly release pressure from the cell after a run is completed. Since it has the largest orfice it is the point at which a vacuum system is normally connected to the unit.

If all valves are closed after pressurizing the cell, the run will be conducted under constant volume conditions. During the run, as the temperature rises, the internal pressure will increase. This is usually negligible since a relatively small volume within and adjacent to the heating block actually reaches the temperature of the experiment. If true constant pressure conditions are required, a back pressure regulator may be connected to the "cool control valve."

A more detailed description of the Pressure DSC cell including a survey of general applications was given by Levy, et.al.[8]

CALIBRATION

The baseline performance run on two empty aluminum sample pans in air at 800 psig and maximum sensitivity is shown in FIGURE 3.

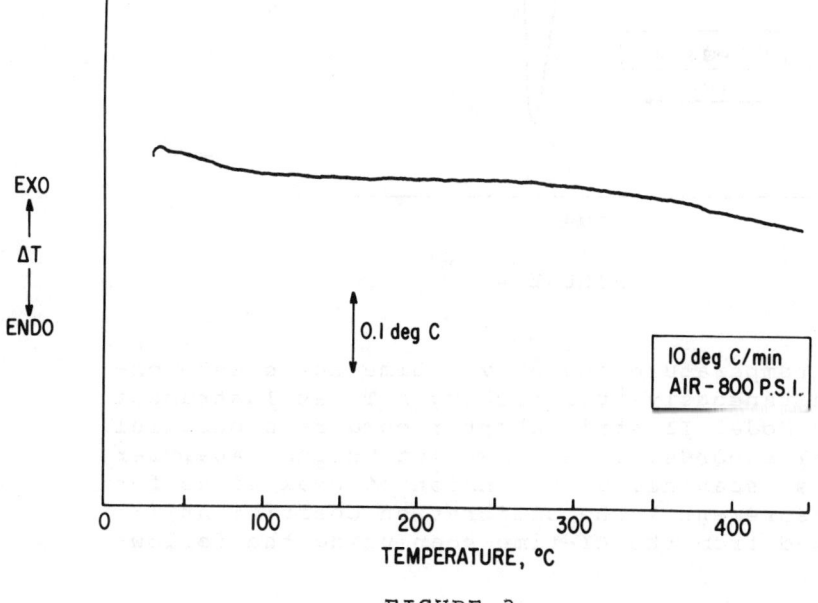

FIGURE 3

At ambient pressure, performance of the Pressure DSC cell is identical to that of a regular Du Pont DSC cell. At elevated pressures there is some change (less than 10% over the entire temperature and pressure range) in the calibration coefficient (E) used to calculate the energies involved in a thermal transition.

For this work the cell was calibrated using an indium metal standard over the pressure range of interest in both helium and hydrogen atmospheres. FIGURE 4 shows the melting endotherm of indium under H_2 at 200 psig.

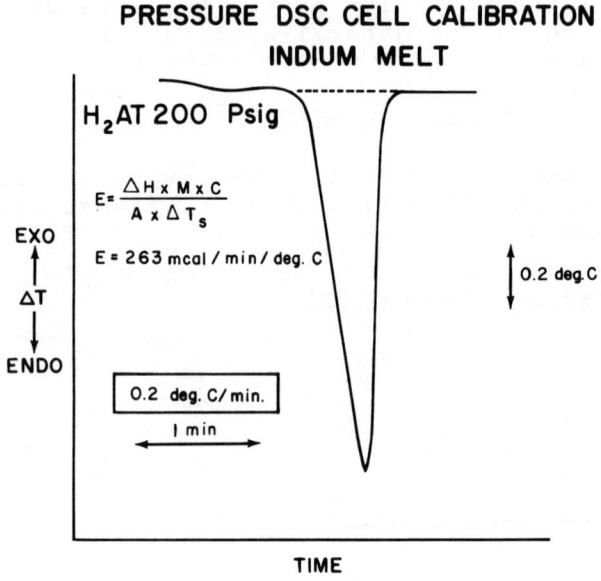

FIGURE 4

ΔT vs. temperature and ΔT vs. time scans were obtained simultaneously by attaching a Texas Instrument Servo/riter Model II strip chart recorder in parallel with the X-Y recorder in the Du Pont Thermal Analyzer. The time-base scan allows expansion of peak areas for ease of measurement. The calibration coefficient (E) is calculated from the ΔT-time scan using the following equation:

$$E = \frac{\Delta H \times M \times C}{A \times \Delta Ts} = 263 \text{ m cal/min/deg C}$$

where: ΔH = 6.79 cal/gm (from Handbook of Chem. & Physics, Vol.   ).
  M = 14.93 mg (indium sample weight)
  C = 2 in/min (chart speed)
  A = 3.86 sq.in. (measured from ΔT-time curve)
  ΔTs= 0.2 deg/in. (Y axis or ΔT sensitivity)

EXPERIMENTAL

Materials

With the exception of the two mixed metal catalysts, the samples are commercial catalysts manufactured by Engelhard Industries, Inc. Samples of aromatic nitro compounds were supplied by the Du Pont Organic Chemicals Department. Cylinders of Matheson Spectro Grade helium and hydrogen were used throughout the investigation.

TEST PROCEDURE I

Catalyst Evaluation

1. Catalyst samples, vacuum dried below 50°C, were weighed in tared aluminum sample pans. Samples weighing approximately 5 mg were found to just cover the bottom of the pan, allowing rapid diffusion of H_2 throughout the sample.

2. Weighed portions of catalyst substrate (e.g. Carbon) may be used as a reference material[3] to distinguish between the heat of physical adsorption on the substrate and that of chemisorption on the metal. This step is not necessary where the only object is to compare close weights of metals on the same substrate. Heat capacity differences using an empty reference pan versus a reference pan containing an inert reference material were not significant in relation to the total heat being measured. An empty pan was used for convenience.

3. The sample and reference pans were placed on their respective platforms in the DSC sample chamber.

4. The cell was flushed with helium for 2 min; the exhaust valve closed; and the cell pressurized to 150 psig while allowing a slight helium bleed (10 ml/min) through the Outlet Valve.

5. A programmed heating rate of 20°/min. brought the cell from room temperature to 75°C where the purge gas was switched from helium at 150 psig, to hydrogen at 200 psig. An endothermic "blip" on the time-base recorder indicated the pressure surge of H_2 into the cell.

6. The experiment was terminated when the ΔT trace returned to the original baseline. The cell was flushed with helium; cooled to room temperature and was then ready for the next run.

TEST PROCEDURE II

Reduction of Test Samples (e.g. aromatic nitro compounds) On Activated Catalysts.

1. A weighed portion of test sample was added to the activated catalyst from Procedure I.

2. Same as Procedure I.

3. Same as Procedure I.

4. The cell temperature was programmed at 20°/min to 100°C then held isothermally at that temperature for 2 min. This allowed the sample and reference temperatures to equilibrate.

5. An auxiliary time-base recorder (T.I. Servo/riter II) was turned on at a chart speed of 4"/min.

6. The purge gas was switched from He at 150 psig to H_2 at 200 psig. The atmosphere change, detected on the recorder by a sharp endothermic spike was usually followed by an immediate exotherm due to reduction of the sample.

RESULTS

I. Catalyst Activities

A typical ΔT-Time curve for the reduction of 5% Pd/Carbon is shown in FIGURE 5. No attempt was made to calculate separately the heat of chemisorption and that of physical adsorption on the carbon. A shoulder on the leading edge of the curve is due to physical adsorption. Six replicas of this sample gave an average ΔH of adsorption of 19.5 $\pm$ 0.25 cal/gm.

The heats of reaction of a series of commercial catalysts using Procedure I, are shown in Table 1.

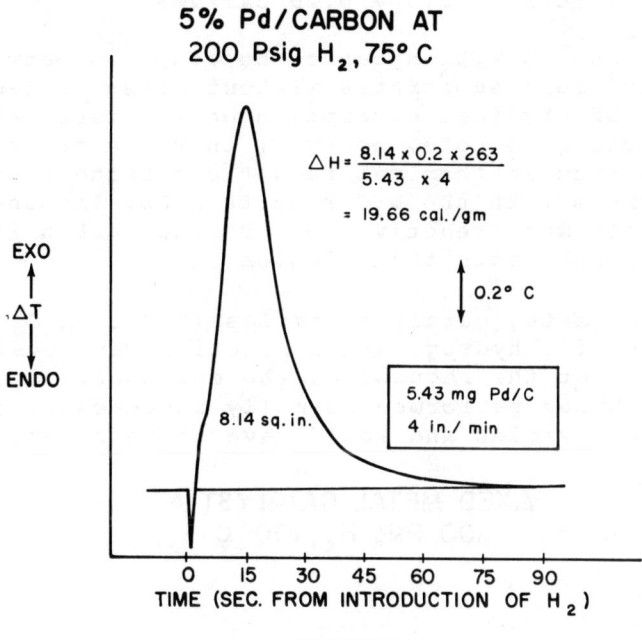

FIGURE 5

Evaluation of Catalyst Activities By DSC
At 200 PSIG H_2, 75°C

	Cal/gm
5% Pd/Carbon	19.5*
5% Pd/Calcium Carbonate	10.6
5% Pd/Alumina	9.4
5% Pt/Alumina	7.1
5% Pt/Carbon	14.0
59% Ruthenium Oxide	298.0

*(Average of 6 runs: 19.5 ± 0.25 cal/gm)

The heats of chemisorption cannot be compared between metals on different substrates without first correcting for the heat of physical adsorption on the substrate. It is interesting to note, however, that the order of activity observed in chemical reductions using these catalysts agrees with the DSC results. For instance, Pd is generally more reactive than Pt and Carbon is a more reactive substrate than Alumina.

Two mixed metal catalyst samples (Pd/Pt on Carbon) used in a specific hydrogenation reaction were evaluated by Procedure I at the request of the end user. Catalyst from one batch had performed normally whereas the second batch gave poor yields and would have to be reworked to

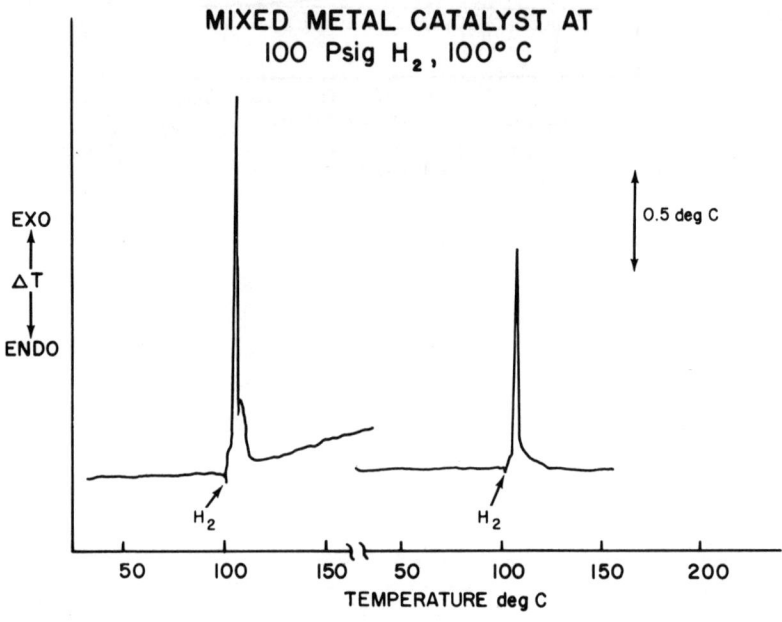

FIGURE 6

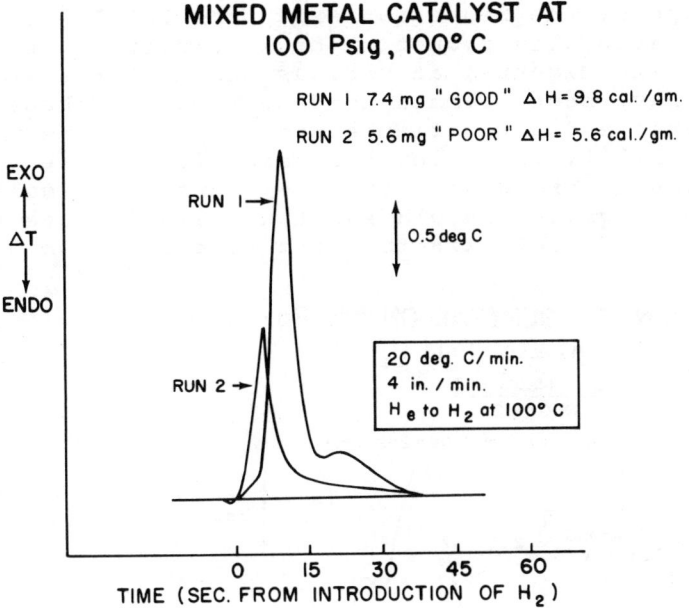

FIGURE 7

recover the precious metals. Reduction of both samples in the Pressure DSC cell are shown in FIGURE 6 (ΔT vs. Temperature) and FIGURE 7 (ΔT vs. Time).

The double peaks in the "good" sample (i.e. Run 1) are believed to represent the reduction of both metals in the catalyst. The missing peak in the "poor" catalyst sample is attributed to poisoning of the more susceptible metal species. Furthermore, the decreased reactivity correlates with the loss of synergism by the secondary metal as discussed by Bond and Webster.[10] On the basis of this data some of the secondary metal was added to the batch of "poor" catalyst and subsequent yields returned almost to normal.

II. Heats of Reduction

m-Dinitrobenzene (DNB) - Weighed portions of DNB were added, in roughly 1:1 ratios, to samples of activated Pd/Carbon catalysts. This excess of catalyst, compared to a commercial process, eliminates catalyst concentration as a rate determining parameter but introduces the possibility of reducing the benzene ring.

The shape of the reduction exotherm, FIGURE 8, indicates a step-wise reduction of each nitro group. A series of experiments, at various temperatures and pressures, have been planned to determine the reaction rate constants and activation energies of each reduction step. The ability to follow the formation of one or more reaction intermediates from reactants with multiple functional groups should substantially improve the understanding of their kinetics and reaction mechanism.

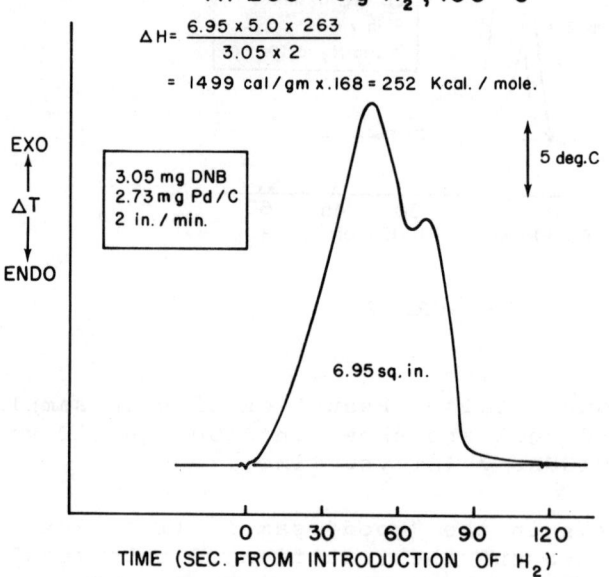

FIGURE 8

Standard reaction enthalpies for the reduction of DNB, as well as the remaining nitro compounds discussed in this section, were calculated from published Enthalpy for Formation ($\Delta H°_f$) data.[10] A sample calculation of $\Delta H°_R$ for the reduction of dinitrobenzene to diaminobenzene is shown below:

$$C_6H_4N_2O_4 \text{ (s)} + 6 H_2 \text{ (g)} \xrightarrow{-\Delta H} C_6H_8N_2 \text{ (s)} + 4H_2O \text{ (g)}$$
$$-6.2 + 0 \qquad\qquad +0.8 \quad 4(-57.80)$$
$$\Delta H°_R = -224.2 \text{ Kcal/mole at } 25°C$$

The measured heat of reduction, $\Delta H_R = 252$ Kcal/mole, as shown on FIGURE 8, is in fairly good agreement with published data.

p-Nitrotoluene (PNT) - The use of 5% Pd/Carbon to reduce PNT, FIGURE 9, did not appear to give complete conversion to p-Toluidine since the measured ΔH_R of 92 Kcal/mole was lower than the 114 Kcal/mole calculated from $\Delta H°_f$ data.[10] Three repeated runs using a mixed metal catalyst of Pd and Pt on Carbon gave values of 117, 113 and 114 Kcal/mole.

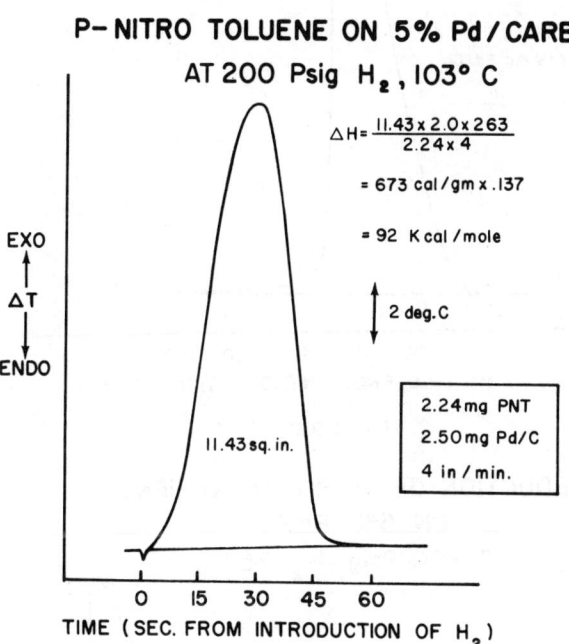

FIGURE 9

o-Nitrotoluene (ONT) - The reduction of o-Nitrotoluene to o-Toluidine, FIGURE 10, gives an experimental ΔH_R of 117 Kcal/mole compared to 116.5 Kcal/mole calculated from $\Delta H°_f$ data.[10]

The ΔT-Time plot, FIGURE 11, of an ONT reaction with 5% Pd/Carbon was run at increased ΔT and time sensitivity (i.e. 0.2°C/in, and at 4"/min chart speed). Since ΔT is directly related to the extent the reaction has proceeded at any time, t, the curve provides all the data necessary to graphically determine the specific rate constant and apparent activation energy as discussed by Bohon.[11]

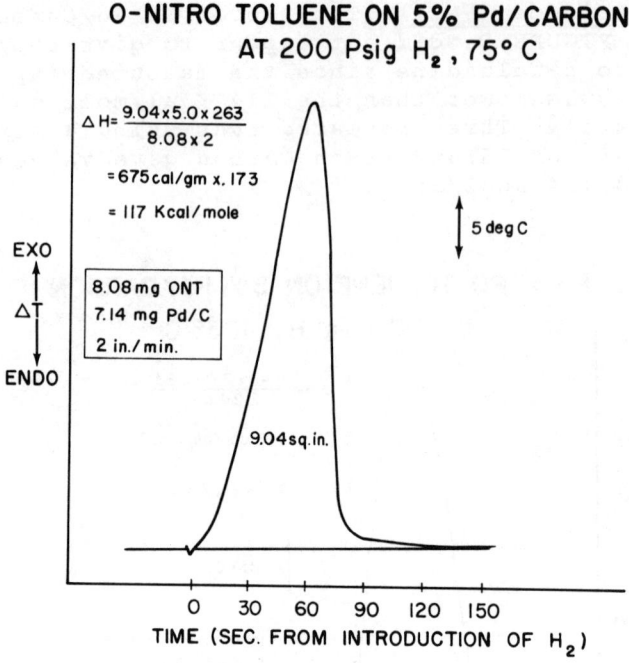

FIGURE 10

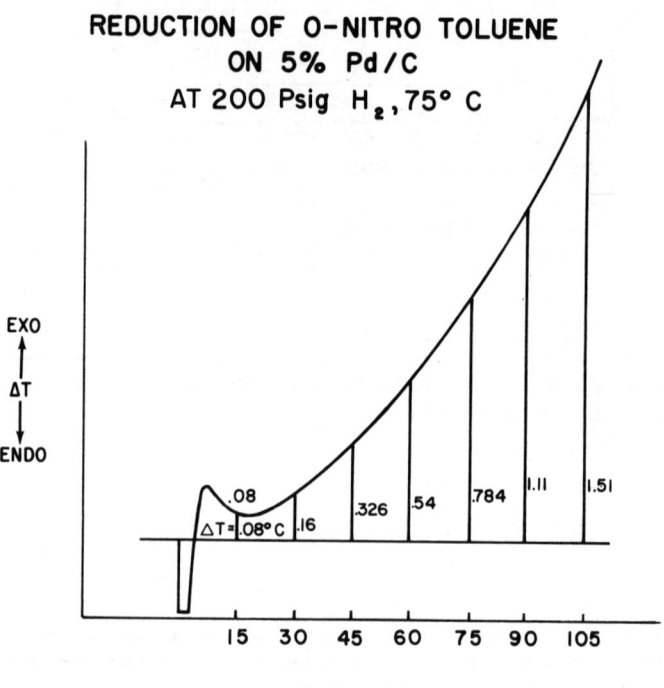

FIGURE 11

α-Nitronaphthalene - FIGURE 12 shows that increasing the H_2 pressure in the DSC cell caused a significant increase in the rate of reduction of αNitronaphthalene. The ΔH_R of 108 Kcal/mole measured at 200 psig agrees very closely with the calculated value of ΔH°_R = 109.8 Kcal/mole.[10] Since ΔH is independent of the reaction path it should not change with pressure. Therefore, the larger values measured at 20 and 30 atmospheres of H_2, are probably due to increased H_2 adsorption on the catalyst and possibly to some reduction of the aromatic ring to tetrahydronaphthalene.

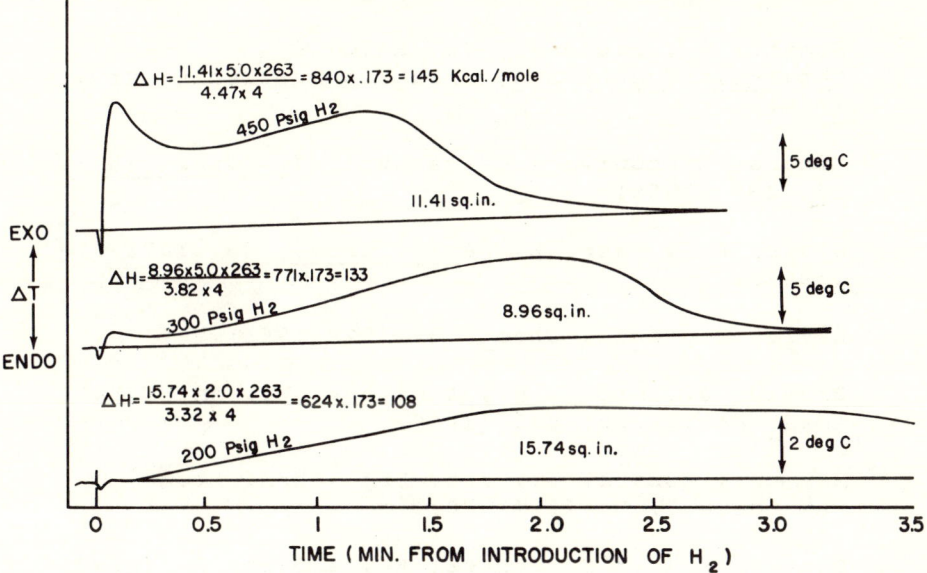

FIGURE 12

CONCLUSIONS

A rapid, economical procedure has been outlined for evaluating the activities of metal catalysts by their direct reaction with reactive gases, such as hydrogen, in a Pressure Differential Scanning Calorimeter.

A second procedure was developed for further screening the reduction properties of catalysts with test compounds directly in the Pressure DSC cell. The measured heat of reduction of four test compounds is in good agreement with literature values.

Future studies will include the calculation of specific rate constants and activation energies for similar reactions at various temperatures and pressures.

ACKNOWLEDGMENT

The author wishes to thank Dr. John R. Kosak for many test samples and helpful advice and Dr. Paul F. Levy for constructive criticism in the preparation of this paper.

REFERENCES

1. Emmett, P., Sabatier, P., Reid, E.: <u>Catalysis Then & Now</u>, Franklin Publishing Co., 1965.

2. Komers, Amenomiya & Cvetanovic, <u>J. Catalysis</u>, 15, 293 (1969).

3. Stone, R. & Rase, H., <u>Anal. Chem.</u>, 29, 1273 (1957).

4. David, D., <u>Anal. Chem.</u>, 37, 82 (1965).

5. Brusie, J., Keely, W., & Mathes, W., Am. N.Y. Acad. Sci., 158, 443 (1969).

6. May, W., Bshara, L., Ind. Eng. Chem. Prod. Res. Devel., 9, 73 (1970).

7. Baxter, R., in Schwenker, R., & Garn, P. (Eds.), Thermal Analysis, Academic Press, N.Y., pp. 65 - 84 (1969); or Du Pont Instrument Reprint #RL-31.

8. Levy, P., Nieuweboer, G., Semanski, L., Submitted to Thermochimica Acta (Feb.,1970). Presented at the American Chemical Society Meeting, Houston, Texas, Feb., 1970.

9. Bond, G.C. & Webster, D., Ann. N.Y. Acad. Sci., 158, pp. 541 (1969).

10. Stull, D., Chem. Thermodynamics of Orig. Cmpds., Wiley, N.Y., (1969).

11. Bohon, R., Proc. First Toronto Sym. on Thermal Anal., pp. 63 (1965).

THE EFFECT OF ENVIRONMENT ON QUANTITATIVE MEASUREMENTS BY

DIFFERENTIAL SCANNING CALORIMETRY

D. J. David

R. L. Stone Company, Analytical and Industrial Division, Columbia Scientific Industries, P. O. Box 6190, Austin, Texas

For many years, investigators have been interested in the quantitative amount of heat energy that accompanies a physical or chemical change in a material under investigation. The need for quantitative enthalpy data prompted the development of a number of dynamic calorimetric techniques.

In the ensuing years, a number of different approaches to dynamic calorimetry followed which have resulted in a variety of instrumentation and permitted the measurement of specific heat, heat of fusion, heat of reaction, etc. (1-17). Recent publications have treated theoretical considerations and approaches to the quantitative measurement of enthalpic effects (18, 19).

Much of the literature is confusing and contradictory on such important facets as the effect of sample geometry, sample packing, heating rate, thermocouple placement, thermocouple design, and sample holder design. These considerations, in many instances, determine the accuracy of the quantitative heat measurement or in fact if such a measurement can be done at all.

These factors, along with the basic instrumental approaches, account for the differing statements in the literature. The basic instrumental approaches which have been used to obtain quantitative enthalpy data are: (1) the use of differential temperature as the ordinate readout, and (2) the use of differential power as the ordinate readout. This latter method uses differential temperature as the basic sensing method which determines the power requirements.

Dosch (17) has reviewed the principles and types of dynamic adiabatic calorimeters. Many dynamic calorimeters employ temperature-controlled calorimetry where the sample temperature is the independent reproducible variable. This latter type of device need not be adiabatic in the classical sense of the term and nevertheless be capable of providing quantitative results.

The temperature-controlled calorimeters provide ordinate readout directly in calories or in differential temperature. Both types must be calibrated with standard materials that evolve or absorb known quantities of heat at their transition temperatures.

Considering the calorimeter that provides ordinate readout in differential temperature as in conventional differential thermal analysis, the question arises as to what design features are required to provide qualitative and quantitative capability as opposed to only qualitative capability.

The design criteria that are necessary for calorimetric work are: (1) the thermocouples must be fixed in position in relation to the sample container as well as the heat source, (2) the differential thermocouple must be external to the sample, (3) the heat flow reaching the sample and reference materials must be reproducible between runs, that is, the scanning rates must be reproducible.

The performance criteria that are inherent in a calorimeter cell are: (1) small amounts of sample material can be used directly without dilution, (2) heat capacity measurements can be performed, (3) enthalpic measurements are free from the effects of scanning rate, (4) sample packing and density exert negligible effects on enthalpic measurements, and (5) repeatable, predictable quantitative response can be achieved.

The effects of design and performance parameters and their effects upon calorimetric measurements have been reported (7, 13, 12). These investigations are in agreement with regard to performance criteria and the presence of these criteria was demonstrated for some systems (12, 13).

The literature confusion enters because some investigators attempt to lump all DTA devices in a category of qualitative (10) which they may or may not be, while other investigators have presented arguments and/or data based entirely on geometry of cylindrical cups which are simplest to define mathematically (18, 19), and which may or may not be applicable to other systems.

While the system of interest here has been studied previously (12), it was not studied over a wide temperature range. Data, in general, are lacking on the effects of vacuum, pressure, and other environmental parameters that determine performance under these conditions. These considerations prompted the present work.

Experimental

Apparatus. The equipment used in this study was an R. L. Stone Thermal System. This consisted of an LB 202F recorder-controller which has a time base recorder and a JP-202 furnace platform. Stone calorimeter cells of the following types were employed: (1) SH-11BR2, which is either nickel or aluminum to cover the temperature range from ambient to 500°C or ambient to 1000°C and is used for static or dynamic flooding atmosphere control, and is equipped with platinel II differential and system thermocouples; (2) H-5 Sub-ambient unit which has a copper cell and operates over the temperature range of -160°C to 300°C and is equipped with iron-constantan differential and system thermocouples; (3) SH-15BR2 which has a nickel inner cell and a stainless steel cap and is used for high pressures up to 3000 psig at 500°C, and is equipped with platinel II differential and system thermocouples; (4) SH-14BR2 which has a nickel cell and is used for high vacuum work down to 10^{-6} Torr., and is equipped with platinel II differential and system thermocouples.

This instrumentation has been described previously (12, 20, 21, 22). All of the above mentioned sample holders have one important aspect in common, which is that they all contain the patented ring type differential thermocouple (23).

Results and Discussion

Ambient Pressure Effects - Static and dynamic gas flow flooding calorimeter cell. We wished to determine the calorimetric response of the nickel version of this cell (SH-11BRNi) over a wide range of temperatures since it is usable to 1100°C.

For this purpose we utilized a number of ACS Reagent Grade metals whose fusion points covered a wide range of temperatures. These materials represent the best calorimetric standards available and they are often used as calibration standards for transition temperature measurements as well. Scanning rates of 10°C/min. were used for all runs unless otherwise specified, and all runs were made in duplicate.

The response of this cell was determined by measuring the area produced by the fusion of each metal from which the heat transfer coefficient of the cell was calculated according to the following:

$$K = \frac{\Delta Hf \text{ of Std (cal/g)} \times \text{chart speed (in minute)}}{\frac{\text{area}}{\text{sample wt.}} \quad \frac{\text{in}^2}{g} \quad \times \text{Attenuation} \frac{(^\circ C)}{(\text{in})}} = \frac{\text{cal}}{\text{minute} \times {}^\circ C}$$

All measurements were corrected by the actual thermocouple output ($^\circ$C/in) in the transition temperature region. The derivations, theoretical considerations and applicability to this system has been presented previously where it was shown that $\Delta H \times K_t \int \text{to } \Delta T dt$ (12, 20). This K factor takes into account the size of the thermocouple bead, length and diameter of the thermocouple wire, geometry of the sample holder, design of the furnace, heat capacity of the thermocouples and sample containers, etc., which is acceptable in a fixed system.

The heat transfer coefficients (K values) for each metal were then plotted as a function of temperature using the first deviation from the baseline which is the transition temperature in this system (24). These results, shown in Curve B of Figure 1, were obtained using undiluted samples of approximately 10 mg. and approximate amounts of undiluted alumina in the reference pan that would balance the heat capacity of the sample. Nitrogen was used as the dynamic gas at a flow rate of 0.05 SCFH.

Ideally, a completely flat response or a constant K is desired as one obtains in adiabatic calorimetry. The increase in K as temperature is increased shows that heat transfer is not constant and that less area is produced by the same number of calories at higher temperatures as compared to the response at lower temperatures. This is the normal type of response curve obtained (25, 26) and permits differential scanning calorimetry to be performed since the area is reproducibly proportional to the total enthalpy of the transformation. Over limited temperature ranges good results have been obtained by considering K constant (12). It has been demonstrated that precision can be improved over narrow temperature ranges by correcting for the non-linear thermocouple output as a function of temperature (27). The present results show that this is not adequate if wide temperature ranges are to be covered.

The calorimeter cell response was then evaluated for ability to provide specific heat measurements. The response of this cell to changing heat capacity as well as fusion is shown in Figure 2. Figure 2 shows a blank run, that is, one without sample pans, sample or reference material; a run with pans and 10 mg. alumina as the sample; and a run with pans and 10 mg. of aluminum as the sample.

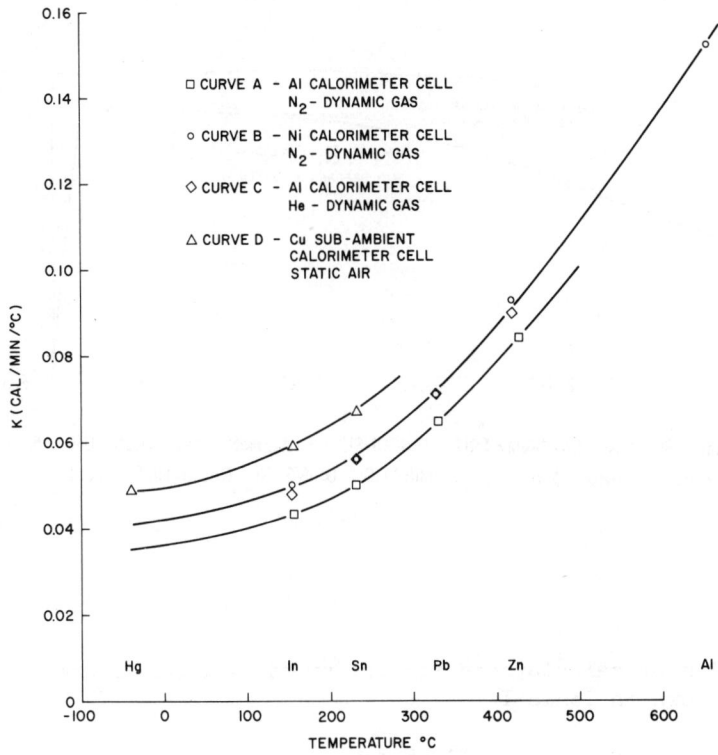

FIG. I HEAT TRANSFER COEFFICIENT (K) VS. TEMPERATURE FOR VARIOUS CALORIMETER CELLS AT AMBIENT PRESSURE AND DYNAMIC GAS ENVIRONMENTS

A number of other metals were run but attention was given only to the areas which were subsequently used for obtaining the K factors. The thermograms in Figure 2 have been included because alumina was used as the calibration standard for specific heat measurements.

The specific heat of the aluminum was calculated at temperatures of interest according to the following:

$$\frac{\text{Wt. of } Al_2O_3}{\text{Amplitude of } Al_2O_3} \times \frac{\text{Amplitude of Al}}{\text{Wt. of Al}} \times Cp Al_2O_3 = Cp\ Al$$

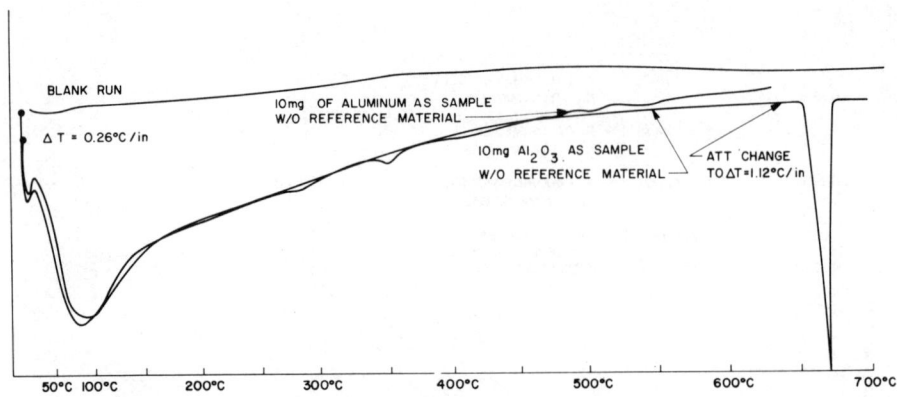

FIG.2 THERMOGRAMS SHOWING CALORIMETRIC RESPONSE OF ALUMINA AND ALUMINUM AS SAMPLES FOR C_p AND ΔH_f DETERMINATION USING Ni CALORIMETER CELL

The results of calculations for the specific heat of the aluminum are shown in Table I.

TABLE I

Comparison of Specific Heat Data

SH-11BR2Al Calorimeter Cell			SH-11BR2Ni Calorimeter Cell	
Cp of Polycarbonate			Cp of Aluminum	
Determined	Reported (29)	T	Determined	Reported (28)
0.25	0.31	50°C	0.21	0.21
0.28	0.35	100°C	0.22	0.23
0.34	0.40	150°C	--	--
0.40	0.47	200°C	0.23	0.24
--		300°C	0.25	0.25
--		400°C	0.25	0.25
--		500°C	0.26	0.26
--		600°C	0.25	0.27

The same basic procedure as described above was used to evaluate the aluminum calorimeter cell. In this case, a narrower temperature range was used for calibration because of the upper temperature limit of 500°C. The specific heat of polycarbonate was measured as the aluminum was, and the results are also shown in Table I.

Curve A in Figure 1 exhibits the same shape as Curve B over the temperature range studied. The lower K values indicate more efficient heat transfer and correspondingly greater sensitivity as we would expect with aluminum due to the better thermal conductivity which also decreases response time.

There is a curve that overlays Curve B that is defined by the parallelograms, i.e., Curve C. This curve was obtained using helium in place of nitrogen as the dynamic gas in the aluminum calorimetric cell. The calorimetric response is lowered because of the thermal conductivity of the helium as we might expect. The shape of the curve matches the other curves in Figure 1.

The response of the aluminum calorimeter cell can be seen in Figure 3. These thermograms show (A) a blank run, that is one without sample, reference, or pans, (B) a run with alumina in the reference pan and an empty pan in the sample side, (C) a run with polycarbonate in the sample side and an empty pan in the reference side, and (D) a run with polycarbonate in the sample pan and alumina in the reference pan.

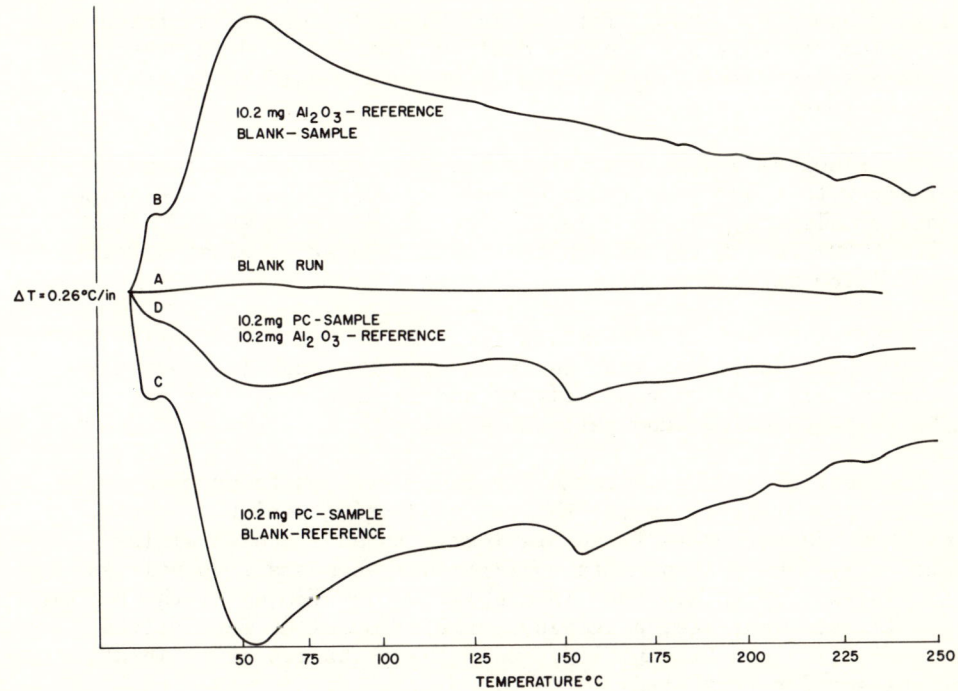

FIG. 3 THERMOGRAMS SHOWING CALORIMETRIC RESPONSE OF ALUMINA AND POLYCARBONATE USED FOR C_p DETERMINATIONS USING Al CALORIMETER CELL

Curves A, B, and C of Figure 3 illustrate the excellent thermocouple matching and uniform heat transfer characteristics of this cell. Curve B was used as the calibration curve for calculating the specific heat values shown in Table I from Curve C even though opposite sides of the differential thermocouple were used. Curve D was used as a cross-check on these values since the heat capacity of the polycarbonate was more closely matched in this instance. Thermogram D gave specific heat values in excellent agreement with those obtained from the previous thermograms.

It will be noted that each of the specific heat curves tends to return to the baseline. The reason for this is that the K values (heat transfer coefficients) become larger as temperature increases. This means that as Figure 1 shows a greater number of calories are required to produce an equivalent response at high temperatures as compared to low temperatures. Nevertheless, heat capacity measurements may be made routinely even at high temperatures as Table I shows. In Thermogram D of Figure 3, the total number of calories required per degree is somewhat less. This is because the sample heat capacity is more closely matched and the differential temperature is not as great as seen by the differential thermocouple. The uniformity and thermal mass of the furnace also influences this and a newly designed furnace used in these experiments increases sample holder temperature uniformity and decreases response time.

The aluminum and polycarbonate specific heat values compare favorably with those reported previously. The polycarbonate values may differ slightly due to molecular weight differences and other factors. All fusion values and the specific heat of alumina were obtained from reference 30.

Recent theoretical treatments (18, 19) as well as experimental data (10) have indicated that the area of a DTA peak is inversely proportional to thermal conductivity and is also dependent upon sample shape, packing, and particle size.

The determined K factors in Figure 1 were not found to be inversely proportional to the thermal conductivity of the sample. These above factors were found previously to have a negligible effect (12). The present data confirm this when small quantities of undiluted samples are used in a sensitive system as in the present case. In this instance, a lumped parameter model as was derived previously (12) can be employed. Only the mechanics of calibration were changed for this study.

This points out the serious error that can be introduced in choosing a particular system based upon DTA with a specified geometry, because that particular geometry can be described in more

Heating Rate Effects

The aluminum calorimeter cell (SH-11BR Al) was chosen to study these effects and thought to be typical of the effects demonstrated by other cells of different materials but of similar designs.

Garn has stated that the problem of increasing area with higher heating rates is due to the problem of heat transfer (31). This statement is not substantiated by the theoretical treatment of Melling et. al. (18) or the work by other authors (12, 32).

For these reasons the area response was studied as a function of heating rate. The area response to a fixed enthalpic input was found to be constant over the range studied, that is, from $2°C/min.$ to $40°C/min.$ scanning rates. This effect is a consequence of good heat transfer and negligible thermal gradients within the sample.

In cases where an X-Y in place of a strip chart recorder is used, an effect on the area as a function of heating rate was observed. This is true because the transition must absorb the same quantity of heat regardless of speed.

Since ΔT increases as the heating rate increases, a shorter amount of time is required for the same number of calories to be evolved or absorbed. When the abscissa represents temperature, the increased heating of the sample (absolute temperature through the sample) causes a correspondingly greater thermal lag of return of the sample temperature to its surroundings. This is manifested as an increased area and is not the case when time base recording is employed with this system.

Sub-Ambient Temperature Effects

<u>Ambient Pressure, H-5 Sub-Ambient Unit.</u> This particular unit has a calorimeter cell consisting of a copper body and an integral heater. This unit has been described previously (33).

The heat transfer coefficients (K values) were calculated as above over the useful temperature range of the cell ($-160°C + 300°C$) and are shown in Curve D of Figure 1.

This curve has the same general shape of those curves run in the previous calorimeter cells except that the K values are higher. This is true in spite of higher output thermocouples (iron-constantan) and the use of copper as the sample holder, which is more thermally conductive and thus should allow more efficient transfer of heat from the heating element which is adjacent to the sample holder.

This is due to the fact that there are slightly greater heat losses in this design calorimeter cell because there is less temperature uniformity since the hot or cool zone does not uniformly surround the cell cavity as in the previous designs.

High Pressure Effects

<u>SH-15BR2SS Cell</u>. This cell is similar in construction to the previous cells in that the cell itself is nickel with a stainless steel cap that contains the gas used for pressurizing.

We wished to examine the calorimetric response of this cell because it was felt that this cell might be useful in studying polymer crystallizations and oxidations under high pressure.

The effect of heat transfer was initially studied as a function of pressure with a constant calorimetric input; i.e., using tin as the calibration standard. These results are shown in Figure 4. The K factor determined for this cell at 0 psig is only slightly higher than that of the ambient pressure calorimeter cell. This is to be

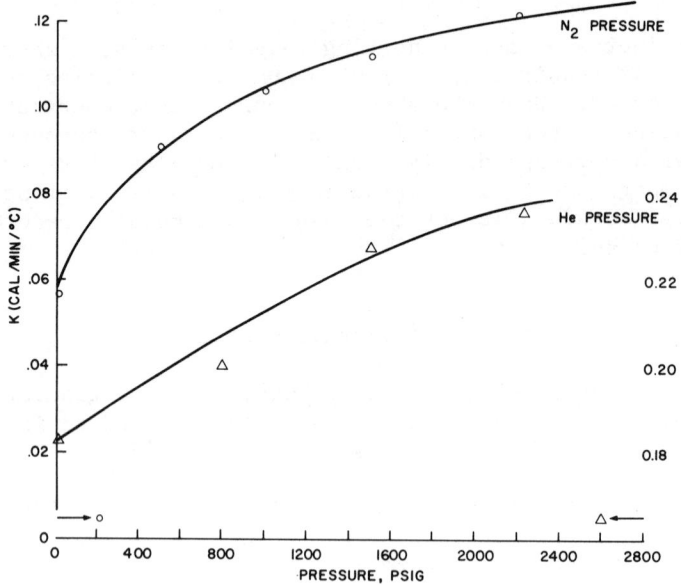

FIG. 4 HEAT TRANSFER COEFFICIENT (K) VS PRESSURE USING A CONSTANT CALORIMETRIC INPUT AND PRESSURE ENVIRONMENT

expected since the heat transfer through the stainless steel cap should be less efficient. At higher pressures, i.e., as the density of the gas increases there is a gradual increase in heat loss as is reflected by the higher K values. This effect is most marked in the case when helium was used to pressurize the cell. Due to the high thermal conductivity of the helium, heat is readily pumped away from the thermocouples.

Calibration curves were then run using the same standards as before except with 2200 psig N_2 and 2200 psig air as the pressurizing agents. These data are shown in Figure 5.

It can be seen that under pressure, the K value response is convex rather than concave and tends to increase very slowly at higher temperatures as opposed to the response at ambient pressures. Each separate pressure setting will define a separate calibration curve. Thus, a family of curves exists that defines the heat transfer coefficients for each pressure and each type of gas.

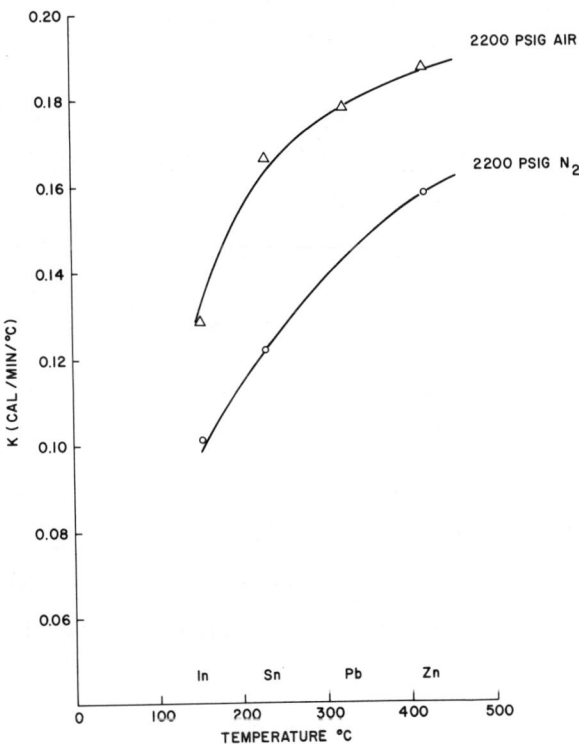

FIG. 5 HEAT TRANSFER COEFFICIENT (K) VS TEMPERATURE FOR THE HIGH PRESSURE CALORIMETER CELL USING DIFFERENT PRESSURE ENVIRONMENTS.

Both as a function of pressure (at constant temperature) and temperature (at constant pressure) the enthalpic effects approach limiting values that are dependent upon exact conditions.

The effect of pressure upon the fusion and crystallization and oxidation of polyethylene was then studied, using the heat transfer coefficients determined from the calibration curves.

Samples of low and high density polyethylene as received were scanned at a rate of 10°C/minute at 0 and 2200 psig N_2. These samples were annealed for 30 minutes at 145°C and cooled at 2°C/minute and then rescanned through the melting point at the same pressures as before. All areas were measured and converted to calorimetric values by means of the following equation:

$$K(cal/min/°C) \times Area\ (in^2) \times Att \frac{(°C)}{in} \times Cht\ Speed\ \frac{(min)}{in} = cal.$$

$$\frac{cal.}{sample\ weight} = \frac{cal.}{g.}$$

These data are shown in Table II.

TABLE II
Heats of Fusion of Low and High Density Polyethylene Under Various Nitrogen Environments

Sample	Gas	Pressure	Sample Size, G.	ΔHf, cal/g
Low Density PE	N_2	Ambient	0.00690	27.6
Low Density PE (Annealed)	N_2	Ambient	0.00690	35.8
High Density PE	N_2	Ambient	0.00695	41.8
High Density PE (Annealed)	N_2	Ambient	0.00695	56.1
Low Density PE	N_2	2200 psig	0.00700	11.1
Low Density PE (Annealed)	N_2	2200 psig	0.00700	15.9
High Density PE	N_2	2200 psig	0.00720	28.0
High Density PE (Annealed)	N_2	2200 psig	0.00720	35.5

Lower ΔHf values were experienced in every case for those samples run at 2200 psig of N_2 but treated in every other respect identically with those samples run at 0 psig. This reflects that lower crystallinities were obtained under pressure than were obtained at ambient pressure. Based on a value of 65 cal/g. for 100% crystalline polyethylene the value for the HD polyethylene (annealed) is about 87% at ambient pressure as compared to 55% at 2200 psig.

Although additional experiments are required, these data strongly suggest that pressures in the ranges utilized in these experiments inhibit crystallization, perhaps by restricting segmental motion and inhibiting perfection of the lamellae. The thermograms of the high density polyethylene under varying conditions are shown in Figure 6.

One of the problems encountered in studying the oxidation and/or thermal stability of polymers is that at the relatively slow scanning rates employed, oxidation produces considerable sample charring. This effect is quite often non-reproducible since it depends on the surface area exposed, use of dynamic gas, etc.

It was felt that one approach to this problem might be the use of air at high pressures to produce quantitative, reproducible oxidations and perhaps serve as a basis for a rapid screening procedure in defining differences in flammability of materials.

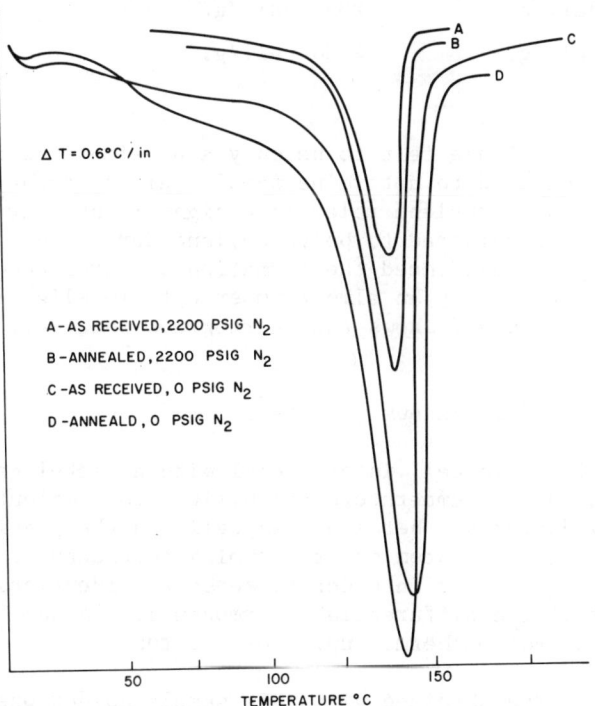

FIG. 6 THERMOGRAMS OF HIGH DENSITY POLYETHYLENE USING THE HIGH PRESSURE CALORIMETER CELL AND N_2 PRESSURE ENVIRONMENTS

In order to determine the effect of air pressure on polymers, polypropylene and polyethylene were run using 2150 psig of air. The heat of fusion was also measured and these data are shown in Table III.

TABLE III

Heats of Fusion and Oxidation of Polypropylene and Polyethylene in Air Environments

Sample	Gas	Pressure	Sample Size	Temperature Ranges of Decomposition	ΔHf	ΔH Decomposition
Polypropylene	Air	2150 psig	0.00056 g.	183°C-217°C	11.7 cal./g.	2330 cal./g.
HD Polyethylene	Air	2150 psig	0.00061 g.	190°C-250°C	40.7 cal./g.	4050 cal./g.

The data in Table III are felt to be only approximate since a micro balance was not used to determine sample weights accurately. It will be noted that polyethylene releases a significantly greater amount of heat energy as compared to polypropylene during the burning. Neither polymer evidenced the formation of carbonaceous residue in the sample pan; only an almost imperceptible slightly brown film remained. Figure 7 shows the thermograms obtained.

High Vacuum Effects

SH 14BR2-Ni Cell. This cell is of nickel with a nickel cap to contain the vacuum and a temperature capability from ambient to 1000° C. The only difference between this cell and the previous cells described is that the system or programming thermocouple is also made in the form of a ring in order to sense the identical heat history as that of the differential thermocouple. A sample pan was placed in the system thermocouple for all runs.

The calibration curve obtained with this sample holder operating at 5×10^{-6} Torr is shown in Figure 8. It follows the general shape of the pressure calibration curves in that the K values tend to approach constant values rather than continue to increase rapidly.

ENVIRONMENTAL EFFECT ON QUANTITATIVE MEASUREMENTS 383

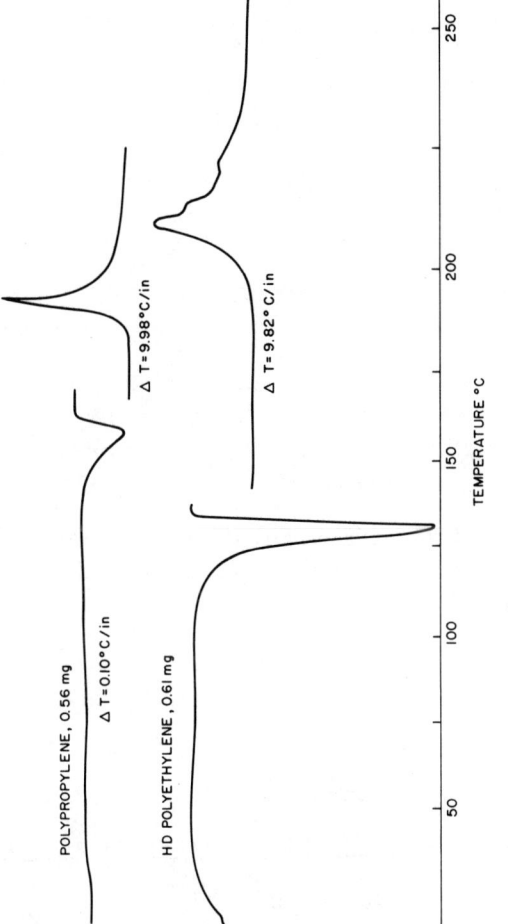

FIG. 7 THERMOGRAMS OF POLYPROPYLENE AND HIGH DENSITY POLYETHYLENE USING THE HIGH PRESSURE CALORIMETER CELL AND AIR PRESSURE ENVIRONMENTS

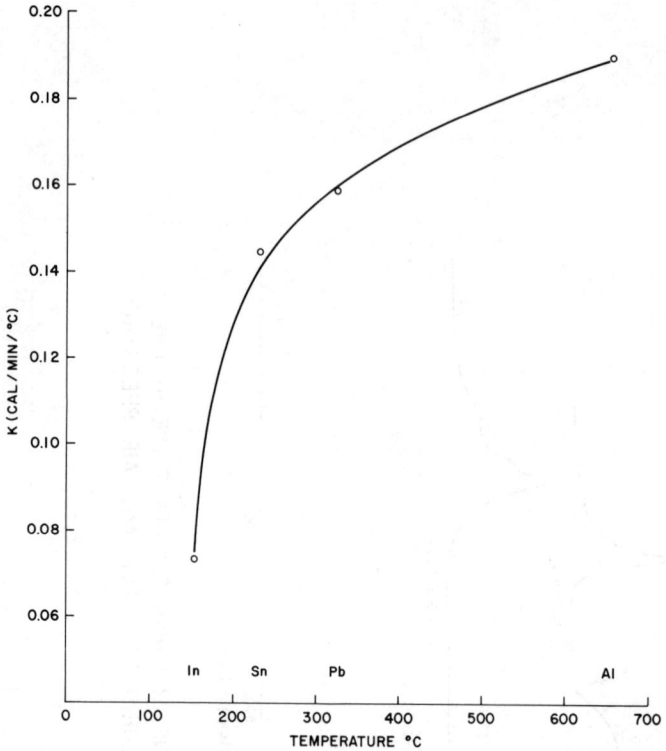

FIG. 8 HEAT TRANSFER COEFFICIENT (K) VS TEMPERATURE FOR THE HIGH VACUUM CALORIMETER CELL AT 5×10^{-6} TORR

TABLE IV

Transition Temperature of Calibration Standards at 5×10^{-6} Torr

Sample	Measured	Reported
Indium	140.5° C	156.6° C
Tin	214.6° C	231.8° C
Lead	296.8° C	327.4° C
Aluminum	628.5° C	660.1° C

In the high vacuum runs, although the design is similar to the nickel cell, the heat transfer coefficients are much higher indicating significantly poorer heat transfer because of the vacuum. The peak shape often obtained with both standards and sample materials is quite distorted from that obtained in any of the other runs. The transition temperature of the standards were significantly lowered, as Table IV shows.

One explanation of this is that the heat uniformity within the cell is poor, which results in a different temperature zone for the system thermocouple as compared to the differential thermocouple. This may also be accentuated by the different heat transfer mechanism that occurs in vacuum. However, it should be noted that: (1) the heat history of the system and differential thermocouple have been designed to be as identical as possible in all respects, (2) small undiluted samples were used, (3) the measured temperatures are consistently lower and show a linear relationship with true transition temperature, and (4) the shape of the curve indicates melting over a broad range.

For the purpose of peak shape comparison, Figure 9 shows the shape of the tin peak under the environment of dynamic nitrogen flow and vacuum.

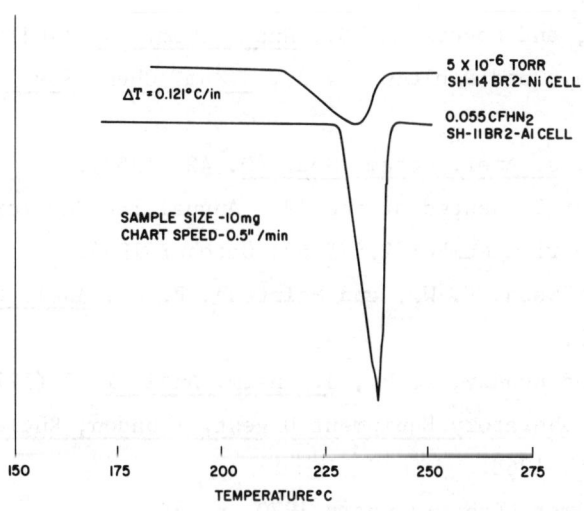

FIG. 9 THERMOGRAMS OF TIN SHOWING THE CALORIMETRIC RESPONSE USING DYNAMIC GAS FLOW AND HIGH VACUUM ENVIRONMENTS

Literature Cited

1. Sykes, C., Proc. Roy. Soc. 148, 422 (1935).
2. Moser, H., Physik Z. 36, (1937).
3. Kumanin, K. G., Zh. Prikl. Khim. 20, 1242 (1947).
4. Eyraud, C., et. al., Compt. Rend. 240, 862 (1955).
5. Solomons, C., and Cummings, J. P., Rev. Sci. Instr. 38, (1964).
6. Eyraud, C., Compt. Rend. 238, 1511 (1954).
7. Vold, M. J., Anal. Chem. 21, 683 (1949).
8. Nagasaki, S., and Takagi, T., Applied Physics (Japanese) 1, (1948).
9. Clarebrough, M. E., Hargreaves, M. E., Mitchell, D., and West, G. W., Proc. Roy. Soc. 215 (1952).
10. Watson, E. S., O'Neill, M. J., Justin, J., and Brenner, N., Anal. Chem. 36, 1233 (1964).
11. O'Neill, M. J., Anal. Chem. 36, 1239 (1964).
12. David, D. J., Anal. Chem. 36, 2162 (1964).
13. Barrall, E. M., Porter, R. S., and Johnson, J. F., Anal. Chem. 36, 2172 (1964).
14. Barrall, E. M., and Rogers, L. B., Anal. Chem. 34, 1101 (1962).
15. Borchardt, H. J., and Daniels, F. J., J. Am. Chem. Soc. 79, 41 (1957).
16. Boersma, S. L., J. Amer. Ceram. Soc. 40, 42 (1957).
17. Dosch, E., Paper Presented at the 19th Annual ISA New York Conference, Preprint Number 2.6-5-64, October 1964.
18. Melling, R., Wilburn, F. W., and McIntosh, R. M., Anal. Chem. 41, 1275 (1969).
19. Berg, L. G., and Egunov, V. P., J. Therm. Anal. 1, 5 (1970).
20. David, D. J., Laboratory Equipment Digest, (London, England), June and August, 1968.
21. David, D. J., Amer. Lab., January 1970, p. 35.
22. Stone-Tracor Bulletins, ST-109, ST-107, ST-101, ST-106.
23. Stone, R. L., and Burress, G. T., U. S. Pat. 3, 298, 220 (to Tracor, Inc.) (Jan. 17, 1967)

24. David, D. J., "Transition Temperatures in Differential Thermal Analysis", Chap. 2, Polymer Thermal Analysis, Slade, P., and Jenkins, L., Editors, Marcel Dekker, N.Y., 1966.
25. Foldvari-Vogl, M., and Klikerszky, B., Acta Geol. (Budapest) 5, 187 (1958).
26. Sabatier, G., Bull. Soc. Franc. Mineral. 77, 953, and 1077 (1954).
27. Smith, M. R., Stone Thermoscope 1, March 1968.
28. Hodgman, C. D., "Handbook of Chemistry and Physics", 30th Ed. Chemical Rubber Publishing Co., Cleveland, Ohio, 1947.
29. O'Reilly, J. M., Karacz, F. E., and Baer, H. E., J. Polymer Sci. Pt. C No. 6, 109 (1964).
30. Kubaschewski, O., Evans, E. L., and Alcock, C. B., "Metallurgical Thermochemistry", Pergamon Press, London 1967.
31. Garn, P. D., "Thermoanalytical Methods of Investigation", Academic Press, New York 1965.
32. Sewell, E. C., Chap. 2, in "DTA of Clays", Min. Soc., London, 1967.
33. David, D. J., Mostyn, W. T., and Stone, R. L., "Design and Applications of a Sub-Ambient Differential Thermal Instrument", paper presented at the 19th Pittsburgh Conference on Analytical Chemistry and Applied Spectroscopy, March, 1968.

A GENERAL METHOD FOR CHARACTERIZING THERMOANALYTICAL DATA

Stuart M. Ellerstein

Thiokol Chemical Corporation

Trenton, New Jersey 08607

INTRODUCTION

The primary objective of this investigation was to find a general method that could be used for the quantitative characterization of thermoanalytical data. It seemed desirable to obtain a method which was not dependent upon any mathematical model, and which was only dependent upon the intrinsic properties of the thermogram itself. On the other hand, if a good model was available then a physical interpretation of the intrinsic parameters might be obtained.

METHOD

The approach that was finally chosen had been developed in the field of statistics in order to characterize a frequency distribution[1]. More recently this type of analysis has been used in the field of chromatography to elucidate details about peak shape and to aid in the resolution of unresolved peaks[2,3].

The basis for this method lies in the fact that a peak may in principle be totally described by its complete set of mathematical moments. These may be calculated by dividing the thermogram (in some cases, e.g. thermogravimetric analysis, the derivative of the thermogram is used) into i equal finite intervals of width d. The n moments are then given by

$$M_n = \sum_i T_i^n f_i \Big/ \sum_i f_i \qquad (1)$$

where T_i is the mid-value of the ith interval along the temperature (or time) axis, and f_i is the corresponding displacement of the peak from the base line. Letting $N = \sum_i f_i$ it is seen that $M_0 = 1$, and that the area under the peak is approximated by Nd (the approximation may be improved by using numerical integration by quadratures).

The centroid of the peak $\bar{T}$ is simply M_1. It is one of the three common measures of central tendency of the peak, the other two being the mode (peak maximum) and the median (which divides the peak into two equal areas). In a symmetrical peak the three values coincide. As a peak becomes more asymmetric the centroid (or arithmetic mean) is displaced further than the median from the mode in the direction of skewness.

It is convenient to take all of the higher moments μ_n for $n \geq 2$ about $\bar{T}$. μ_2 is now seen to be the variance, i.e., the square of the standard deviation σ, which gives a measure of dispersion. Another quantity which is of value is the coefficient of variation or relative standard deviation, i.e., $\sigma/\bar{T}$.

The next moment μ_3 gives the direction and magnitude of skewness (a negative sign indicating skewness to the left). In practice, however, a dimensionless quantity is preferred. One that is often used is $C_3 = \mu_3 / \mu_2^{3/2}$.

The kurtosis of the peak is related to μ_4. This gives information about how broad the modal region of the peak is, as well as, how the tails taper off. Once again a relative value of the kurtosis is preferred. One that is often used is $\beta_2 = \mu_4 / \mu_2^2$.

Higher moments of the peak are also easily calculated with a digital computer, but are usually unneccessary. Odd moments reflect asymmetry, while even moments extend the concept of kurtosis.

CALCULATIONS

The numerical calculations were performed using an IBM 360/50 time-sharing computer. All programs were written in BASIC, and are available to interested parties.

In order to illustrate this method peaks were generated using the assumed normalized rate law.

$$(1/H_o) \, dH/dT = k/q \, (H_r/H_o)^n \quad (2)$$

where as usual dH/dT is the instantaneous change of enthalpy with absolute temperature, $k = A\exp(-E^*/RT)$ is the Arrhenius rate constant, H_r is that part of the enthalpy remaining at temperature T, q is the constant heating rate dT/dt, n is the apparent order of reaction, and H_o is the total enthalpy under the peak. This type of relation has been widely used by many workers, e.g. references 4 and 5. (On the other hand, in many experimental investigations equation (2) is often criticized with respect to its validity, which was a motivating factor in this present endeavor.)

Equation (2) may be integrated by parts making use of the non-elementary integral Ei (-x) where $x = E^*/RT$. In the present work up to eleven terms of the asymptotic expansion were used in order to minimize the error over the complete range of the reaction. Other workers have used two and three term approximations, which can lead to large errors.

RESULTS

It may be seen that equation (2) has the four parameters E^*, n, A, and q with the last two appearing as the ratio A/q. Calculations were performed using the following values of the above parameters:

E^* 30, 40, 50, 60, 80, and 100 k.cal./mole

n   0, .5, 1, and 2

A/q 10^{15}, 10^{16}, 10^{17}, and 10^{18}

where the ratio was obtained by fixing A at 10^{17} and varying q from 100 to 0.1. For each case the temperature range had to be first located. This was done by having the computer search for the interval using a value of 20°K for the increment. Once the range had been roughly established it was more precisely determined by using a smaller increment, e.g., 1°K, or for the case of n = 0 even 0.1°K at its high temperature end. The entire peak was then generated using a 1°K increment, and its intrinsic parameters were calculated. It is noted that each peak was tested by the Method of Freeman and Carroll[6] to see if the peak output was consistent with the input. If agreement was not within the round off error, then the run was rechecked, e.g., input: $E^* = 40,000$, n = 1.0, $A/q = 10^{17}$; output: $E^* = 40,000.7$, n = 0.99978, $A/q = 1.00058 \times 10^{17}$, which was considered to be good. It is noted that all calculations were performed using a single precision mode.

Some typical examples of the accrued data are now presented. In Table I is seen the effect of keeping n and A/q constant, while allowing E^* to vary.

Table I

The Effect of Varying the Activation Energy Upon the Intrinsic Parameters. (n = 1 and $A/q = 10^{17}$)

E^* k.cal/mole	$\bar{T}$ °K	σ °K	$\sigma/\bar{T}$	C_3	β_2
30	361.2	10.3	.0285	-.898	4.38
40	478.5	13.5	.0283	-.899	4.38
50	595.1	16.7	.0281	-.899	4.38
60	711.1	19.9	.0280	-.898	4.37
80	942.1	26.3	.0279	-.900	4.38
100	1171.8	32.5	.0277	-.901	4.37

It is seen that E^* is approximately proportional to both $\bar{T}$ and σ. This is borne out by the almost constant value of the coefficient of variation. The negative sign of C_3 tells that the peaks are all skewed to the left, which is characteristic of a first order reaction. Also, the value of β_2 being greater than three reveals that the peaks are all leptokurtic, i.e., having a narrower modal portion and higher tails than the corresponding Gaussian curve, cf., $C_3 = 0$ and $\beta_2 = 3$.

In the next table both the apparent order of reaction and the heating rate are varied.

Table II

The Effect of Varying Both n and q upon The Intrinsic Parameters ($E^* = 40,000$ k.cal./mole and $A = 10^{17}$)

q °k/min.	$\bar{T}$ °k	σ °k	$\sigma/\bar{T}$	C_3	β_2
n = 2					
0.1	460.8	18.4	.0406	.386	4.61
1.0	485.3	20.4	.0420	.407	4.64
10.0	512.5	22.7	.0443	.426	4.68
100.0	542.8	25.4	.0468	.440	4.66
n = 1					
0.1	454.7	12.3	.0270	-.911	4.43
1.0	478.5	13.5	.0283	-.899	4.38
10.0	504.9	15.0	.0297	-.890	4.36
100.0	534.3	16.7	.0313	-.878	4.32
n = .5					
0.1	452.3	10.5	.0232	-1.37	5.59
1.0	475.9	11.6	.0243	-1.38	5.69
10.0	502.0	12.8	.0256	-1.37	5.67
100.0	531.1	14.3	.0269	-1.36	5.61
n = 0					
0.1	450.8	9.3	.0206	-1.75	7.11
1.0	474.2	10.2	.0215	-1.77	7.27
10.0	500.0	11.3	.0227	-1.76	7.25
100.0	528.9	12.6	.0239	-1.75	7.18

Looking at Table II it is seen that $\bar{T}$ and σ both increase with reaction order and heating rate. Moreover, the coefficient of variation also increases with these kinetic parameters. The positive sign of C_3 shows that a second order reaction is skewed to the right, while the others are skewed to the left. This therefore predicts that for this set of kinetic parameters a symmetrical peak will occur between n = 1 and n = 2. C_3 also shows that for n = 2 the skewness increases slightly with heating rate, while for n = 1 there is a slight decrease (n = 0 and n = .5 are apparently unaffected). Finally, β_2 appears to go through a minimum value, although all peaks calculated to date have been leptokurtic.

DISCUSSION

The example used to illustrate the method shows that a peak may be characterized by its intrinsic parameters, which are related to its mathematical moments. Like other methods of calculation used it will be subject to base line placement error. Also, since the numerical integration accuracy is dependent upon the width of the class interval, this will have some effect on the calculated parameters. A value of one degree Kelvin was arbitrarily chosen for these calculations. A ten degree interval would not have made too much difference since these peaks are rather broad. On the other hand, a narrower peak as might be encountered during an impurity determination would require a smaller interval. The availability of digitizers permit great flexibility in selecting an interval which is suitable for a given problem. In comparison studies it is best to use the same interval throughout.

At present we are studying the trajectory of the centroid $\bar{T}$ vs. temperature. The results are incomplete at this writing, but will be reported on at a later date.

CONCLUSION

It is concluded that the salient features of a thermoanalytical peak may be characterized using its mathematical moments. Since they are not dependent upon a particular model they have been termed the intrinsic parameters. They are very useful since at the very least they can quantitatively supplement other more specific methods that are in use.

ACKNOWLEDGEMENT

Thanks are given to Mrs. M. Conte for helping to prepare this manuscript.

BIBLIOGRAPHY

1. Croxton, F. E., "Elementary Statistics", Dover, New York, 1959.

2. Grushka, E., Myers, M. N., Schettler, P. D., and Giddings, J. C., Anal. Chem., 41, 889 (1969).

3. Grushka, E., Myers, M. N., and Giddings, J. C., Anal. Chem. 42, 21 (1970).

4. Flynn, J. H., and Wall, L. A., J. Res. Nat. Bur. Stand., A70, 487(1966).

5. Ellerstein, S. M., "Analytical Calorimetry", Plenum, New York, 1968, p. 279.

6. Freeman, B. S., and Carroll, B., J. Phys. Chem., 62, 394 (1958).

THERMAL ANALYSES OF POLYMERS. VI.

THERMAL DEPOLARIZATION ANALYSIS (TDA)

Gerald W. Miller

E. I. du Pont de Nemours & Company (Inc.)
Instrument & Equipment Division
Wilmington, Delaware 19898

INTRODUCTION

Magill published some of the first work on the use of the hot-stage microscope for investigating the crystallization of polymers, particularly Nylon 66 (1,2). These isothermal crystallization studies were carried out in the temperature range of 160° to 250°C, after initially melting the polyamide at 300°C. His kinetic data agreed with that which had been published by other authors using other techniques for the study of the crystallization of this polymer.

A major contribution to the use of fusion microscopy in polymer characterization can be illustrated by the work of Barrall, Johnson and Porter (3), wherein they describe the modification of a polarizing microscope for coupling it to a Thermal Analyzer for temperature programming and signal recording. Their work shows excellent agreement with other thermal methods, e.g., differential thermal analysis. From their scan of cast polypropylene, it would seem that the depolarized light measurement may illustrate morphological changes that are not easily observed by differential thermal analysis or other thermal methods. This is also shown to be true for the changes in depolarized light intensity as a function of temperature for poly(3-methyl-1-pentene). Isothermal crystallization curves have also been observed by this technique for polyethylene terphthalate, Penton polyether (4) and polypropylene (5).

This report illustrates the extension of uses of a polarizing microscope similar to that of Barrall, et al. (3) for the illustration of specific morphological changes in polymers which are either not readily resolved or are undetected by other thermal techniques. This technique can detect changes in molecular order as observed by x-ray techniques as well as those which lack this extended three-dimensional structure, e.g., short range order. Furthermore, glassy transitions for simple molecules are not observed by this measurement. The technique of TDA is also shown to correlate well with conventional thermal techniques of DTA, DSC, linear expansion and dilatometry.

EXPERIMENTAL

All of the data contained herein were run in duplicate and, in certain cases, in triplicate. The microscope set-up and operation were patterned after that of Barrall, et al. (3) and are illustrated in Figure 1. The hot-stage attachment was changed to accommodate programmed cooling with the addition of liquid nitrogen. Proper insulation and the introduction of liquid nitrogen directly on the heater pedestal allowed cooling to -150°C, and temperature programming and signal readout were achieved with a Du Pont 900 Thermal Analyzer. DSC and DTA heating rates were 10°/minute, and TDA and expansion, 5°/minute. Employing heating rates greater than 5°/minute for TDA induces a significant temperature error.

RESULTS AND DISCUSSION

Solid crystalline materials are known to depolarize light in that they convert plane polarized light into eliptically polarized light. This is caused by changes in birefringence, and as the birefringence of the material changes at a specific temperature, there will be a change in the depolarized light intensity, giving rise to the name, "Thermal Depolarization Analysis".

The measurement of depolarized light transmission is also affected by the scattering capabilities of the material in that a material upon melting may decrease its scattering intensity and appear that it is depolarizing further. This effect is particularly inherent in powdered materials. Furthermore, a change in birefringence may be detected according to the ratio of amorphous to crystalline areas within a given substance, and this refractive index change according to this ratio has been termed form birefringence. Consequently, while the measurement apparently hinges on the nature of the crystalline or molecularly ordered state and particularly since it is influenced by the size of the crystallites, their orientation, and the configuration of the unit cell, it would appear initially that measurements by TDA produce a quantity which expresses the summation of a number of physical contributions to changing the light intensity, including light scattering, form birefringence, orientation, configuration of the sample and temperature.

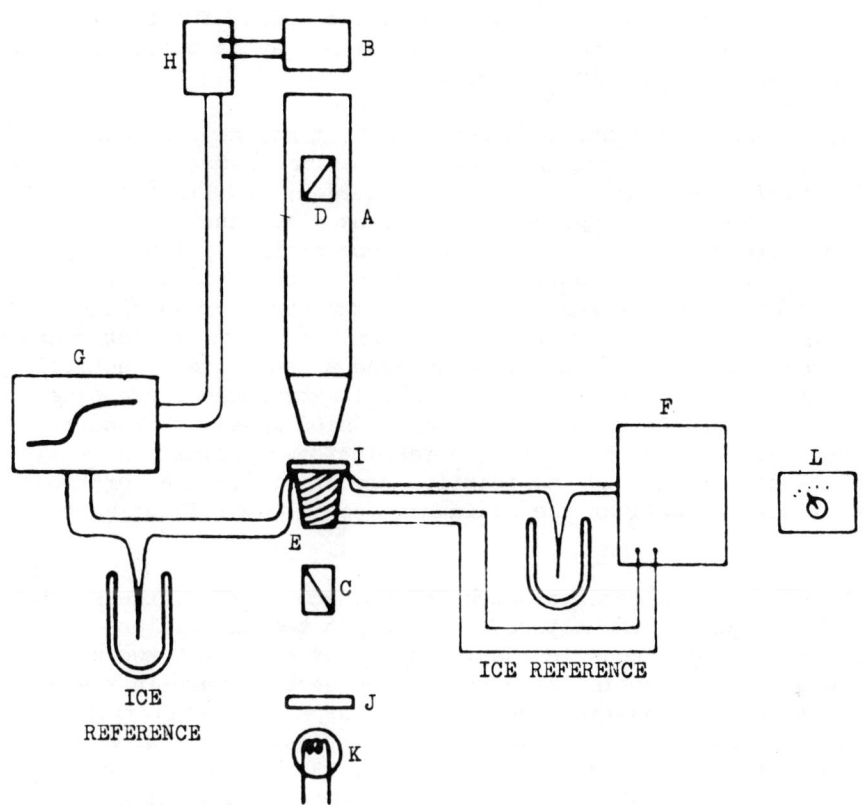

A	MICROSCOPE OPTICS	G	RECORDER
B	PHOTOCELL, CADMIUM TELLURIDE	H	PHOTOCELL AMPLIFIER
C	POLARIZER	I	SAMPLE
D	ANALYZER	J	FILTER
E	HOT STAGE	K	LIGHT SOURCE
F	TEMPERATURE REGULATOR	L	PREMELTER

Fig. 1. Modification of a polarizing for TDA measurements.[3]

In some cases, it may be desirable to keep the physical conformation of the sample for measuring its changes in depolarization, whereas in other situations the sample can be either melt pressed or solution cast in order to have a thin film between two cover slips for the microscope stage. The proper preparation of very thin films helps to eliminate much of the contribution due to scattering unless the film is quite translucent. The size of the samples for this measurement is usually of the order of a few milligrams. In some cases one or two cover slips may be used, depending upon the desire of the investigator.

1. <u>Ammonium Nitrate</u>. In some cases it may be desirable to isolate a particular wavelength band in order to measure the changes in birefringent color of a given material, and, initially, it would seem that each compound and the reason for the investigation would dictate whether or not and what type of interference filter should be used. For example, in using no filter for dotriacontane, maximum light transmission is observed, whereas a green filter slightly decreases the transmission level, a yellow filter decreases the transmission level somewhat more, and a red filter practically nulls out the transmission level when all measurements are made with the same sample at the same transmission level. For dotriacontane, there is no change in the resolution capability as well. These changes are due to the relative amount of light absorption by the filter as well as the maximum response peak of the photocell at 530 nm.

However, for a material such as ammonium nitrate, there are definite changes in the nature of the first two transitions as well as some degree of change in the intensity of the first two peaks (Figure 2). For example, with no interference filters in the beam, the first three transitions occur at the expected temperatures of 56°, 83° and 132°C. However, the light intensity increases at 56°, decreases at 83° and further decreases at 132°C. When a green filter is used, a peak begins at 56°, reaching a maximum at 60°. There is then an increase in light intensity at 83° and as with no filter, a decrease at 132°C.

A yellow filter shows a small decrease in light intensity at 56°, followed by an increase in light intensity at 57°. At 78°, the light intensity increases, plateauing at 82° and decreasing at 92°C. The usual decrease at 132°C occurs as well. The runs with no filter or the green or yellow filters were run at the same sensitivities.

On using a red filter at five times the sensitivity of the other curves in Figure 2, the light intensity at 56° increases, and there is no evidence of the 83° light change. The third peak again decreases to lower light levels.

The cubic to isotropic transition is observed at the highest sensitivities of the TDA instrument at 172°C. The transition apparently only weakly affects polarized light. In every scan, the transformation to cubic form at 133°C (third transition) is readily apparent, independent of the filter used.

The four transitions identified with ammonium nitrate are, in sequence (6):

a. β-rhombic ⟶ α-rhombic

b. α-rhombic ⟶ rhombahedral

c. rhombahedral ⟶ cubic, 132°C

d. cubic ⟶ isotropic liquid, 172°C

The latter two transitions can occur at the indicated temperatures, but the first two have been reported to be dependent on pretreatment, thermal history and other environmental effects (7). The variation in direction of light change as well as the presence of the transition appear to indicate the first two transitions are also wavelength dependent.

2. <u>Benzoic Acid</u>. In the elementary course in organic chemistry, benzoic acid is one of the standard substances used for melting point determinations, and it is often used as a standard in DTA measurements for both the determination of thermal and calorimetric accuracy. The DTA response to the melting of benzoic acid is shown in Figure 3 in which the onset of the endotherm occurs at 119° with total melting at 121°C. Since this measurement is carried out with the thermocouple in the sample, there is likely no more accurate method than this for the determination of the melting temperature. An examination of the melting behavior of benzoic acid by use of Thermal Depolarization Analysis (TDA), shown also in Figure 3, indicates that premelting begins as low as 100°, with the premelting peak at 120° and final fusion at 125°C. It would appear from the correlation of the results in DTA and TDA that while the DTA measurement averages the melting phenomenon, the TDA measurement is actually able to separate two types of disordering for benzoic acid. Naturally, as the light intensity increases, there is an increase in molecular order, and as it decreases, the molecular order is melted out. In comparing the two scans of benzoic acid by TDA and DTA measurements, the final melting and return to baseline of the DTA scan coincide well in temperature at 125° with the fusion as indicated by the TDA measurement.

The TDA curve in Figure 3, labeled "initial run", shows the response of benzoic acid powder between two cover slips, whereas the second melt, which is much more intense in its depolarization,

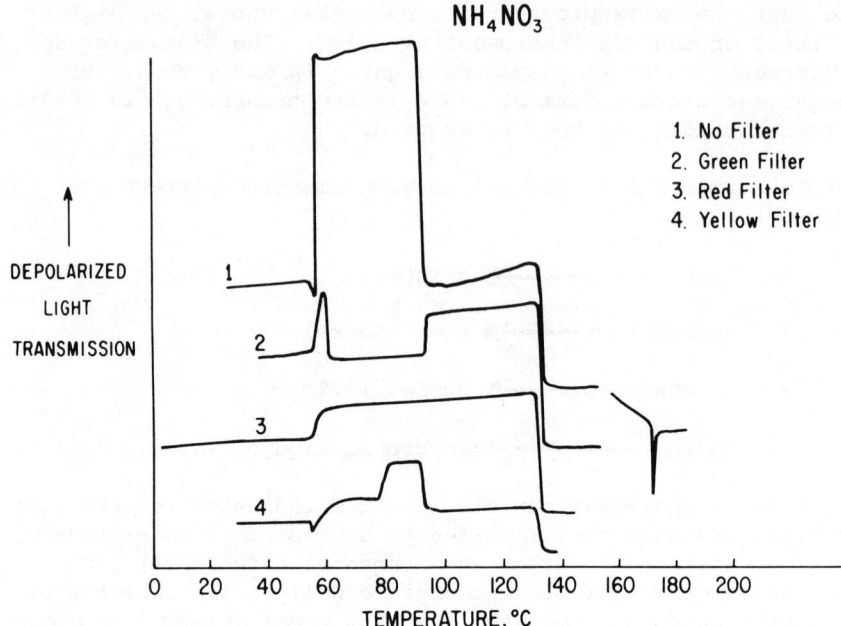

Fig. 2. Filter effects on the TDA measurement of NH_4NO_3.

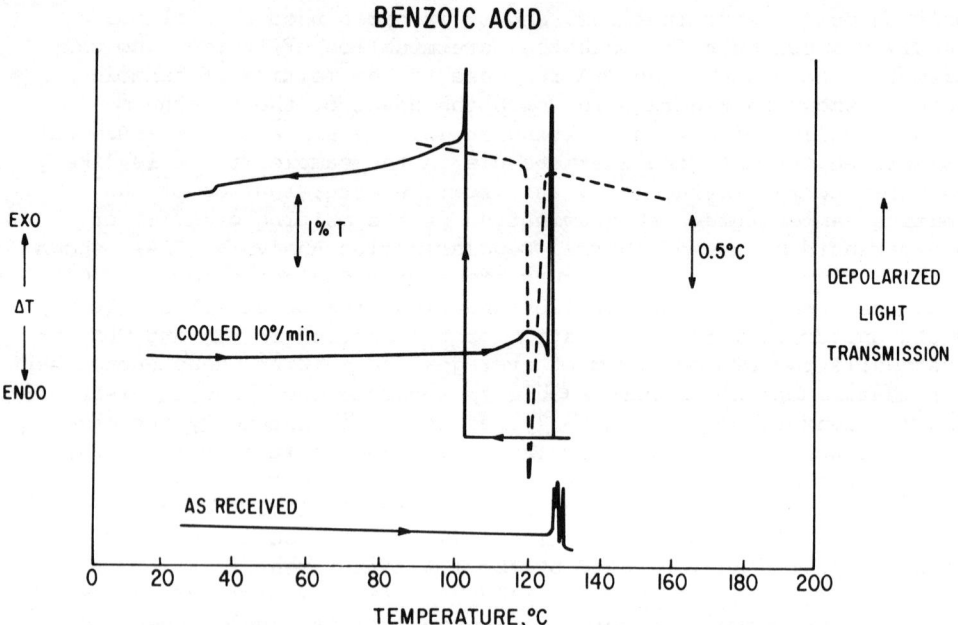

Fig. 3. TDA and DTA response of benzoic acid.

is the remelt between two cover slips. The lower intensity in the depolarization peak for the first melt is due to the large amount of scattering of light from the surface of the benzoic acid powder, and, hence, the change in depolarization from the crystalline state to the isotropic liquid is very small. This change is quite large in the second melt. Hence, it would appear that while one peak shows the thermal response of benzoic acid by DTA measurements, a change in the lattice structure occurs prior to final fusion when measuring the melting phenomena by TDA.

3. <u>Tristearin</u>. Barrall and Guffy (8) have illustrated the polymorphic behavior of tristearin using DTA, NMR, and DLI or TDA measurements. The DTA data in Figure 4 agree well with their data for the endothermal maximum at 56°, the exothermal maximum at 61°, and the second endothermal maximum at 72°C. TDA data also agree quite well in that the onset of fusion begins at 56° with recrystallization beginning at 59°, ceasing at the temperature at which fusion begins (68°), with final fusion terminating at 75°C. A comparison of the DTA and TDA characteristics of tristearin illustrates the increasing depolarization of light as the exothermal reordering occurs, with consequent fusion indicated by a decrease in light intensity to zero light level. For a clear-cut case like tristearin, it's easy to recognize this correlation. However, for very subtle changes in reordering for which the heat of exotherm is extremely small and not perceivable by DTA, TDA measurements can tell quite easily something about this reordering process.

It has been reported in the literature of Charbonnet and Singleton (9) that the melting phenomenon occurring at 56° is the transition of the tristearin from the α to the β form. The exotherm at 61° represents the conversion of the β to the β' form, with the fusion at 72°, indicating the conversion of the β' form to the isotropic liquid. The thermal characterization of polyglycerides will be treated in a forthcoming publication.

4. <u>Cholesteryl Proprionate</u>. The observation of the mesomorphic character of cholesteryl proprionate has been made with the techniques of fusion microscopy, differential thermal analysis, depolarized light intensity, dilatometry and pressure calorimetry (10). Measurements of changes in depolarization in Figure 5 reflect basically the same changes observed by Barrall, Porter and Johnson (11). However, Figure 5 serves to illustrate more the effect of the physical state of the sample rather than the chemical makeup. After the cholesteryl proprionate has been melted between two cover slips, the material was heated at 5° per minute. The curve appears rather flat until the crystal to cholesteric phase transition occurs at 103°C followed by a change from the cholesteric phase to the isotropic melt at 119°C. The sample is then program-cooled at 5° per minute, illustrating in Figure 5 that by the time the sample had reached 75°C, all crystallization was complete. Between 20° and

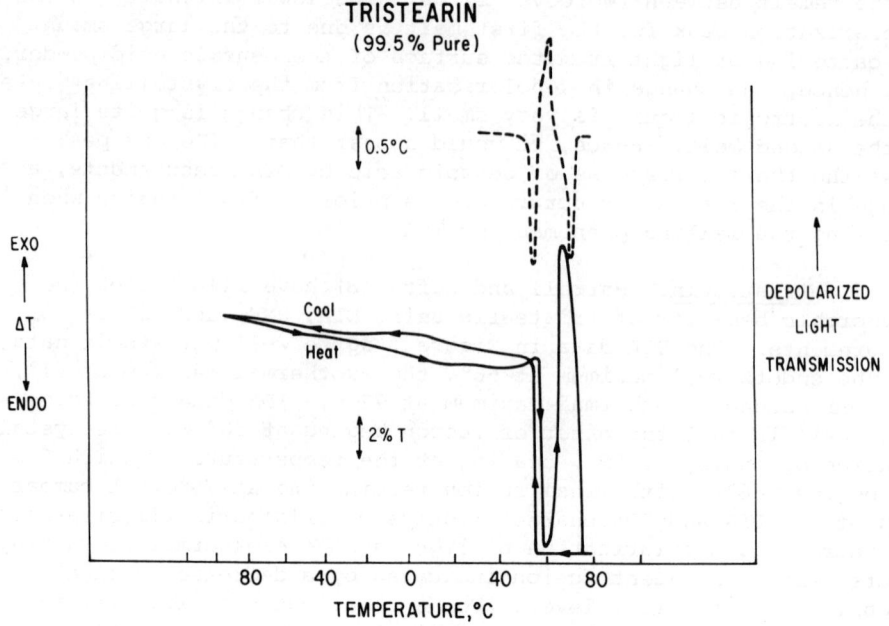

Fig. 4. TDA and DTA response of tristearin.

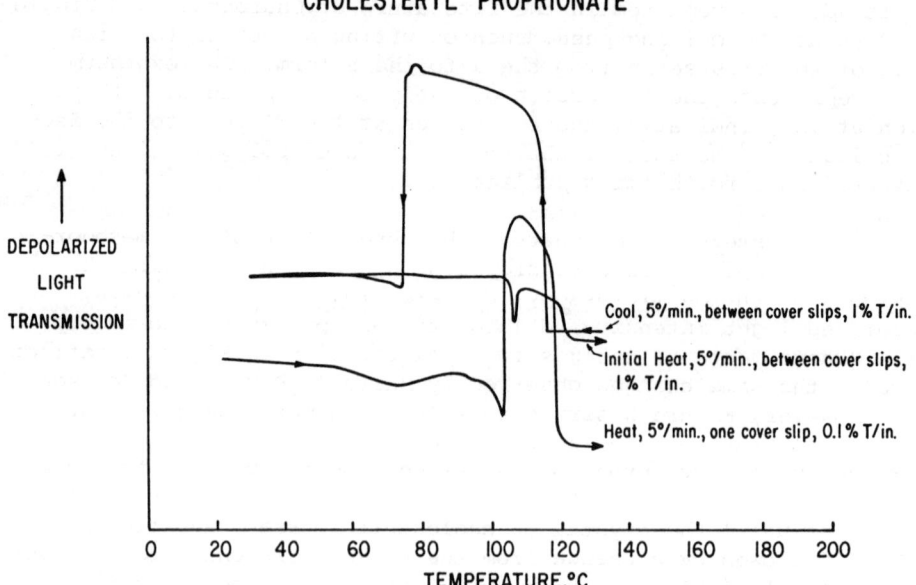

Fig. 5. TDA curves for cholesteryl proprionate.

60°C, the level of depolarized light is the same for the heating and cooling cycles. This is the normal behavior that one gets when using similar slow heating and cooling rates. When one of the cover slips is taken from the sample and only one-half of the sample rerun at ten times the sensitivity of the first runs, the difference exists that there appears to be some type of fusion or molecular disordering occurring at 56°C as well as the detection of melting, beginning at 99°C with final fusion at 104°C. At this point there is a rapid rise in the light intensity as the conversion is essentially carried through the whole temperature regime from 99° to 107°C, finally reaching the total average crystalline depolarization necessary to represent the cholesteric state. At 110°C fusion or disordering of the cholesteric phase begins and becomes very rapid at 119°C, prior to the light field becoming black at 120°C. The differences in the two heating curves here reflect the ability of the crystals to orient themselves on the surface of one cover slip, whereas running the material between two cover slips can apparently change the light level and decrease the total amount of ordering possible. More than that, for phase transitions or phases which may be pressure sensitive, subjecting a small particle of a material to the weight of a cover slip can cause distortion or changes in intensity as well as preferred surface ordering of the material as it goes through a given transition. Controlled cooling of materials, such as cholesteryl proprionate, is essential for the reproducibility of the scan.

While the measurement of depolarized light is also accompanied by scattering at the surface of the material being measured and while the scattering is more abundant in a sample resting on top of one cover slip, there are two measurement parameters which favor one cover slip:

1. The thermal gradient and the heat transfer to the material should be much more rapid than with two cover slips, in that the second cover slip acts as another thermal insulator to the material. Consequently, it may appear to cause the material to melt at a higher temperature.

2. With the sample placed on top of only one cover slip, any small changes in order which may be restricted by a load on the sample under measurement are easily seen as with the change at 56° and again the beginning of melting at 99°, corresponding to the literature value for the melting behavior of cholesteryl proprionate. The melting for cholesteryl proprionate under the same conditions using two cover slips begins at 103°C and is erroneous.

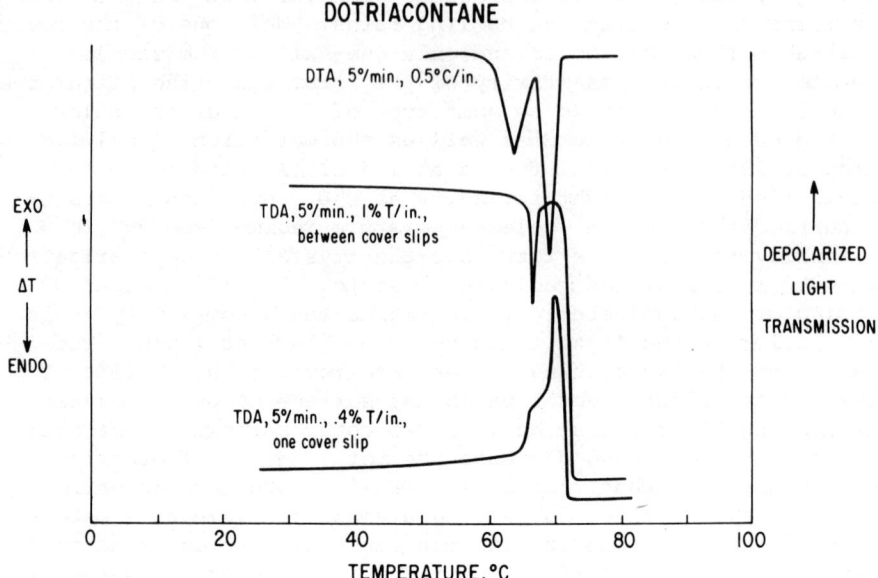

Fig. 6. TDA and DSC response of dotriacontane.

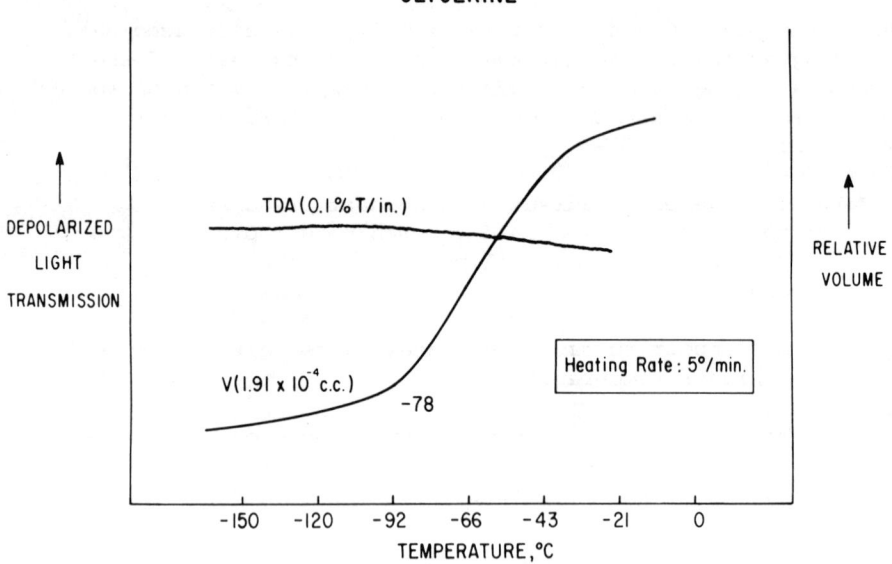

Fig. 7. TDA and dilatometric response of glycerine.

The investigator will have to decide on whether he sandwiches his sample between two cover slips and heats at very slow heating rates to obtain temperature accuracy, or places the sample upon one cover slip and tolerates changes due to scattering at reasonable heating rates for temperature accuracy.

5. <u>Dotriacontane</u>. Another standard commonly used by people in thermal analysis is the C_{32} hydrocarbon, dotriacontane, whose premelting doublet has been recognized as a standard. In contrasting the TDA behavior where the dotriacontane is between two cover slips and resting on top of one cover slip (Figure 6), we again see the concurrence of the peak temperatures for DTA with the changes in depolarized light using a sample on top of a cover slip. However, as the sample is meshed between two cover slips, the temperature at which fusion occurs has shifted up slightly, again indicating the insulation of the sample by two cover slips. The insertion of a thermocouple directly in the sample between the cover slips as was done in the work of Barrall, Porter and Johnson (3) obviates this difficulty.

6. <u>Glycerine</u>. The literature abounds with reports of temperatures at which phase transitions have occurred, and many are termed as due to glass transitions. Glycerine has been shown to have a glass transition from the glassy phase to the amorphouse state around -80°C (12), and this transition is shown in Figure 7 as a broad, sweeping volume change when measured dilatometrically. Thermal Depolarization Analyses, in measuring changes in molecular ordering, apparently does not respond to glass transition temperatures. Consequently, we see in Figure 7 that there is no change in depolarized light intensity as a function of temperature for glycerine in the temperature range of -80°C at the highest sensitivity of the instrument. A correlation in measurements by the differential scanning calorimeter, thermal penetration measurement and dilatometric change have been observed near -80°C (21). Hence, we might initially postulate that glass transitions cannot be seen by Thermal Depolarization Analyses.

7. <u>Glucose</u>. Crystalline glucose exhibits a melting endotherm at 163°C by DSC, with the peak occurring at 167°C (Figure 8). The TDA trace shows the onset of melting at 168°C, exhibited by an increase in light transmission with two peaks at 170° and 172°C. The depolarized light intensity peaks at 170° and 172°C indicate an intermediate crystalline form whose order is more perfect than the original crystal. This completion of fusion coincides with the temperature at which the DSC endotherm returns to the baseline. Comparison of the DSC and TDA behavior indicates that the onset of fusion could occur at a slightly lower temperature with DSC than TDA. Changes in specific volume for a given polymer occur at a higher temperature than the onset of fusion by a DSC measurement, and it would appear that both TDA and dilatometry, only at very

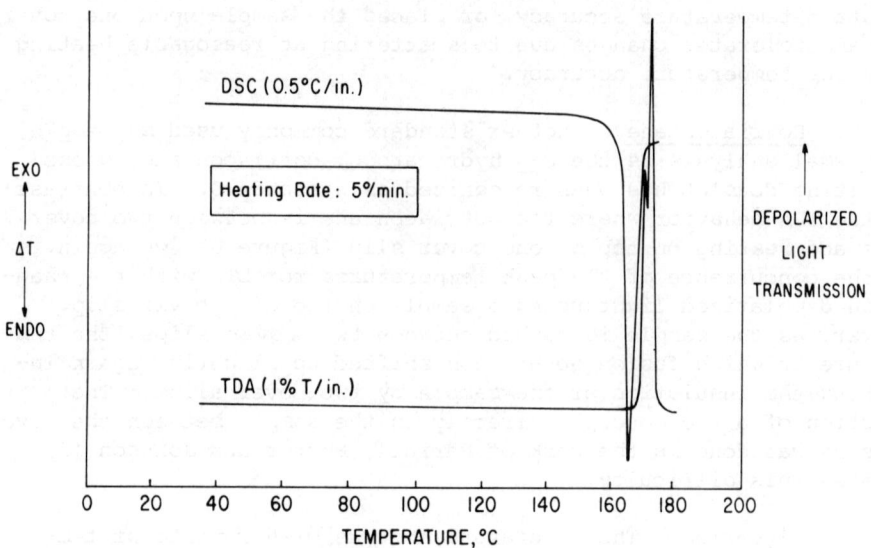

Fig. 8. TDA and DSC response of crystalline glucose.

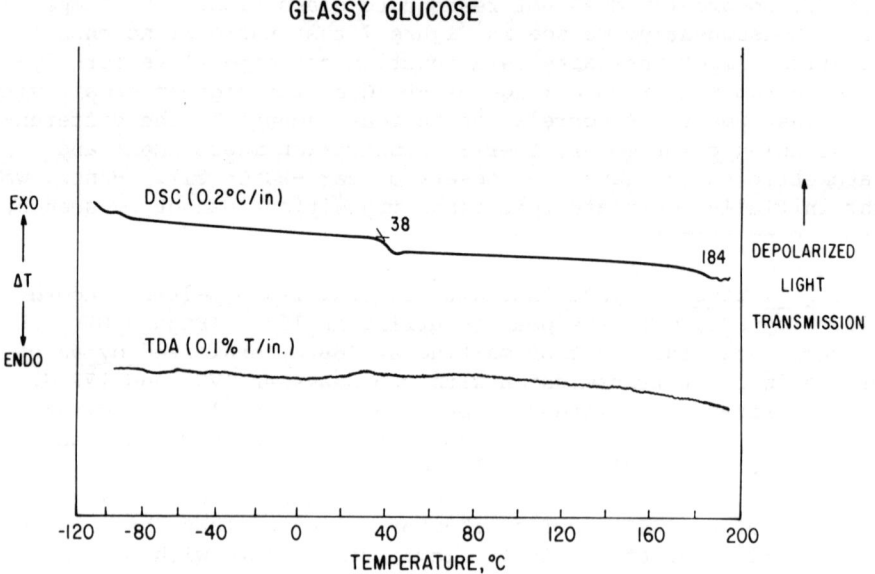

Fig. 9. TDA and DSC response of glassy glucose.

THERMAL ANALYSIS OF POLYMERS

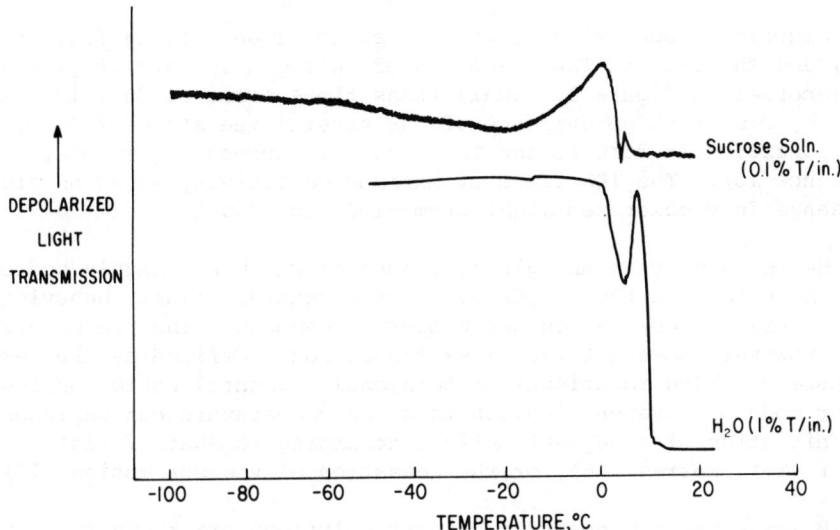

Fig. 10. TDA response of water and a concentrated solution of sucrose.

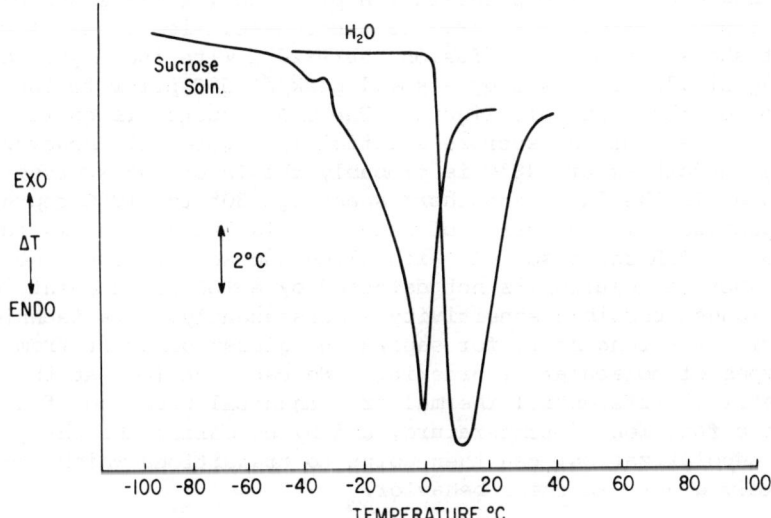

Fig. 11. DSC response of water and a concentrated solution of sucrose.

slow heating rates, may coincide with DSC data. On the other hand, it appears that TDA can resolve the molecular order/disorder processes much better than the heat averaging effects of DSC. More work is needed to confirm this initial indication.

Glucose is converted to the glassy state on cooling from the melt, and the DSC and TDA responses of this glassy form of glucose are recorded in Figure 9. Small transitions occur at 36°, 183° and 190°C by DSC, with no evidence of the crystalline state of glucose. The transition at 36°C is the Tg of glassy glucose (Tg = 32°C, Reference 13). The TDA trace at maximum sensitivity shows no sign of change in depolarized light from -100° to +200°C.

Hence, glycerine and glucose, both of which are highly hydrogen bonded through their hydroxyl groups, exhibit glassy behavior both by visual observation and property response. The TDA measurement, however, does not see glass transitions, defined as the temperature at which rotational or torsional segmental motion begins. Either mode of movement implies that the temperature can represent the initiation of a segment motion, according to Shatzki (14), Miller (15), Meares (16), or the cessation of viscous motion (17).

8. <u>Sucrose</u>. Concentrated sugar solutions are known to exhibit glassy behavior at subambient temperatures (15). In Figure 10 are shown the TDA curves for a concentrated solution of sucrose in water (10 g/10 ml) and for water alone. The TDA trace for water shows a high degree of depolarization prior to its two modes of fusion at 0° and 8°C. However, the sucrose solution by a TDA measurement shows the onset of fusion near -20°C with the major melting occurring at 1°C, followed by a small peak at 5°C prior to the formation of the isotropic liquid. The TDA response is confirmed by the DSC trace of the sucrose solution in Figure 11. However, the small endotherm at -48°C is probably the Tg of the sucrose solution while the large endotherm spanning -30° to +10°C represents the suppression of the fusion of water. This latter peak is identical to the TDA response. As with glycerine and glucose, the Tg of this sucrose solution is not detected by a TDA measurement, even at the highest possible sensitivity. Consequently, this technique can provide a strong basis for separating glassy behavior from other types of molecular disordering. We can then look at the dilatometric, differential thermal or mechanical response of a material as a function of temperature, and by comparing the changes in thermal depolarization, can then point to transitions which are most likely glassy in their behavior.

9. <u>Polyethylene</u>. Low density polyethylene is known to have a broad endothermal characteristic when measured by differential scanning calorimetry, and the dilatometric reponse shown in Figure 12 is illustrative of the broad melting range of the crystallites present in low density polyethylene. Examination of this

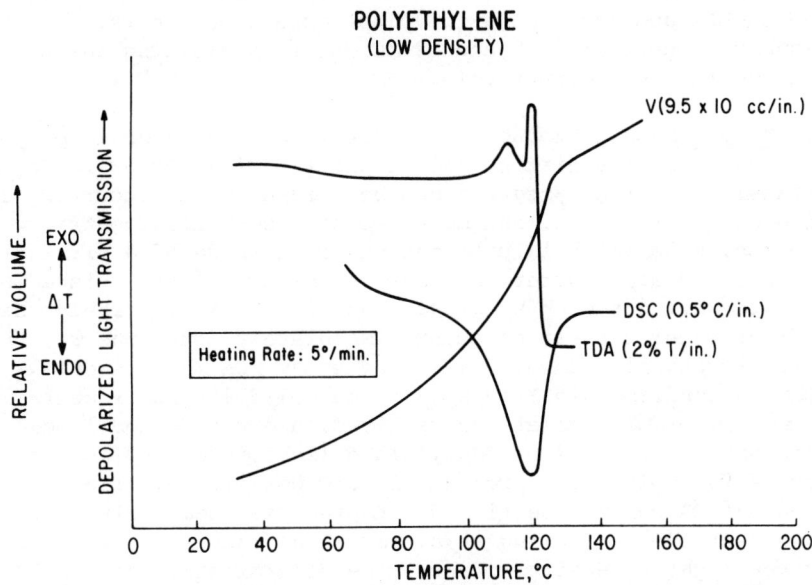

Fig. 12. TDA, DSC and dilatometric correlation of low density polyethylene.

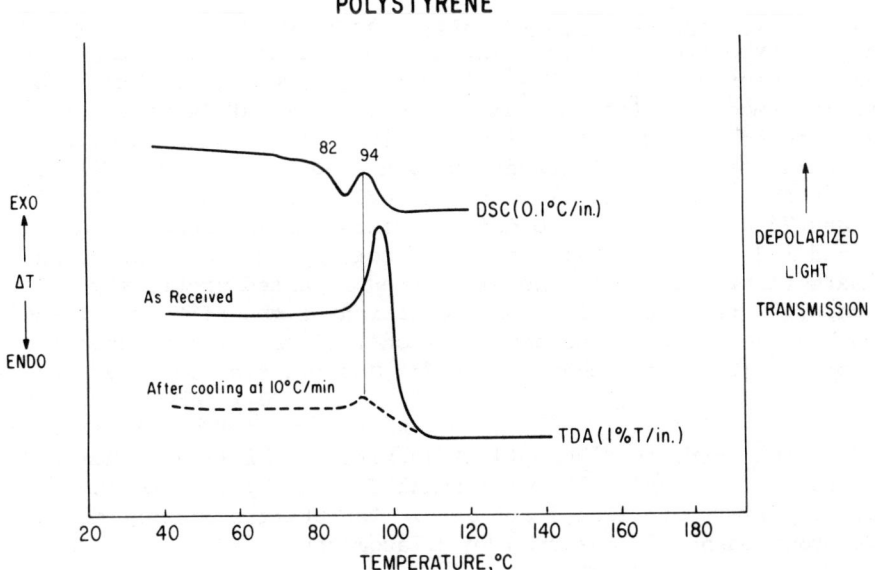

Fig. 13. TDA and dilatometric response of polystyrene.

material by means of Thermal Depolarization Analyses shows that there is a change around 45°C and that there are two melting phenomena occurring in the temperature range in which the endotherm is observed in DTA and the change in volume expansion occurs. The TDA peak responses occur at 113° and 120°C due to premelting chain rotation and fusion, respectively.

10. **Polystyrene.** Probably the most studied polymer is polystyrene because of its history and simplicity of structure. The glass transition for polystyrene has been reported as occurring in the range of 75° to 105°C, and more recently some dilatometric work has been published which illustrates the occurrence of a first order transition immediately after the glass transition (18). Since TDA measurements will not pick up glass transitions, it is likely that if the first order transition occurs immediately after the Tg, the first order transition should be picked up by TDA measurements, and this is confirmed in Figure 13. The solid line represents the first run on the pellets as received, and the dotted line indicates the lower level of depolarized light intensity as the polystyrene is cooled at a program of 10°C per minute. The onset of fusion by TDA can be identified with the peak temperature of the DSC curve near 89°C. Cooling at slower rates raises the level of depolarized light intensity. Hence, the determination of small cooling rate effects on the degree of molecular order are possible with TDA. It should also be possible to measure these small changes in molecular order isothermally.

11. **Poly(Bisphenol-A-Carbonate).** X-ray measurements have shown that polycarbonate is essentially non-crystalline, i.e., there is no significant amount of three-dimensional order (19). Its response to various modes of thermal analysis have shown at least two major phase changes to occur near -120° and 150°C by torsion pendulum measurements (20) and others at 55° and -140°C (21). The dilatometric scan in Figure 14 illustrates small volume changes at -137°, -47° and 74°C as well as two definite first order changes near 148° and 163°C. The change at -137°C corresponds to a change in ultimate tensile strength and elongation reported previously (21), and the 74°C transition is a manifestation of the 55°C transition, which has been shown to be dependent on the annealing treatment of the polymer (21). The change at -47°C has not been reported previously.

The TDA curve in Figure 14 illustrates two fusion phenomena occurring at 148° and 152°C with final fusion (by TDA) to the apparently isotropic liquid at 165°C. The TDA trace confirms the first order phenomena observed by dilatometry.

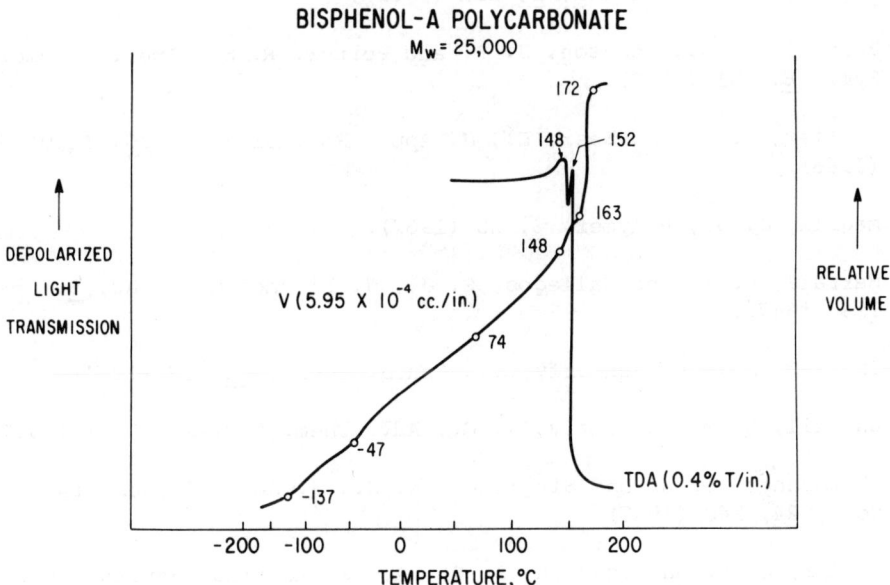

Fig. 14. TDA and dilatometric response of polycarbonate.

ACKNOWLEDGMENT

The author is grateful to E. M. Barrall and J. F. Johnson for an initial introduction to and for continuing discussions of this technique. Permission by the Du Pont Company to publish this work is appreciated.

REFERENCES

1. Magill, J. H., Nature, 187, 770 (1960).

2. Magill, J. H., Polymer, 2, 221 (1961).

3. Barrall, E. M., Johnson, J. F. and Porter, R. S., Appl. Polymer Sym., 8, 191 (1969).

4. Collier, J. R. and Baer, E., J. Appl. Polymer Sci., 10, 1409 (1966).

5. Magill, J. H., Polymer, 3, 35 (1962).

6. Barrall, E. M. and Gallegos, E. J., J. Polymer Sci., A2, 5, 113 (1967).

7. Theoret, A., and Sandorfy, C., Can.J. Chem., 42, 57 (1964).

8. Barrall, E. M. and Guffy, J. C., Adv. Chem. Series, 63, 1 (1967).

9. Charbonnet, G. H. and Singleton, W. S., J. Am. Oil Chemists Soc., 24, 140 (1947).

10. Miller, G. W. and Johnson, J. F., A.C.S. Meeting, Houston, February, 1970.

11. Barrall, E. M., Johnson, J. F. and Porter R. S., Mol. Cryst., 3, 103 (1967).

12. Miller, G. W., "Thermal Analysis", R. F. Schwenker and P. D. Garn, Eds., Academic Press, New York, 1969, p. 435.

13. Kauzman, W., Chem. Rev., 43, 219 (1948).

14. Schatzki, T. F., J. Polymer Sci., C14, 139 (1966).

15. Miller, R. L., "The Structure of Polymers", Reinhold, New York, 1966, p. 281-286.

16. Meares, P., "Polymers: Bulk Structure and Properties", Van Nostrand, New York, 1965, Chapter 10.

17. Wunderlich, B., private communication.

18. Miller, G. W., "Treatise on Adhesives and Adhesion", Volume III, R. L. Patrick, Ed., Marcel Dekker, New York, in press.

19. Schnell, H., "Physics and Chemistry of Polycarbonate", Interscience, New York, 1962.

20. Nielsen, L., "Mechanical Properties of Polymers", Reinhold, New York, 1962, p. 187.

21. Miller, G. W., "Analytical Calorimetry", R. S. Porter and J. F. Johnson, Eds., Plenum, New York, 1969, p. 71.

THE APPLICATION OF COMBINED DIFFERENTIAL SCANNING CALORIMETRY-MASS SPECTROMETRY (DSC-MS) TO THE STUDY OF THERMAL AND OXIDATIVE DECOMPOSITIONS

George Dugan, J. D. McCarty, and R. J. Friant

Research Center

Hercules Incorporated, Wilmington, Del. 19899

INTRODUCTION

As shown by the classic work of Borchardt and Daniels (1), differential scanning calorimetry (DSC) can be of considerable value in the study of thermal decomposition reactions. Heats of reaction and rates of reaction are readily calculated directly from the thermal analysis record, while other information concerning the thermal stability of a substance can often be inferred using DSC.

However, for definitive study of complex reactions involving more than one reaction step, additional analytical data may be required, such as identification of intermediate chemical species and determination of the reaction stoichiometry of each step.

Perhaps the most powerful technique which might be used for this purpose is mass spectrometry (MS), since both qualitative and quantitative work may be carried out in the same analysis. Mass spectrometry does, however, impose certain limitations of its own which must be taken into account. These limitations are associated with the introduction of a sample gas into the evacuated spectrometer inlet.

One approach that might be used is that of heating the sample by means of a furnace within the mass spectrometer inlet itself. This immediately restricts one to the study of non-volatile materials and to reactions in which atmospheric or pressure effects play no important role.

An alternative method is that of leading the effluent gases from an external furnace to the MS inlet via a molecular leak or sampling valve. Here the sample is flushed continuously by a flowing atmosphere of inert or reactive gas of up to 2 atmospheres pressure, but if the volatile reaction products are condensable, the effluent path to the MS inlet must be kept heated. Depending on the length of this path, there may be a significant delay between the thermal event and the measurement of off-gases by the mass spectrometer. Only a portion of the effluent gases is sampled for analysis by MS, which may impose some limitations on the quantitative aspect of analyzing the off-gases.

The least attractive method would appear to be the classic one of trapping off-gases in a separate experiment, then transferring the trapped products to the MS inlet. The usual difficulties associated with this method are magnified when the starting material is of the order of 1-10 milligrams.

A practical consideration in any case is the time and money spent on an instrument dedicated to monitoring thermal decomposition reactions. Usually the cost of such an instrument turns out to be prohibitive.

In our laboratory we have recently combined the advantages of differential scanning calorimetry and mass spectrometry without requiring a dedicated hybrid instrument. The technique is simple and requires relatively little time, yet permits both qualitative and quantitative analysis of off-gases from thermal decomposition reactions on a small (1-10 mg) sample. At the same time, a DSC record of the reaction being studied is obtained.

Three steps are required for the analysis:

1) The weighed sample, contained in a hermetically sealed metal capsule, is placed in a DSC or equivalent instrument and given a programmed or isothermal heat treatment. The DSC curve is recorded simultaneously if desired. A blank sample is sealed in a similar capsule but is not given the thermal treatment.

2) The treated sample capsule is transferred to a special sampling device which in turn is coupled to the MS inlet. After evacuation of the inlet the sample capsule is punctured to permit sampling of the volatile reaction products. The blank is sampled similarly.

3) Finally, the products are identified and analyzed quantitatively by mass spectrometry.

EXPERIMENTAL

Sample Preparation

The sample (1-10 mg) is weighed into a special aluminum sample pan obtainable from either the Perkin-Elmer Corp., Norwalk, Conn., or the Instrument Products Division of E. I. du Pont de Nemours & Co. (Inc.), Wilmington, Del. The pan is then sealed by means of a cold-weld pressure sealing device available from the manufacturer. If properly sealed, these aluminum pans (capacity about 20 microliters) are capable of containing an internal pressure of approximately 2 atmospheres. Similar pans made of gold foil, obtainable from Perkin-Elmer, are said to withstand an even higher internal pressure.

Should a special atmosphere be desired (e.g., pure oxygen or dry nitrogen), the sample may be sealed while inside a dry bag or dry box. Samples which react with the sample pan material should not be analyzed by this technique, but most organic materials appear to be compatible with the aluminum pans.

Treatment of Sample in the DSC

Thermal treatment of the sample is carried out by heating at some programmed rate or holding isothermally in a Perkin-Elmer DSC-1B Differential Scanning Calorimeter or similar instrument, in order to generate the volatile species to be analyzed subsequently by mass spectrometry, as well as to obtain the thermal analysis curve from which calorimetric information is derived. The sample heating can be controlled so as to obtain specimens representing different stages of a reaction, either by holding isothermally for different periods of time, or by programming up to a given temperature, then cooling the sample quickly to quench the reaction. Partioning the reaction products in this way permits the study

Figure 1

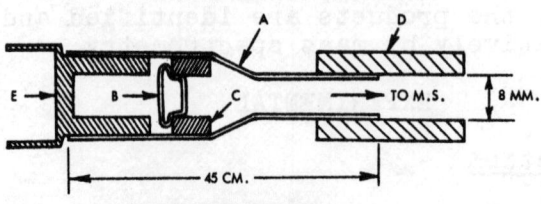

GLASS SAMPLE HOLDER FOR DSC-MS ANALYSIS

A. GLASS SAMPLE HOLDER BODY
B. ALUMINUM DSC SAMPLE PAN
C. SLOTTED TEFLON SPACER
D. 1/4" I.D. HEAVY WALL TYGON TUBING
E. NO. 14 RED RUBBER STOPPULE

Figure 2

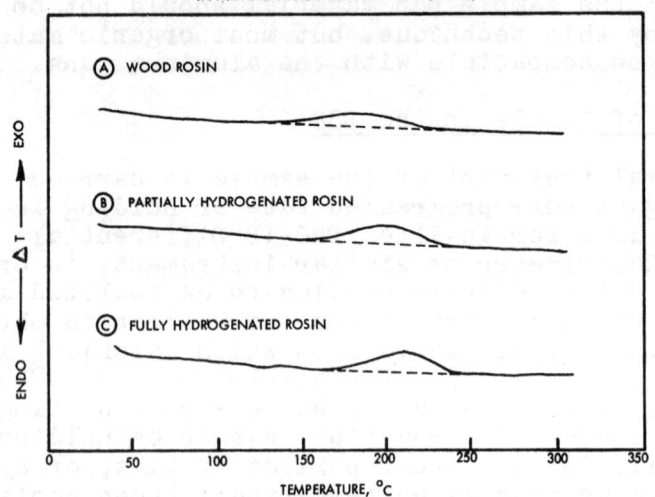

DIFFERENTIAL THERMAL ANALYSIS CURVES SHOWING THE EFFECT
OF HYDROGENATION ON THE OXIDATIVE STABILITY OF ROSINS
(SAMPLES WEIGHING ONE MILLIGRAM WERE SEALED
IN AIR AND PROGRAMMED AT 20°C./MIN.)

of multistep reactions.

Analysis of Off-gases by Mass Spectrometry

After thermal treatment, the encapsulated sample is transferred to the glass sample holder shown assembled in Figure 1. The assembled holder containing the sample pan (B) is then attached to the gas inlet of a C.E.C. Model 21-103C mass spectrometer by means of the Tygon tubing. After evacuation of the MS inlet a solid stainless steel needle is inserted through the rubber stoppule to puncture the sample pan. The total pressure of the volatiles is then measured with a micromanometer. If no pressure is detected, it is assumed that the sample pan leaked and the sample is discarded. If a pressure reading is obtained, the mass spectrometer scan is begun.

RESULTS AND DISCUSSION

The combined DSC-MS technique was successfully applied to the identification of volatile products from the decomposition of rosin and its derivatives, dicumyl peroxide, methyl parathion, and a variety of other research samples. Some of these results will be described here.

Thermal and Oxidative Decomposition of Modified Rosins

The air oxidation of rosins and rosin derivatives which occurs at elevated temperatures may in some uses have a deleterious effect on the properties of the product. Oxygen absorption measurements at room temperature and high pressure indicate that some rosins, such as the hydrogenated and esterified derivatives, are less susceptible to oxidation than unmodified rosins (2). Programmed DSC runs of treated rosins also demonstrate differences in oxidative stability of rosin and hydrogenated rosins. (Figure 2). The apparent onset of the exothermic oxidation peak occurs at a lower temperature for the wood rosin than for the hydrogenated rosins, indicating greater oxidative stability in the latter.

In a study of the effect of hydrogenation and dehydrogenation on the thermal and oxidative stability of rosins, the DSC-MS technique was used to analyze off-gases from the oxidation of a hydrogenated rosin and a pure resin acid, dehydroabietic acid. Table I summarizes some of the data from these experiments.

Table I - Effect of Thermal Treatment and Reaction Atmosphere on the CO_2 Yield of Dehydroabietic Acid

Sample Wt., mg	Temp. Program	Original Atmosphere	Moles $CO_2 \times 10^8$
0.112	50-200°C	Oxygen	10.2
0.104	50-240	Oxygen	15.5
0.096	50-300	Oxygen	23.5
1.004	50-200	Nitrogen	0.2
1.016	50-240	Nitrogen	0.3
1.020	50-300	Nitrogen	0.5

Dehydroabietic acid samples were encapsulated in an oxygen atmosphere using a dry bag, then heated up at 20°C/min. The amount of carbon dioxide produced in the reaction is clearly a function of the upper temperature limit of the run. When dehydroabietic acid was heated in the absence of air (i.e., when the sample was encapsulated in a nitrogen atmosphere), a much smaller quantity of carbon dioxide was obtained. The carbon dioxide liberated under these conditions can be attributed to either decarboxylation or to oxidation by surface-absorbed oxygen.

Other products detected were acetone, water, methane, and hydrogen, all of which might be expected from the free-radical oxidation of this material. Volatile products from the oxidation of hydrogenated rosins were found to be similar to those observed for dehydroabietic acid.

Free-Radical Decomposition of Dicumyl Peroxide

Polymer cross-linking agents must possess sufficient thermal stability to be incorporated into the polymer at milling temperatures without the risk of premature cross-linking, or "scorching". To aid in predicting the stability of such materials, DSC data has often been used in our laboratory to determine reaction kinetics and heats of decomposition.

Early in the development of the combined DSC-MS technique, we examined the thermal behavior of dicumyl peroxide, which is used as a free-radical initiator and polymer cross-linking agent. While little new information regarding this material was learned in the process, the results demonstrate the advantages of combined DSC-MS over conventional methods: small sample size, elimination of difficult trapping techniques, complete recovery of products, and results obtained in a matter of hours.

The main decomposition products of dicumyl peroxide were identified as acetophenone, methane and dimethylbenzyl alcohol. Lesser amounts of methanol and the methyl ether of dimethylbenzyl alcohol were also detected. As shown in the following reactions, this decomposition is already well understood in terms of a mechanism involving the intiial formation of two cumyloxy radicals by thermal dissociation:

$$(CH_3)_2\underset{\phi}{C}\text{-O-O-}\underset{\phi}{C}(CH_3)_2 \xrightarrow{\Delta} 2\ (CH_3)_2\underset{\phi}{C}\text{-O}\cdot \quad (1)$$

$$(CH_3)_2\underset{\phi}{C}\text{-O}\cdot \longrightarrow \phi\text{-}\underset{}{\overset{O}{\overset{\|}{C}}}\text{-}CH_3 + CH_3\cdot \quad (2)$$

$$CH_3\cdot + RH \longrightarrow CH_4 + R\cdot \quad (3)$$

$$(CH_3)_2\underset{\phi}{C}\text{-O}\cdot + RH \longrightarrow (CH_3)_2\underset{\phi}{C}\text{-OH} + R\cdot \quad (4)$$

Rearrangement of the cumyloxy radical accounts for the acetophenone, while products such as methane and dimethylbenzyl alcohol are formed when hydrogen atoms are abstracted from a substrate by free radical species. If the substrate is a polymer chain, secondary radicals can be generated. A cross-link results when two such secondary radicals combine:

$$\{CH_2CH_2CH_2\} + CH_3\cdot \longrightarrow \{CH_2\dot{C}HCH_2\} + CH_4 \quad (5)$$

$$2\ \{CH_2\dot{C}HCH_2\} \longrightarrow \begin{array}{c}\{CH_2CHCH_2\}\\|\\\{CH_2CHCH_2\}\end{array} \quad (6)$$

Figure 3

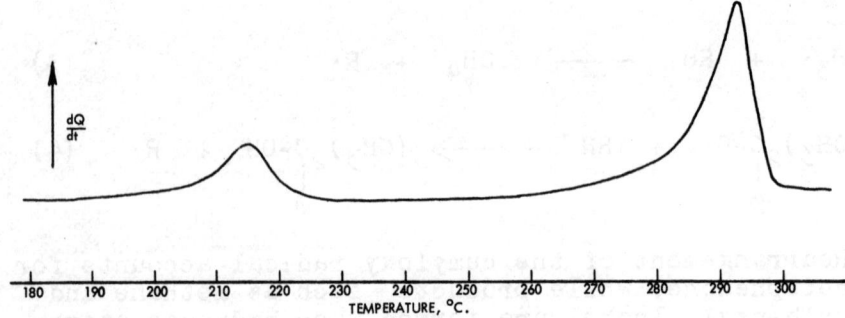

DIFFERENTIAL SCANNING CALORIMETRY CURVE SHOWING EXOTHERMIC DECOMPOSITION PEAKS OF METHYL PARATHION

(PROGRAMMED AT 5°C./MIN. IN SEALED PANS)

Thermal Decomposition of Methyl Parathion

Phosphorothioate insecticides, such as the various homologs of parathion, have been known to decompose vigorously while being stripped of solvent by distillation at abnormally high temperatures. We have studied the decomposition by DSC to examine the thermal effects involved, and by MS to identify and analyze the volatile products of the reaction.

The DSC curve shown in Figure 3 indicates that methyl parathion decomposes in two exothermic steps. To resolve the products of each step, two different methyl parathion samples were thermally treated in the DSC. Sample A was programmed from 170°C to 220°C and therefore included only volatile products from the first decomposition peak. Sample B was programmed from 170°C to 300°C and included decomposition gases from both reaction peaks. Less than two milligrams of this toxic material were required for analysis of volatiles from both reaction steps.

The data shown in Table II indicate that the first decomposition step yields dimethyl sulfide as the main volatile product, along with a much smaller amount of dimethyl disulfide. Note the absence of SO_2 in this sample. In the second reaction step oxidation is clearly indicated by the products SO_2, CO_2, and H_2O.

Table II - Mass Spectral Analysis of
Gases from Methyl Parathion Decomposition

	Sample A	Sample B
Dimethyl disulfide	0.7%	3.8%
Dimethyl sulfide	99.3	21.6
Sulfur dioxide	-	4.5
Carbon dioxide	-	10.0
Water	-	60.1 *

* Oxygen in sample is lower than background.

These results are similar to those of McPherson and Johnson (3), who carried out extensive work on several parathion homologs.

The first step in the decomposition of methyl parathion is probably best explained by the following reaction scheme:

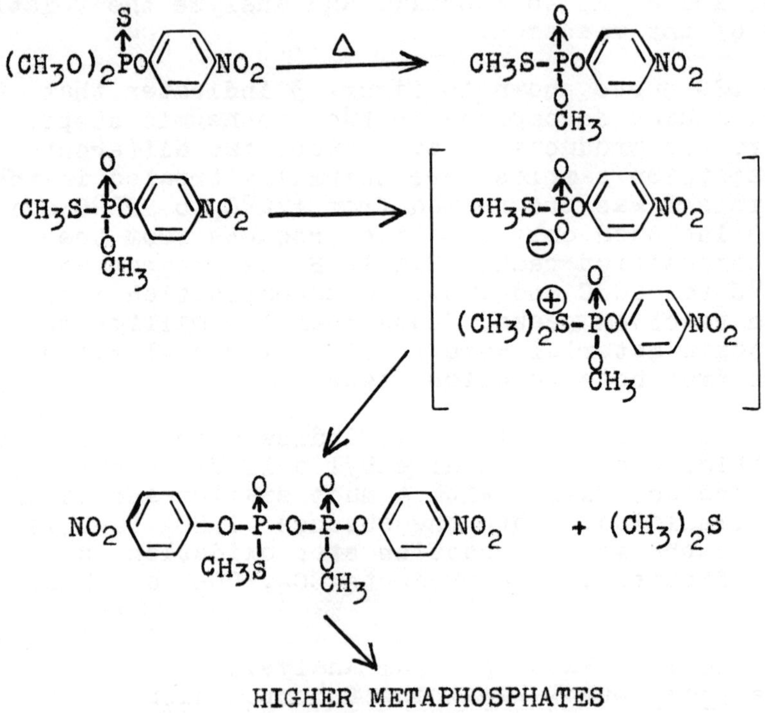

$$(CH_3O)_2\overset{S}{P}O\text{-}C_6H_4\text{-}NO_2 \xrightarrow{\Delta} CH_3S\text{-}\overset{O}{P}(OCH_3)O\text{-}C_6H_4\text{-}NO_2 \quad (7)$$

$$CH_3S\text{-}\overset{O}{P}(OCH_3)O\text{-}C_6H_4\text{-}NO_2 \longrightarrow \left[\begin{array}{c} CH_3S^{\ominus}\text{-}\overset{O}{P}(O)O\text{-}C_6H_4\text{-}NO_2 \\ (CH_3)_2\overset{\oplus}{S}\text{-}\overset{O}{P}(O)(OCH_3)O\text{-}C_6H_4\text{-}NO_2 \end{array} \right] \quad (8)$$

$$NO_2\text{-}C_6H_4\text{-}O\text{-}\overset{O}{\underset{CH_3S}{P}}\text{-}O\text{-}\overset{O}{\underset{CH_3}{P}}\text{-}O\text{-}C_6H_4\text{-}NO_2 + (CH_3)_2S \quad (9)$$

↓

HIGHER METAPHOSPHATES

Methyl parathion first isomerizes upon being heated as in Equation 7, followed by formation of an ion pair upon further heating of the S-methyl isomer. The ion pair then reacts as shown in the next equation to yield dimethyl sulfide and a poly(arylmetaphosphate). Higher metaphosphates are formed by further condensation.

Support for this mechanism, which was suggested by the work of Hilgetag, *et. al.*, include the following facts:

1) McPherson and Johnson (3) found that the dark, oily residue left after the first reaction step has a higher molecular weight that the starting material, and is insoluble in benzene and hexane but mostly soluble in acetone. They also gave

evidence that the residue hydrolyzes to yield phosphoric acid, phenol, and p-nitrophenol.

2) Disappearance of thiono sulfur by heating methyl parathion at 120°C is essentially complete after 18 hours. By contrast, only about 5% of the material has been converted to dimethyl sulfide in this length of time.

3) The only volatile products of the first reaction step are dimethyl sulfide and dimethyl disulfide.

4) Phosphate esters are known to be excellent alkylating agents, and many sulfonium salts are known to be thermally unstable.

The tarry residue from the second reaction step in the decomposition of methyl parathion is not at all well characterized; however, the evidence suggests that the second exothermic reaction step is associated with oxidation by the nitro group.

CONCLUSIONS

In summary, we have combined the advantages of differential scanning calorimetry and mass spectrometry to obtain a technique which offers a rapid and convenient means of studying complex thermal and oxidative reactions. Cost is relatively low, and only a few milligrams of sample are required for analysis, minimizing the hazard in handling unstable or toxic materials.

REFERENCES

1. H. J. Borchardt and F. Daniels, J. Amer. Chem. Soc. 79, 41 (1957).

2. H. I. Enos, Jr., et. al., "Rosin and Rosin Derivatives", Encyclopedia of Chemical Technology, Vol. 17, 2nd Ed., John Wiley and Sons, Inc. (1968).

3. J. B. McPherson, Jr. and G. A. Johnson, J. Agr. and Food Chem. 7, 42 (1956).

4. Gunter Hilgetag, et. al., J. prakt. Chem. (4) 8, 73-89 (1959).

THE REDUCTION OF DSC CURVES AND THE CALCULATION OF REFINED AUTOCATALYTIC RATE CONSTANTS BY A COMPUTER TECHNIQUE

Robert W. Crossley, Ernest A. Dorko and Ronald L. Diggs

Dept. of Aero-Mechanical Engineering, Air Force

Institute of Technology, Wright-Patterson AF Base, Ohio

Introduction

A previous report[1] gave the results of a study in which the kinetic scheme could be determined for a reacting system initially in the solid state. The instrument utilized was a differential scanning calorimeter which was operated in the isothermal mode. The present report describes two refinements of the previously described analytical procedure and also extends the studies to decomposition in matrix.

A computer reduction technique for the DSC isothermal thermogram has been developed which replaces the planimeter reduction. The new technique is much less time consuming and eliminates the problem of errors due to operator fatigue.

A refinement in the determination of the rate constants for a solid state decomposition which can be described by an autocatalytic mechanism has also been achieved.

An autocatalytic reaction can be described by eq 1 and 2

$$A \xrightarrow{k_1} B \qquad (1)$$

$$A + B \xrightarrow{k_2} B \qquad (2)$$

where A represents the initial reactant and B represents product. The rate expression which corresponds to this kinetic scheme is given in eq 3[1].

$$\frac{-da}{dt} = -\dot{a} = k_1 a + k_2 a(1-a) \qquad (3)$$

where a is the reactant fraction remaining at time t. Dividing through by a gives eq 4

$$-\frac{d\ln a}{dt} = k_1 + k_2(1 - a) \tag{4}$$

A plot of ln a versus t produces a curve whose slope equals k_1 as a→1 (at the start of reaction) and $k_1 + k_2$ as a→0 (toward the end of reaction). The constant, k_2, can then be determined by subtracting the values of the initial and final slopes.

In the previous analysis[1], the above method was utilized to obtain rate constants. These constants were designated k_A (constant for the acceleratory period) and k_D (constant for the deceleratory period). However, it was found that k_D was two orders of magnitude larger than k_A. This means that the initial slope of the ln a vs t curve, if obtained at even 1% reaction, has a contribution from k_2 as large as the value of k_1. An attempt to obtain the slope of the curve by either graphical or computer techniques below 1% reaction is not feasible. Therefore, a different approach had to be taken.

Values of k_1 and k_2 were obtained by a simultaneous solution of eq 3 and its integral, eq 5.

$$\frac{1}{a} = \frac{1}{k_1 + k_2}(k_1 \exp(k_1 + k_2)t + k_2) \tag{5}$$

(The solution technique will be discussed in the next section.) Eq 3 and 5 were solved at each value of a and average values of k_1 and k_2 were taken over appropriate time intervals to give "true" values of k_1 and k_2.

Mathematics of Thermogram Reduction

The data reduction is divided into two phases, mechanism independent solutions for the reactant fraction, a, and various functions of a, and solutions for mechanism dependent rate constants. Each phase is performed by a separate computer program.

Program DATAR[2]

Ordinal points, referred to as "coarse data" and evenly spaced in time over the time span of the thermogram, are read into the computer. Up to 1000 points may be read, but 40-50 are usually sufficient for acceptable accuracy. The reactant fraction remaining at time, t, is calculated by eq 6

$$a(t) = \frac{\int_{t}^{t_{max}} (ORD) \, dt}{\int_{0}^{t_{max}} (ORD) \, dt} \tag{6}$$

A Simpson's Rule procedure modified to handle odd numbers of intervals is used to calculate the integrals. To preserve accuracy as t→t_{max} and when there is substantial noise in the thermogram data, additional ordinal data points, spaced closer than the coarse data, are used to calculate a(t) during the last 10-15% of the time span.

The program calculates and prints the following at specified times: time in seconds and in minutes, a, ln a, log 100a, 1/a, $1/a^2$, (1-a)/a, and log 100[(1-a)/a].

Program PARACT[3]

This program is used to determine the true constants, k_1 and k_2. First, k_1 is found by numerical iteration of eq 7.

$$k_1 = \frac{1-a}{at} \ln\left(\frac{k_1}{-\dot{a}/a^2}\right) + \left(\frac{-\dot{a}}{a^2}\right) \quad (7)$$

Eq 7 results from eliminating k_2 between eq 3 and 5. The reactant fraction, a, is calculated as in Program DATAR, and a Lagrangian numerical differentiation procedure[4] is used to obtain å from a. When k_1 is known, k_2 is calculated from eq 3 re-arranged to the form of eq 8

$$k_2 = \left(k_1 + \frac{-\dot{a}}{a}\right) \bigg/ (1-a) \quad (8)$$

As a check of consistancy and accuracy, the sum, $k_1 + k_2$, is calculated from eq 9

$$k_1 + k_2 = \frac{1}{t} \ln\left(\frac{-\dot{a}/a^2}{k_1}\right) \quad (9)$$

which is the result of taking the logarithmic form of eq 5 and inserting eq 3.

When the calculations have been completed, the program prints the following at specified times: time, t, in seconds and in minutes a, k_1, k_2, $k_1 + k_2$, and $-\dot{a}/a$. In addition, the program averages and prints the values of k_1, k_2, and $k_1 + k_2$ for designated reaction intervals.

Results

The data for the solid state decomposition of pure, crystalline cupferron tosylate (N-phenyl-N' tosyloxydi-imide N-oxide) and of its p-chloro and p-methyl derivatives were reduced using the computer technique. In addition, results obtained with the tosylate substrate imbedded in potassium bromide matrix were also reduced with the computer technique.

Table I reports the values for the reactant fraction at time, t, as obtained with the use of a planimeter and from the program DATAR. The data are for pure, crystalline p-chloro-cupferron tosylate decomposed at 118°C.

In Table II are listed the results of the rate constant calculations for the decomposition of the pure, crystalline cupferron

TABLE I

Comparison of the Values for Reactant Fraction Obtained by Planimeter Reduction and by Computer Reduction of a Typical DSC Thermogram. The Thermogram was Obtained During the Decomposition at $118°C$ of Pure, Crystalline p-Chlorocupferron Tosylate.

TIME (sec)	REACTANT FRACTION Planimeter Reduction	Computer Reduction
0	1.000	1.000
60	0.999	0.999
120	0.995	0.996
240	0.986	0.987
360	0.971	0.971
480	0.952	0.948
600	0.923	0.920
720	0.884	0.882
840	0.836	0.834
960	0.771	0.772
1080	0.686	0.689
1200	0.558	0.559
1320	0.390	0.391
1440	0.208	0.209
1560	0.092	0.095
1680	0.031	0.033
1800	0.005	0.003
1860	0.000	0.000

tosylates and for the tosylates imbedded in KBr matrix respectively.

In Table III and IV are listed the Arrhenius parameters for the systems studied based on the true values for k_1 and k_2.

Discussion

Program DATAR

An example comparison of reactant fraction, a, obtained by mechanical integration (planimeter) and by numerical integration (computer) appears in Table I. Using the planimeter, reactant fraction could be determined to within 0.01 units of the area under the DSC curve. This limit of accuracy results in a minimum relative error of 1% when a is close to 1.0, with increasing relative error at lower values of a, and this is the accuracy limit of the planimeter technique. In contrast, the accuracy of the computer technique is in essence limited only by the number of points at which the thermogram is read. The number of points used gave a comparable relative accuracy range of 0.1% to 1%. This was confirmed by doubling the number of points in selected test cases.

An important advantage of the computer technique is that it indicates regions on each end of the thermogram for which the chosen base line effectively excludes reaction, i.e., $a \geq 1.0$ near $t = 0$ and $a \leq 0.0$ near $t = t_{max}$. This indication is especially important for thermograms with significant scatter near the ends of reaction as such regions are not visually observable *a priori*.

Program PARACT

The only meaningful limitation to the accuracy of the solution of eq 7 for k_1 is the accuracy of a and $\dot{a}$. The accuracy of a was discussed in the previous section. A numerical differentiation algorithm based on a seven point Lagrangian interpolation formula gave approximately the same accuracy in $\dot{a}$ as in a. The seven point formula was chosen for curve smoothing as well as for point-by-point accuracy.

The actual solution of eq 7 was performed in 17 digit arithematic. This was necessary to insure rapid, essentially monatonic convergence in spite of large variations (order of 1 to order of 10^{20}) in the rate of change of the right side of eq 7 with respect to k_1. Convergence to a minimum of seven digits was always obtained within 15 iterations. Calculations for a complete thermogram take less than three seconds on an IBM 7090.

The calculation of k_2 and $k_1 + k_2$ by eq 8 and 9 gives a semi-independent check on accuracy when $(k_1 + k_2) - k_2$ is compared to the originally calculated k_1. The two determinations of k_1 agreed to at least five digits.

From the above, it is concluded that the calculated values of k_1 and k_2 are as accurate as the values of a and $\dot{a}$ being used. This is a significant advance over visual determination of k_A and

TABLE II
Reaction Rate Constants (sec^{-1})

Temperature T°C	Acceleratory Period		Deceleratory Period	
	k_1 (x10^5)	k_2 (x10^3)	k_1' (x10^5)	k_2' (x10^3)

100% Cupferron Tosylate

Temperature T°C	k_1 (x10^5)	k_2 (x10^3)	k_1' (x10^5)	k_2' (x10^3)
121	7.56	12.10	27.26	8.89
118	5.87	7.37	10.40	7.03
115	4.70	4.62	1.07	6.85
114	5.40	3.67	1.06	5.78
112	3.50	2.82	0.93	4.18
109	5.44	1.47	0.34	3.30
106	2.27	1.27	0.13	2.61
103	2.17	0.93	0.73	1.50
100	2.80	0.72	0.68	1.18

8% Cupferron Tosylate in KBr

Temperature T°C	k_1 (x10^5)	k_2 (x10^3)	k_1' (x10^5)	k_2' (x10^3)
115	5.73	4.01	3.90	4.80
113	5.41	3.34	3.23	4.16
111	4.61	2.09	1.03	3.27
109	3.60	1.52	0.43	2.64
107	3.13	1.25	0.19	2.37
105	2.60	0.95	0.28	1.65
103	1.92	0.73	0.04	1.62
101	1.45	0.62	0.06	1.22

4% Cupferron Tosylate in KBr

Temperature T°C	k_1 (x10^5)	k_2 (x10^3)	k_1' (x10^5)	k_2' (x10^3)
115	11.00	7.34	21.11	5.80
113	11.40	6.05	8.40	6.57
111	5.12	3.33	1.61	4.39
111	6.48	3.00	2.60	3.83
109	4.28	2.20	2.50	2.66
107	2.59	2.09	6.03	1.70
105	2.68	1.30	1.64	1.57
103	2.05	0.99	0.50	1.46
101	2.61	0.60	0.05	1.52

TABLE II (CON'T)

Temperature T°C	Acceleratory Period		Deceleratory Period	
	k_1 (x10^5)	k_2 (x10^3)	k_1' (x10^5)	k_2' (x10^3)

2% Cupferron Tosylate in KBr

115	5.65	4.34	3.43	5.09
113	5.06	3.27	2.07	3.32
111	6.07	2.96	5.87	3.17
109	4.84	1.93	1.74	2.64
107	3.40	1.64	0.80	2.47
105	2.33	1.32	0.32	2.11
103	3.05	0.73	0.10	1.75

100% p-Chlorocupferron Tosylate

130	3.59	10.84	0.63	15.22
127	3.36	6.90	0.16	11.86
124	5.22	3.60	0.10	10.48
121	2.70	2.47	0.00	7.87
118	4.55	1.18	0.00	6.32
115	7.05	0.63	0.00	4.91
112	4.63	0.47	0.00	3.45

100% p-Methylcupferron Tosylate

118	2.31	10.49	15.23	6.92
115	1.69	4.59	5.86	3.89
112	1.91	2.61	2.15	3.00
109	1.04	2.34	1.55	2.43
106	1.26	1.42	1.26	1.65
103	1.87	0.55	1.64	1.61

k_D from plots of functions of a vs t.

Results

The rate constants as determined from the solution of the Bernoulli equation and its integral are referred to as the true rate constants. Solution of the equations at small fixed time interval throughout the reaction gives results that can be averaged for the initial, or acceleratory, and decay, or deceleratory, periods of reaction. The two periods are separated by the maximum point in the thermogram which occurs at 40-70% of reaction, depending on composition and temperature.

There is a distinct difference in the values of the constants for the acceleratory and deceleratory periods. This result is consistent with the observed bend in the Ostwald plot. The indication is that a relatively subtle change in the reaction process is occurring. (see below) The distinction between the acceleratory and deceleratory periods is observed during decomposition of all the substrates whether decomposed in the pure state or when imbedded in KBr matrix.

The "KBr matrix effect" (i.e., the over-all effect on the rate constants due to imbedding the substrate in KBr matrix) appears to be relatively minor. However, there does seem to be a small effect on the activation energy associated with the uncatalyzed reaction (eq 1).

Insights into the mechanism of reaction are provided by the activation energies obtained from Arrhenius plots of the rate constants. The activation energy, E_1, for the uncatalyzed reaction is quite low for the initial decomposition of cupferron tosylate. Arrhenius plots for the two derivatives of cupferron tosylate show that the rate constants (k_1), are not exponential functions of the reciprocal of the temperature. Indeed when a plot was prepared of the value of the rate constant, k_1, versus temperature a straight line relationship was shown to exist. Similar plots of k_2, k_1', and k_2' showed non-linear relationships between the rate constants and the temperature. (See Figure 1 for representative plots). These observations indicate that the rate of the initial, uncatalyzed reaction is controlled by a process different than the chemical reaction or at least that chemical control is not exclusive. On the other hand, Arrhenius plots of k_2, k_1', and k_2' show the expected straight-line relationships, so the limiting rates in these processes can be the rates of the chemical reactions. The fact that the rates of the chemical reactions are not exclusively controlling is indicated by the fact that the activation energies for the catalyzed reactions are less during the deceleratory portions than during the acceleratory portions of the processes.

A final observation can be made by comparing the activation energies, E_2, E_1', and E_2'. In every case, the value of E_1' is very large compared to the values of E_2 and E_2'. This indicates that in the deceleratory period the uncatalyzed reaction makes essentially

TABLE III
Activation Energies (kcal/mol)

System	Acceleratory Period E_1	E_2	Deceleratory Period E_1'	E_2'
100% Cupferron Tosylate	15.71	39.96	51.09	29.25
8% Cupferron Tosylate	28.58	39.55	96.60	28.21
4% Cupferron Tosylate	37.10	49.66	67.82	34.82
2% Cupferron Tosylate	20.20	40.32	85.96	22.60
100% p-cl-Cupferron Tosylate	-----*	56.33	74.22	24.71
100% p-CH$_3$-Cupferron tosylate	-----*	50.84	44.15	27.81

*The rate constants do not obey an Arrhenius temperature dependence.

TABLE IV
Log Arrhenius Frequency Factors

System	Acceleratory Period A_1	A_2	Deceleratory Period A_1'	A_2'
100% Cupferron Tosylate	4.55	20.18	24.19	14.23
8% Cupferron Tosylate	11.90	19.86	50.01	13.57
4% Cupferron Tosylate	16.93	25.82	34.32	17.41
2% Cupferron Tosylate	7.18	20.37	44.20	10.37
100% p-cl-Cupferron Tosylate	-----*	28.55	34.75	11.59
100% p-CH$_3$-Cupferron Tosylate	-----*	26.38	20.63	13.30

*The rate constants do not obey an Arrhenius temperature dependence.

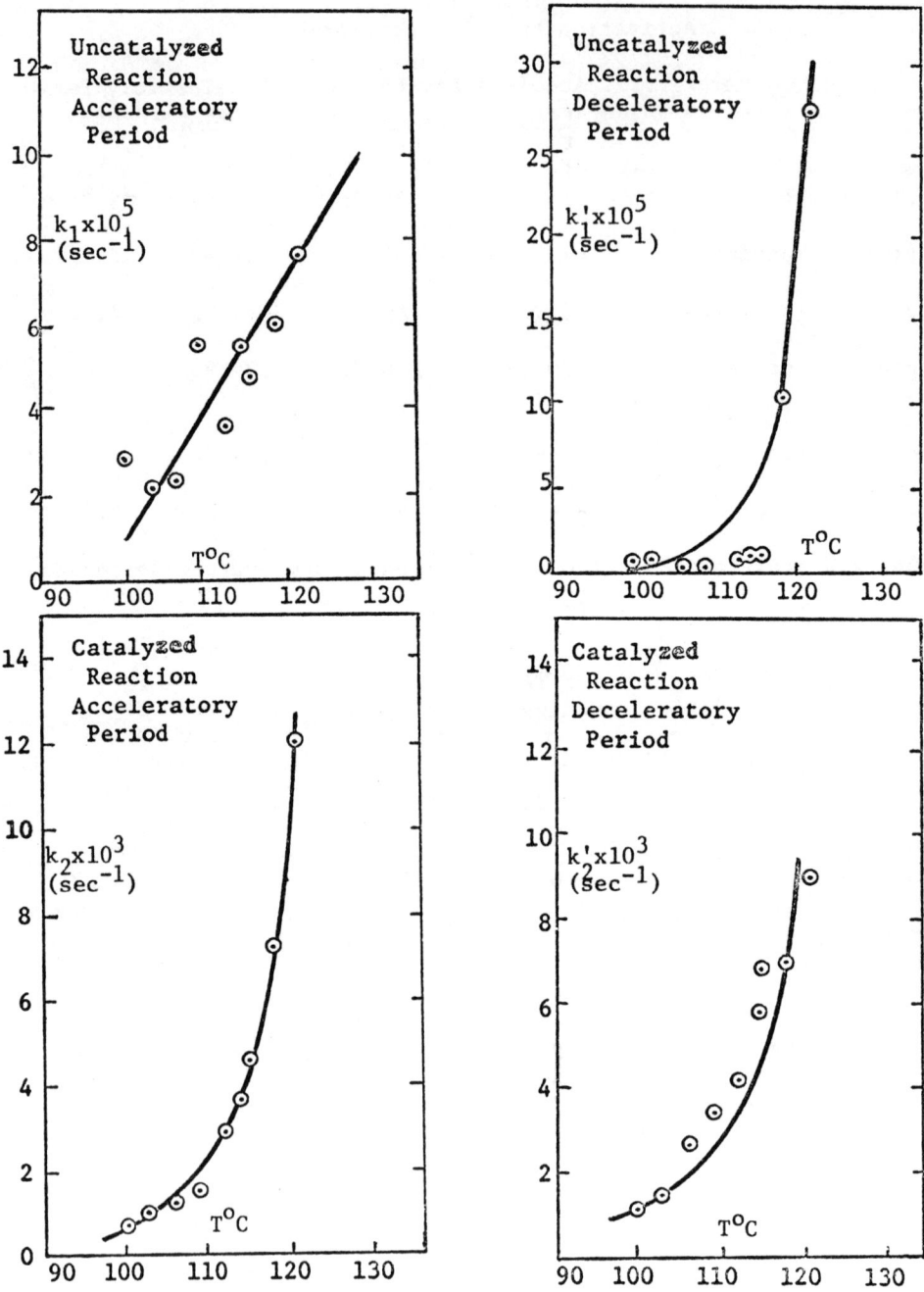

Figure 1. Graphs of the rate constants for the decomposition of pure, crystalline cupferron tosylate.

no contribution to the kinetic mechanism.

The conclusion which can be drawn from the preceeding discussion is that the decomposition of the cupferron tosylates, whether pure or in KBr matrix, proceeds in three phases. The decomposition is initiated by an uncatalyzed reaction whose rate does not show an Arrhenius temperature dependence but rather a linear temperature dependence. This initial reaction is rapidly followed by a catalyzed reaction whose rate changes rather abruptly at ca 50% reaction to another catalyzed reaction requiring a somewhat lower activation energy which is rate controlling to the end of reaction.

In a previous report[1], it was shown that the decomposition of cupferron tosylates could be related back to the reactant and product fractions. However, it was also recognized that the melting, or liquefaction, played a substantial role during the course of the reaction. In the present report, it becomes clear that effects other than melting also play prominant roles. Three can reasonably be postulated: the chemical decomposition itself or the chemical effect, the heat transfer effect, and the diffusion effect.

A predominantly chemical effect is expected to exhibit an Arrhenius temperature dependence. This is the case for the catalyzed portion of the reaction. The catalyzed reaction is that occurring in the liquefied state. It is here postulated that the bend in the Ostwald curve and the corresponding increase in the reaction rate occurs at the point where the entire reacting mixture is liquefied. At this point, the chemical effect is maximized relative to heat and mass transfer effects. Prior to complete liquefaction, the heat transfer effect would be expected to be a more important factor in the reaction mechanism. Indeed the rate is slower prior to liquefaction, and the activation process is more difficult. The lack of any significant KBr matrix effect on the decomposition indicates that mass transfer, or diffusion, effects do not play significant roles either before or after complete liquefaction.

In the initial, uncatalyzed reaction, the most striking deviation from chemical control is demonstrated. The linear dependence of the rate constant on temperature suggests a process whose rate is controlled largely by heat transfer effects, in particular, heat capacity, liquefaction, and spacial effects. Analysis of these phenomena is continuing using the heat conduction equation, eq 10

$$\rho C \frac{\partial T}{\partial t} = \nabla \cdot (\lambda \nabla T) + q \qquad (10)$$

where λ is the thermal conductivity and q is the combined volumetric generation rate due to chemical processes and liquefaction. Results will be reported subsequently.

References

1. E.A. Dorko, R.S. Hughes, and C.R. Downs, Anal. Chem., $\underline{42}$, 253 (1970).
2. R.W. Crossley and E.A. Dorko, Program DATAR, available from the Quantum Chemistry Program Exchange, Indiana University, Bloomington, Ind., No. QCPE-162.
3. R.W. Crossley and E.A. Dorko, Program PARACT, in prep. for submittal to the Quantum Chemistry Program Exchange.
4. Z. Kopal, Numerical Analysis, John Wiley and Sons, Inc., New York, N.Y. (1961).

THERMAL RESISTANCE FACTORS IN DIFFERENTIAL SCANNING CALORIMETRY

W. P. Brennan, B. Miller, and J. C. Whitwell

Textile Research Institute and Department of Chemical Engineering, Princeton University, Princeton, New Jersey 08540

GENERAL THEORY

In the analysis of heat flow processes in differential scanning calorimetry it has been generally agreed(O'Neill) that the controlling thermal resistance is between sample container and sample holder, and, consequently, that the sample and sample container are at the same temperature. The thermal equation applying to this situation, usually referred to as Newton's Law, is

$$\frac{dq_s}{dt} = \frac{T_p - T_s}{R} \qquad (1)$$

where dq_s/dt = total rate of flow of thermal energy to the sample, T_p = programmed temperature, T_s = sample temperature, and R = thermal resistance.

The thermal resistance, R, is usually determined by the slope of the line recorded for a sharp temperature dependent transition, e.g. melting. The time derivative of equation (1) is

$$\frac{d\left(\frac{dq_s}{dt}\right)}{dt} = \frac{\frac{dT_p}{dt}}{R} - \frac{\frac{dT_s}{dt}}{R} ; \qquad (2)$$

during melting dT_s/dt is assumed to be zero. The slope of the curve generated during melting, $d\left(\frac{dq_s}{dt}\right)/dt$ in equation (2), is then equal to dT_p/dt, the programmed temperature rate, divided by the thermal resistance, R.

Furthermore, the peak height, $dq_s/dt\big)_{MAX}$, (measured above an extension of the pre-melt base line) is equal to the slope multiplied by the time to the peak, t_{MAX}:

$$\frac{dq_s}{dt}\bigg)_{MAX} = t_{MAX} \; \frac{1}{R} \frac{dT_p}{dt} \quad . \tag{3}$$

In this relationship, t_{MAX} has been shown to be (Gray)

$$t_{MAX} = \left(R^2 C_s^2 - \frac{2\lambda R}{dT_p/dt}\right)^{1/2} - RC_s \quad . \tag{4}$$

where C_s = total heat capacity of sample and container, and λ = total enthalpy of fusion of the sample (a negative value for endothermic transitions).

The theory of DSC-generated peak shapes for sharp endothermic transitions also predicts that area under a DSC peak should be independent of the effective thermal resistance. (This prediction has been discussed by Gray and will not be considered here.) Further, the peak height should be proportional to the inverse square root of R (O'Neill).

The purposes of this paper are: one, to question these basic assumptions and their implications (i.e., the location and the effect of the thermal resistance on the peak height); and, two, to discuss the effect of thermal resistance on kinetic analysis.

PEAK HEIGHT AS A FUNCTION OF RESISTANCE

From equations (3) and (4) it is apparent that dq_s/dt is a function of both R and C_s, when heating rate and total enthalpy of transition are fixed. Only

THERMAL RESISTANCE FACTORS

in the unrealistic case where $C_S = 0$, and therefore,

$$t_{MAX} = \left(\frac{-2\lambda R}{dT_p/dt}\right)^{1/2} \quad (5)$$

is $dq_S/dt)_{MAX}$ proportional to the inverse square root of R.

For $C_S \neq 0$ the relationship is more complex. Theoretical relationships between $dq_S/dt)_{MAX}$ and $1/\sqrt{R}$, for various values of C_S (plotted log-log) are shown in Figure 1.

Clearly the peak height is dependent not only on R, but also on C_S.

FACTORS INFLUENCING THE THERMAL RESISTANCE

Consider the situation where the controlling resistance is not between sample container and heater but is the sum of a series of resistances as illustrated in Figure 2. In this and subsequent examples in this section the sample is not capped in order to eliminate conductive heat transfer to the sample through the cap and thereby to simplify the model.

The total resistance to heat flow, R, between the heater and the sample is the sum $R_1 + R_2 + R_3$, where R_1 = resistance between heater and pan, R_2 = resistance of the aluminum pan (considered negligible due to the high thermal conductivity of aluminum) and R_3 = resistance between the aluminum pan and sample (e.g., indium).

The resistance R is then defined as

$$R = R_1 + R_3 = \frac{1}{A_1 h_1} + \frac{1}{A_3 h_3} \quad (6)$$

where A = an area of contact, and h = a heat transfer coefficient. In the experiments to be described the resistance R_1 will be assumed to be constant since the pans are similar in shape and contact area. It is, however, possible to vary A_3 and hence R_3 systematically by using samples with different areas of contact.

To accomplish this different lengths (L) of indium wire were used and A is then proportional to L or,

$$A = PL \tag{7}$$

where P = an unknown proportionality constant. Consequently a graph of R, calculated from Equation (2) versus $1/L$ should produce a straight line of non-zero slope if the resistance R_3 is not negligible and if h_3 is a constant. A zero slope would indicate that $R_1 >> R_3$ and is consequently the controlling resistance. Figure 3 shows the result of using different lengths of indium and tin wires.

The measured resistances vary considerably with L, the slopes of the lines being proportional to $1/h_3$ and the proportionality constant in each case being $1/P$. Since the proportionality constant changes only the slope of the curve the intercept at $1/L = 0$ (infinite length and area) is the resistance R_1.

These observations have serious implications if one is depending on any one standard metal to establish the basic resistance from which to calculate a thermal lag correction for use with other materials; it is obvious that the estimated resistance depends on the sample contact area (hence on sample size and shape).

To exaggerate the effect of thermal resistance that might be encountered with polymeric materials (e.g. films, fibers, etc.), a disc of teflon 0.089 cm thick X 0.646 cm diameter was placed in an aluminum sample pan with indium on top of the teflon. This arrangement is illustrated in Figure 4.

The potential resistances are: R_1 = resistance between heater and pan, R_2 = resistance of the aluminum pan, R_3 = resistance between pan and teflon, R_4 = resistance of the teflon cylinder, and R_5 = resistance between teflon and indium. The total resistance, R, between heater and sample is the sum $R_1 + R_2 + R_3 + R_4 + R_5$. Again, if R_2 is assumed to be very much smaller than the other resistances,

$$R = R_1 + R_3 + R_4 + R_5 \tag{8}$$

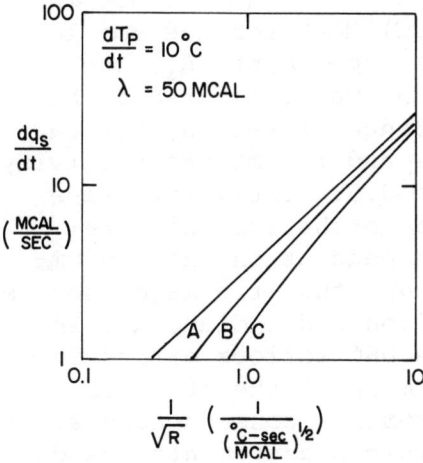

Figure 1. Peak maximum as a function of thermal resistance and heat capacity.

Line	C_S (MCAL/°C)
A	0
B	5
C	10

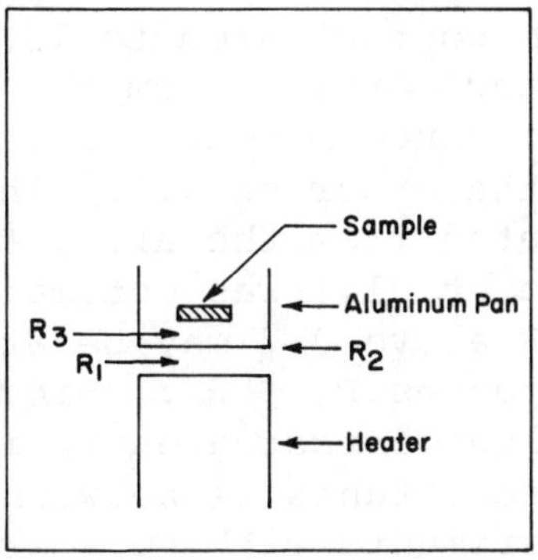

Figure 2. Illustration of experimental arrangement and possible effective thermal resistances.

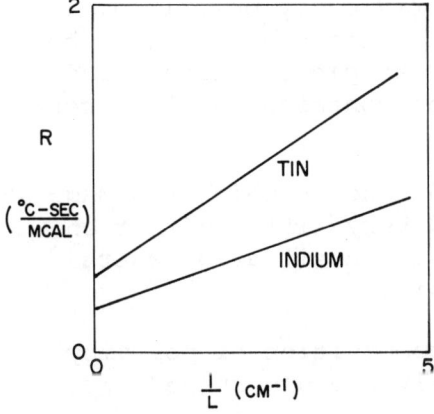

Figure 3. Resistance as a function of length for indium and tin wires.

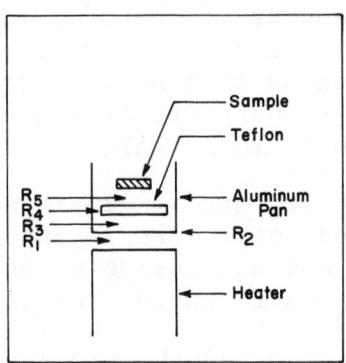

Figure 4. Illustration of experimental arrangement and possible effective thermal resistances.

A series of runs was undertaken using three samples of indium each with a different area of contact. Calculated values of R ranged from 1.8 for the largest indium contact area to 13.0 (°C-SEC)/MCAL for the smallest contact area. From the previous experiment, R_1, the resistance between heater and pan, was estimated to be of the order of 0.2. The resistance of teflon, R_4, calculated from the area, thickness and thermal conductivity (Hsu et.al.) was estimated to be 0.5. Thus, any value of R above 0.7 may be attributed to the remaining resistances $R_3 + R_5$. Since R is dependent on the indium-teflon contact area it is concluded that the major source of resistance is between the teflon and indium, R_5. In experiments with polymers in the DSC serious errors in thermal lag corrections will be made if the standard procedure is followed. Even more serious consequences arise in calorimetric DTA where enthalpic measurements are dependent on the thermal resistance. If there are significant and unknown resistances in the system, other than the "constant" resistance between heater and sample container, calorimetric data from DTA would appear to be of dubious value.

EFFECT OF RESISTANCE ON REACTION KINETICS

This section will consider the consequences produced by a change in the value of R on reaction kinetics independent of the complications previously described. In this and the following example the resistance R will be taken as that between heater and sample container, and the temperature of the sample and container will be considered identical.

Two cases were investigated, one in which the magnitude of the resistance, R, had very little effect on the kinetic analysis, and secondly, a situation where it greatly influenced the analysis.

The endothermic thermal decomposition of polyvinyl alcohol, producing a comparatively broadly shaped reaction trace was the example of case 1. Solid residual product was negligible and its heat capacity was consequently assumed to be zero. The DSC trace for the reaction, corrected to a zero base line, is shown in Figure 5.

The heat balance for any point in the trace CAFBD, is

$$\frac{dq_s}{dt} = C_s \frac{dT_s}{dt} - \frac{dh}{dt} \qquad (9)$$

where dq_s/dt = heat flow from the thermal energy source, C_s = total heat capacity of sample plus container, dT_s/dt = rate of temperature change of the sample and container, and dh/dt = heat adsorbed by sample per unit time. The rate of heat adsorbed is proportional to the rate of reaction through the following relationships,

$$\frac{dh}{dt} = \lambda \frac{dx}{dt} = \lambda A_o e^{-E/\overline{R}T}(1-x)^n \qquad (10)$$

where λ = total enthalpy of reaction, dx/dt = rate of reaction, x = fraction reacted, $A_o e^{-E/\overline{R}T} = k$, the Arrhenius' form of the rate constant, and n = apparent order of the reaction.

A plot of ln k versus 1/T°K will be a straight line when the order of reaction, n, has been correctly chosen. Methods for determining k using differential scanning calorimetry have been discussed (Borchardt) and will not be repeated here.

A determining factor in kinetic analysis is the establishment of a base line between the initiation and termination points of the reaction (Points A and B, Figure 5). Consider first the situation where the resistance, R, equals zero so that, as indicated by equation (2),

$$\frac{dT_s}{dt} = \frac{dT_p}{dt} .$$

For a first approximation, the base line is taken as a straight line AB between the initiation and termination points of the reaction. With this base line the Arrhenius plot for n = 1 is shown in line A, Figure 6.

The activation energy, calculated from the slope of

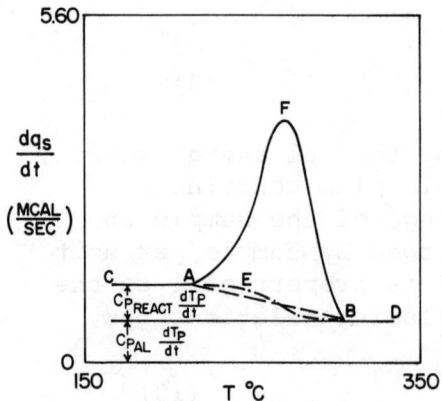

Figure 5. DSC trace for the decomposition of polyvinyl alcohol.

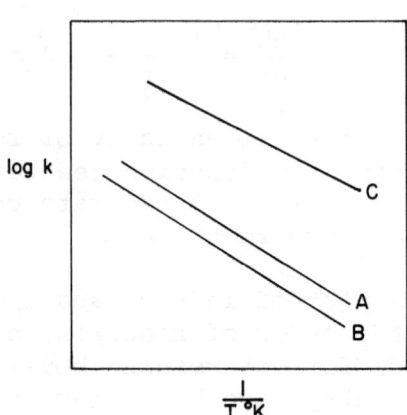

Figure 6. Arrhenius plot for the decomposition of polyvinyl alcohol.

Line	R	Base Line Determined	E(Kcal/Mole)	n
A	0	straight line	40.6	1.0
B	0	calculated-equation(11)	40.6	1.0
C	3	calculated-equation(12)	42.8	1.06

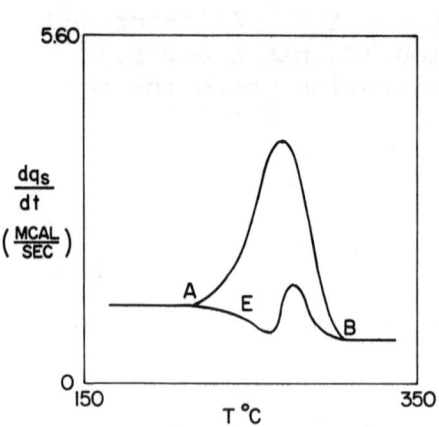

Figure 7. Calculated base line for R = 3.0.

this line, is 40.6 Kcal/mole and equation (10) with n = 1 appears to be a suitable approximation to a realistic expression of the rate.

Since the straight base line method described above is only an approximation the effect of using a more realistic base line is of considerable interest. An iteration method for approximating the true shape of this base line when R = 0 has been described (Brennan et. al.). The equation of this base line, now called AEB has been written

$$[\,C_{P_{AL}} + C_{P_{REACT}}(1-x) + C_{P_{PROD}}(x)\,]\,\frac{dT_p}{dt} \quad (11)$$

assuming no thermal lag. This resulting base line, AEB, is indicated in Figure 5.

Kinetic analysis using base line AEB produces very little change. The Arrhenius plot is line B, Figure 6. Again n = 1, E = 40.6 Kcal/mole; however, the pre-exponential factor A_o is reduced.

A different situation will arise when R ≠ 0. Then the equation for the base line becomes (Heuval)

$$[\,C_{P_{AL}} + C_{P_{REACT}}(1-x) + C_{P_{PROD}}(x)\,]\,\frac{dT_s}{dt} \quad (12)$$

In this instance the term dT_p/dt in equation (11) is replaced by dT_s/dt, and this rate is determined from the slope of the generated trace using equation (2)

$$\frac{dT_s}{dt} = \frac{dT_p}{dt} - R\,d\!\left(\frac{dq_s}{dt}\right) \quad (13)$$

The iterative procedure (when R ≠ 0) for establishment of the base line is essentially the same as when R = 0. In Figure 7 the final, iterated base line AEB is indicated for R = 3.0; line C, Figure 6 represents the Arrhenius plot with the value of n changed to reestablish the straight line. The apparent order of reaction is now 1.06 and the activation energy is 42.8

Kcal/mole. The change in these values over those for R = 0 is very small particularly in view of the large value chosen for R, one which would appear to be unrealistically high based on previous results using indium.

The melting endotherm of polyethylene was chosen as an example of case 2. This process produces a comparatively sharp transition trace. While the heat capacity changes slightly as a result of the transition, the effect is small enough to be considered negligible. The DSC trace for the transition, corrected to a zero base line, is shown in Figure 8.

Equations (9), (11), (12), and (13) also apply to this transition. If melting is considered to be a zero-order process equation (10) becomes

$$\frac{dx}{dt} = A_o e^{-E/RT} \qquad (14)$$

From the sharpness of the transition the total resistance, R, might be expected to have a large effect on the kinetic model. To test this assumption, four values of R were assumed: 0.00, 0.20, 0.50, and 1.00. The corrected base lines for these values of R were determined according to the procedure previously described and are shown in Figure 9.

Arrhenius plots, constructed for n = 0, are shown in Figure 10.

While it may appear that the transition is not zero order (which may well be true), other explanations can account for the shapes of the lines.

First, the calculated value of k is sensitive to the value of R near the beginning and particularly near the end of the transition. If these points are eliminated from the Arrhenius plots in Figure 10 the transition will appear to be zero order and the lines essentially parallel.

Second, the value of R may be greater than 1.00.

THERMAL RESISTANCE FACTORS

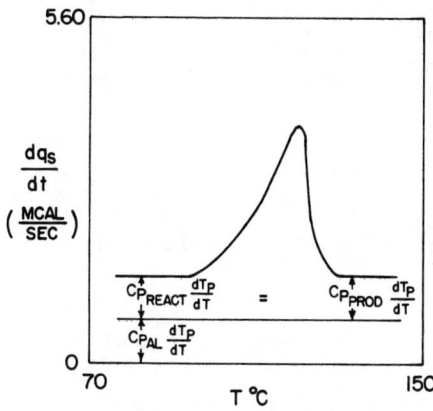

Figure 8. DSC trace for the melting of polyethylene.

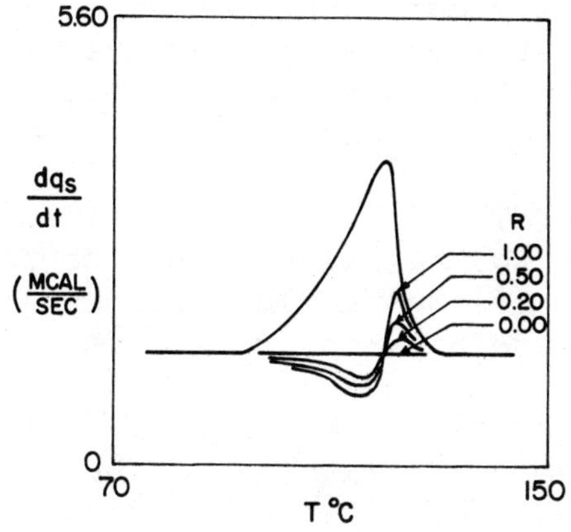

Figure 9. Effect of thermal resistance on base line.

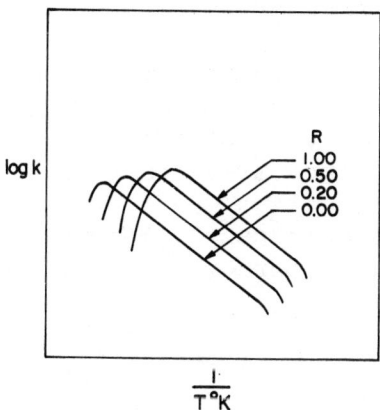

Figure 10. Arrhenius plot for the melting of polyethylene.

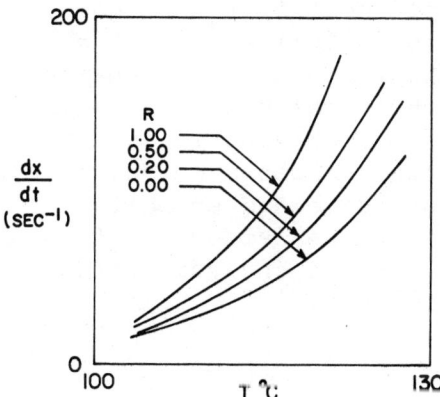

Figure 11. Effect of thermal resistance on the apparent rate of melting of polyethylene.

If the transition were truly zero order, all of the area under the trace after the peak must be due to heat capacity catch-up(Reed et.al.). For this situation to exist a value of R greater than 1.0 must be chosen so that the corrected base line exactly matches the trace of the transition after the peak.

Third, the effective contact area for heat transfer to the polyethylene could be decreasing with shrinkage of the sample during melting, producing a decrease in the apparent rate of transition.

To view the effect of R in a more sensitive manner, Figure 11 illustrates the rate of transition(up to the peak of the trace), dx/dt, as a function of temperature for the four values of R.

CONCLUSIONS

1. For sharp temperature dependent transitions, e.g. melting, the peak maximum is not simply proportional to $1/\sqrt{R}$ but is also a function of the total heat capacity of the system.

2. Experiments indicate that there may be significant resistances in the system other than that between sample container and heater. This observation has serious implications if one is depending on any one standard metal to establish the basic resistance from which to calculate a thermal lag correction for use with other materials; it is obvious that the estimated resistance depends on the sample contact area(hence on sample size and shape).

3. Resistance factors can significantly affect the kinetic analysis of data from differential scanning calorimetry.

REFERENCES

1. Borchardt, H. J., and Daniels, F. J., J. Am. Chem. Soc. $\underline{79}$, 41 (1957)-

2. Brennan, W. P., Miller, B., and Whitwell, J. C., Ind. & Eng. Chem. Fundamentals $\underline{8}$, 314 (1969).

3. Gray, A. P., "Analytical Calorimetry", Plenum Press, New York, 1968, pp. 209-218.

4. Heuval and Lind, Anal. Chem. (in press).

5. Hsu, K. L., Kline, D. E., and Tomlinson, J. N., J. Appl. Polymer Sci., $\underline{9}$, 3567 (1965).

6. O'Neill, M. J., Anal. Chem. $\underline{36}$, 1238 (1964).

7. Reed, R.L. Weber, L., and Gottfried, B. S., Ind. & Eng. Chem. Fundamentals $\underline{4}$, 38 (1965).

INDEX

A

α-helical conformation 147, 218
Acoustic transducers 285
Acrylonitrile 51, 172
Acrylonitrile-butadiene-styrene 51
Activation energy 104, 292, 447
Adiabatic calorimetry 372
  compressibility 79
  dynamic calorimeters 370
Alkali metal tetrafluoroborates 286
Alumina 348
Aluminum 376
Aluminum oxide 342
  powdered 271
p-Aminocinnamic acid esters 121
Ammonium nitrate 400
Amorphous density 9
  layer 13
Analysis of off-gasses 421
Annealing of aqueous gelatin
    solutions 225
  of polyethylene crystals 1, 3, 19
Aromatic nitro compounds 353, 359, 360
Arrhenius analysis 327
  plot 107, 437, 449
  rate constant 391
As-S 309
As-Se glasses 309
Autocatalytic rate constants 429
Autocatalytic reaction 429

B

Baseline shift 111
Beef muscle tissue 137, 142
Benzoic acid 115, 401
Bi-Se vapor quenched glass 310
Born-vonKarman (24)
  lattice 83
Bound water 135
Breccia Rocks 274, 278
Brominated polyethylene
  single crystals 27
Bulk modulus 82
Buna N 157
Butyl cellosolve
  pelargonate 65
Butyl rubber 61

C

Calorimeter cell 370, 372
  static and dynamic gas
    flow flooding of 371
Calorimeters, dynamic
  adiabatic 370
Calorimetric accuracy 339
  measurements on organic
    nitrogen base adducts 239
Calorimetry 9, 11 (see also
  adiabatic calorimetry and
  differential calorimetry
Carbonates 113
Carboxyl end-groups 97
  -terminated polystyrene
    95, 97
Catalytic reduction 353
Cellosolve pelargonate 62
Chalcogenide glasses 309
Chemisorption 359
Cholesteric state 405
Cholesteryl esters 113
  propionate 403

Cocrystallization 41
Coevaporation 311
Cold drawing 12
Complexes of organotin halides 239
Compressive flexural strength 186
Computer technique 429
Configurational entropy 31
Copolymers 41, 33, 220
   ethylene-vinyl acetate 41, 44
Core-tube samples 270
Crosslinking 2
Crystal, dilute solution 2
   phase change 291
   size distribution 24
   thickness 13, 14
Crystalline core 23
   density 9
   phase change 302
   rock 274, 278, 279
Crystallinity 9, 17, 22, 23
Crystallization 306, 314
   effect of pressure on 380
   of polymer 397
Cupferron tosylate and derivatives 429
Cure kinetics, dynamic 209, 201
   epoxy 201
   isothermic 201
   reaction 207
Cyclic scanning 302

D

Debye characteristic temperature 84
Decomposition 255, 292
Deflection temperature 187
Degree of cure 185
Dehydration 291
Dehydrogenation 421
Dehydroxylation 347
Deionized bone gelatin 225
Densities in a gradient column 2
Density 9, 11, 15, 16, 18
Dibenzazepine 127
Dicarboxyl-terminated polystyrene 98
Dicumyl peroxide 421, 422
Diethanolamine 192
Differential enthalpic analysis 107
Differential microcalorimetry 135

Differential scanning calorimetry 1, 28, 41, 61, 71, 104, 113, 130, 137, 148, 171, 187, 225, 255, 256, 310, 353, 369, 372, 398, 410, 429, 441, 452
   computer interface 205
   dynamic 171, 201
   high density 353
   isothermal mode 429
   mass spectrometry 417
Differential thermal analysis 30, 61, 103, 180, 211, 269, 291, 295, 310, 339, 398
   peak temperatures 299
Digitizers 392
Dilatometry 398
Dilute solution crystals 2
m-Dinitrobenzene 363, 364
Dinitrotoluene 353
Di-tert-butyl peroxide 175
Dotriacontane 407
Draw rate 11
   ratio 11, 15, 17, 18, 22, 23, 24
Dynamic cure analysis 201
   cure experiments 209
   differential scanning calorimetry 171, 201
   rate equation 202

E

Egg albumin 137, 142
Elastic modulus 15, 16, 17
Elastomer 61 (see also ethylene-propylene rubbers, butyl rubber, polybutadiene)
Electrical properties 186
Electro-balance 137
Elution fractionation of polyethylene 2
Enthalpic, differential analysis 107
   changes 107
Enthalpy 291 (see also transition enthalpy)

INDEX

Entropy, configurational 31
  of fusion 131
  of transition 113, 125, 131
Epoxy adhesive 155
  cure kinetics 201
  resins 201
Ethylene-propylene rubbers 187
Ethylene-vinyl acetate copolymers 41, 44
Ethylene-vinyl alcohol copolymers 44
Exothermic 299
  peak 111
Exotherms 274
Expansion dilatometer 214 (see also dilatometry)

F

Ferroelectric transition 256
Fiber structure 11, 12, 13, 21, 22, 23
Fire retardant 58
First-order reaction 107
Flexon 62, 65
Form birefringence 398
Free amine 292
Freezing 299
Frequency factor 107
  distribution 389
Fusion, effects of pressure on 380 (see also heat of fusion
  microscopy 397

G

Gallium 115
Gas phase reactions 340
Gelatin 137
  of dionized bone 225
Gelatin 211, 215
  thermodynamics of thermally reversible 212
Gel permeation 96
Gel permeation chromatography 100
Gels, polymeric 211
  thermally reversible 211
Gel-sol transition 225
Glass transition 111, 407
  chalcogenide 309
  phenomena 310
  temperature 52, 61, 310

Glass transition (cont'd)
Glucose 407
Glycerine 407
Grüneisen ratio 77, 78

H

Halogenation 27
Heat:
  balance 447
  capacity 52, 77, 81
  of chemisorption 362
  of crosslinking 214
  of crystallization 310, 314
  flow processes 441
  of fusion 9, 11, 15, 16, 17, 18, 28, 30, 39, 382
  losses 341
  of oxidation 382
  polymerization 171, 172
  of reaction 339
  specific 9, 79, 84, 373
  transfer coefficient 372
  of transition 114, 147, 151, 152, 281
Heating rate 104, 270, 271,
  effects 376
Helical conformation 220
  winding 218
Helix-coil transition 147, 215
Hydrogen bonding 41
Hydrogenation 362
Hydrolysis 307
Hydroxylammonium perchlorate 291
Hypersthene 270, 276, 278

I

Impact modified polyblends 51
resistance 186
Impurity determination 392
Indium 115
Infrared 207
Initiators:
  di-tert-butyl peroxide 175
  tert-butyl peroctoate 175
Instrument calibration 270
Internal standard method 281

Irradiated platelet polyethylene, crystal melting of 5
Irradiation 269
Isothermic cur kinetics 201

K

Kaolinite 347, 348
Karman lattice 77
Kinetics of Dissolution 319
Kurtosis 390

L

Lamella slip 12
Le Chatelier's principle 259
Leptokurtic peaks 392
Light scattering 96, 398
Linear coefficient of thermal
  expansion 80
  expansion 398
  polyethylene 9, 10 (see also polyethylene, high density)
Low temperature dynamic properties 61
Lunar fines 270
  samples 269

M

Magnesium oxide 319
Marker's acid 113
Mass spectrometer 302
  spectroscopy 291
  spectroscopy-Differential Scanning Calorimeter 417
Melting 291
  curves 10
  point of brominated crystals 28, 30
  point by differential thermal analysis 30
  of polyethylene crystals 1
  process 304
  temperatures 13
Mercury 115
Mesomorphism 121
Methacrylate-acrylonitrile-butadiene-styrene system 53
Methacrylate-butadiene-styrene 53
Methyl methacrylate 172
  parathion 421, 425
Microfibril 10, 12, 21, 22, 23
Micronecking 12, 21

Microscopy 114
  fusion 397
  thermal 291
Molecular weight determination 95
Muscle tissue 135

N

Nematic transitions 125
Network polymer structures 185
Nitric acid treated polyethylene 14
Nitrotoluene, ortho and para 365
N.M.R. spectrum 127
Noryl 51
Numerical integration 390
Nylon 82, 85

O

Off-gasses, analysis of 421
Orientation 24
Organotin halides, complexes 239

P

Paracrystalline disorder 22
Particle size 294
Peak area 107
  shape 111
  temperature 104
Penetrometer 214
Perchlorate 291
Perchloric acid 292
Pharmaceuticals 353
Phase separation 314
Phenolic-silica prepregs 155
Plastically deformed polyethylene 10, 11, 12, 14
Plasticizers 62, 65
  endotherms 71
Plastogen 62, 65
Polarized light 302
Polarizing microscope 116
Poly (Aryl ether) 103
Poly (Acrylglycinamide) 216
Poly (Bis-phenol-A-carbonate) 412

Polyblends 51
Polybutadiene 51
Polycarbonate 376
Polyethylene 9, 410, 450
  deformed 10, 11, 12, 14
  elution fractionation of 2
  high density 27, 178, 381, 382
  linear 9, 10
  melting of crystals 1, 5
  oxidation of 380
  radiation of crystals 1
  single crystals of 27, 28, 35
Polyethylene terephthalate 397
Poly-γ-benzyl-1-glutamate 147, 218
Poly-1-glutamic acid 148
Poly (3-methyl-1-pentene) 397
Polypeptide 151
Polyphenylene oxide 55
Polypropylene 382, 397
Polystyrene 55, 82, 85, 412
  carboxyl-terminated 95, 97
Ps-PPO alloys 55
Polysulfone 103
  thermal degradation 183
Poly (vinyl chloride) 51, 220
Prepreg 155
Pressure, ambient effects 371
  effects on crystallization 380
  high 353, 376
  sensitive transitions 405
Protein, soy and milk isolates 137
Pyrolysis product 104

Q

Quadratures 390
Quantitative measurements 270
  thermal analysis 51
Quartz 270
Quenched film 19

R

Radiation: damage 269
  of polyethylene crystals 1
Random-coil conformation 147
Rate constants 429
Reaction rate 107, 447
Refractory Corrosion 319
Rhombic-cubic transition 286

Rochelle salts: deuterated 255
  hydrated 255
  tetrahydrate 258
Rocks 270
Rolled polymer 9, 11
Rosin and its derivatives 421

S

Sample: holder design 111
  loading device 339
  packing 270
  size 270
  weight 270
Schiff's bases 121
Silicon carbide 321
Simpson's rule 429
Single crystals 255, 330
  of polyethylene 27, 28, 35
Smectic transitions 125
Sodium silicate 319
Soy and milk protein isolates 137
Specific heat 9, 79, 84, 373
Spherulite destruction 22
Standard cell 274
Stearic acid 115
Stored energy 269, 274
  artificially induced 276
  radiation induced 278, 279
Stress to break 16
Styrene 172
Styrene-acrylonitrile 51, 172
Styrene-butadiene 56
Sub-ambient temperature 376
Sucrose 410
Surface energy 9
  fines 271
  free energy 37, 39
  layer 23

T

Tartrates, sodium and potassium 258
Teflon 82, 83

Temperature, Debye characteristic 84
  deflection 186
  of drawing 11, 15
  limiting transmission 225
Tensile properties 186
  strength 17
Terrestrial materials 270
Tert-butyl peroctoate 175
Tetrafluoroborate 281
Tetrahydrate of rochelle salt 258
Tetrahydronaphthalene 367
Thermal analysis 61, 225, 291
  quantitative 51
Thermal behavior of aqueous gelatin solutions 225
  conductivity 342, 439
  decomposition 446
    of single crystals 255
  degradation of polysulfone 103
  degradative stability 187
  depolarization analysis 398
  diffusivity 342
  expansion 77
    linear coefficient of 80
  history 225
  microscopy 291
  properties 41
  resistance factors 441, 443
Thermally reversible gels 211
Thermocouple 344, 372
Thermodynamics of thermally reversible gelatin 212
Thermogram 389
  derivative of 389
Thermogravimetric analysis 155, 187, 255, 258, 291, 295, 389
Thermomechanical analyzer 80
Thermometric enthalpy titration 95
Thermosets 185

Thermosetting polymers 201
  Resins 185
Tie molecules 10, 11, 22, 23, 24
Thio analogues 113
Time of flight 302
Toluidene, ortho and para 365
Transition enthalpy 147, 225
  entropy (see entropy of transition and heat of transition)
  ferroelectric 256
  heat of quartz 271
  materials 271
  pressure sensitive 405
  temperature 225
Tristearin 403

U

Ultimate tensile strength 15
Ultrasonic sound 81
  velocity 77

V

Vacuum heating 307
Van't Hoff equation 115, 130, 147, 152
Vapor deposition 309
Vicat softening point 186
Volume resistivity 207

W

Water abstraction 136
Water binding index 135, 142
  of proteins 136

Y

Yagfarov method 286

Z

Zimm-Bragg theory 147

QD
511
A49
v.2

DEC 17 1976